MUSK OXEN
MUSK OXEN
See back endpapers for closeups of Europe and the British Isles.
Canadian Arcott, Outaouais Arcott & Rideau Arcott
Type A Merino
Type B Merino
Montadale
American Tunis
Gulf Coast Native
Hog Island
Merino
ANGORA GOAT
Tunisian Barbary sheep
DROMEDARY CAMEL
DROMEDARY CAMEL
Karakul sheep
YAK & CASHMERE GOAT
SOUTH AMERICAN CAMELIDS
Equator
Falkland Islands
N
W
E
S
1 inch = 1250 mi/2000 km

The Fleece & Fiber Sourcebook

French Angora Rabbit
Musk oxen
Debouillet sheep
Alpaca
Angora goat

SOURCEBOOK

More than 200 Fibers from Animal to Spun Yarn

Deborah Robson & Carol Ekarius

Storey Publishing

The mission of Storey Publishing is to serve our customers by publishing practical information that encourages personal independence in harmony with the environment.

Edited by Gwen Steege, Deborah Burns, Sarah Guare, and Nancy D. Wood
Art direction and book design by Mary Winkelman Velgos
Text production by Jennifer Jepson Smith

Front cover photography by © John Polak except Alpaca © 2009 David C. Phillips/Garden Photo World and Dalesbred Sheep © FLPA/Wayne Hutchinson/agefotostock.com
Back cover photography by © Wayne Hutchinson/Minden Pictures
Author photgraphs on flap by © Kristi Schueler (Deborah Robson) and © Ken Woodard Photography (Carol Ekarius)
Interior photography by © John Polak except as noted on pages 428–429
Photo research and coordination by Mars Vilaubi
Maps by Alison Kolesar
Illustrations on pages 9 and 10 by © Elayne Sears

Indexed by Kathryn Bright
Knit swatches by Diana Foster
Woven swatches by Jennifer Jepson Smith

Storey Publishing
210 MASS MoCA Way
North Adams, MA 01247
www.storey.com

Printed in China by Toppan Leefung Printing Ltd.
10 9 8 7 6 5 4 3 2 1

LIBRARY OF CONGRESS CATALOGING-IN-PUBLICATION DATA

Robson, Deborah.
The fleece and fiber sourcebook / by Deborah Robson and Carol Ekarius.
p. cm.
Includes bibliographical references and index.
ISBN 978-1-60342-711-1 (hardcover : alk. paper)
1. Animal fibers. 2. Wool. I. Ekarius, Carol. II. Title.
TS1545.R66 2011
677'.3—dc22

2010051175

Contents

PART 2

Other Species: The Rest of the Menagerie 333

Yak

Pygmy goats

Preface

Sometime in antiquity (no one is sure exactly when, but we'll tell you what is known on the subject in chapter one) a human discovered that fibers could be twisted and pulled to create a cord. This twisting and pulling of fibers probably occurred quite by happenstance, yet what a profound impact it came to have on humanity. The finest cords, when twisted or braided together, became larger cords — thread, string, yarn, rope. Just think what handy inventions these must have been for cave dwellers, nomads, and other early people. And the

Before and after buddies. A Clun Forest ewe and fingerless mittens from Solitude Wool's Clun Forest/alpaca yarn.

strands of twisted fiber could be woven together to become cloth. Suddenly clothing options expanded from the skin of a dead animal (one imagines, a heavy, smelly covering) to much lighter and more versatile garments.

Over the eons, since our foremothers and -fathers made their discovery, the fibers of animals and plants have served humankind and seemingly insinuated themselves in our DNA, carving out a place in the hearts and minds of fiber buffs everywhere. We are those kind of people. Like so many others who share this passion, we caught the bug as kids, starting to sew before we hit school, and learning the basics of knitting and crocheting by 10 or so. We both dabbled with some kind of loom as kids, and again a bit more seriously in our late teens, but it was in young adulthood when we each became serious about weaving, and soon after that we discovered spinning. Carol's interest in fiber and food led to farming and raising critters, including sheep, and to writing about farming and animals. Deb's home situation kept her from ever owning her own fiber animals, but her fiber addiction led her to pursue fiber professionally, as a writer, editor, and publisher of fiber-related magazines and books.

The two of us met years ago, in a context totally unrelated to fiber or animals, but as we discovered our common interest we became friends, and ultimately that led us to pursue the idea of doing a project together. This is our project, and our goal for it is to look at the animal fibers in a way that hasn't really been done before. We are looking in more depth than we believe anyone has before at the animals that have provided handspinners, knitters, and weavers with the foundation of their craft and artistry for thousands of years. You won't find patterns in this book, but we hope you will learn a great deal about the wool and hair fibers that have clothed and served us for generation upon generation, back to the person who first picked a fluff of wild sheep fibers out of a bush and twisted them together.

If you have ever run your hand over a fleece and sighed, picked up a ball of yarn and groaned with pleasure, run your hand along the light and breezy swath of a fine jersey woolen fabric and felt a tingle run down your spine, or worn a favorite wool jacket season after season, then you are a person who may just lose yourself in the pages that follow. We hope it tantalizes your senses and that it encourages you to venture into projects using fibers you never really thought of before.

Deb Robson

Carol Ekarius

Fiber Fascination

Make stuff. Stay home. Draw. Dress locally. Hand wash. Learn to darn. — Sarah Swett, Fiber Artist

Pick up a ball of yarn and cast stitches onto a needle. Start knitting. Quickly see a form begin to take shape — a sweater for a loved one, perhaps, or fun socks to keep your tootsies warm on a cold winter night. Or pull a stool up to a beautiful loom and begin the rhythmic work of weaving weft into warp. Back and forth, back and forth, throw the shuttle, pull the beater: An intricate pattern begins to reveal itself. What great satisfaction this work brings. It connects us to our past and creates memories for our future. And it all starts with fiber.

This book focuses on animal fibers, but before we move along, we want to call your attention to the other natural sources of fibers: plants and minerals. Cotton, flax, and hemp are the most common plant fibers, but other vegetable-source fibers, such as sisal, coconut, and pineapple fiber, are in use around the globe. In terms of mineral fibers, you wouldn't want to knit a sweater out of asbestos or weave a blanket of "mineral wool" or spun glass (also known as fiberglass), but these natural fibers find their way into a variety of industrial applications and products we routinely use.

What about synthetic fibers? Today they account for over 65 percent of global production, and this share is growing steadily as the market for natural fibers falls. The industrialized world prefers synthetic fibers because they are more standardized, relatively cheap, and easier to manage on a large scale than are natural fibers. Consumers have also been trained to think that the synthetics perform better than their natural equivalents. In some cases they do, but in many others the benefits of the synthetics have been significantly overrated and the drawbacks of becoming dependent on manufactured fibers have been glossed over.

What difference does it make, this shift from natural to synthetic textiles? Why should we care if natural fibers disappear? First of all,

Definitions of Fiber

We cover fiber in great detail in the pages to come. To avoid confusion, let's introduce two definitions that tend to overlap:

The commercial definition. Fiber is a long, narrow, and flexible material that may be of animal, plant, synthetic, or mineral origin. It is used not only in the production of textiles, but also for paper, rope, and a wide array of other useful items, including automobiles, sporting goods, cosmetics, and food.

The zoological definition. Fiber is an external, multicellular structure made up primarily of protein. It grows from the skin, and its primary function is to provide a creature with protection from the elements and from predators.

Slow Food Movement

Slow Food is a movement that started in Italy and has spread around the globe, with more than 80,000 members in the United States. Slow Food USA envisions a future food system that is based on the principles of high quality and taste, environmental sustainability, and social justice — in essence, a food system that is good, clean, and fair. This movement supports a shift away from the destructive effects of an industrial food system and toward the regenerative cultural, social, and economic benefits of a sustainable food system, regional food traditions, the pleasures of the table, and a slower and more harmonious rhythm of life. Learn more at www.slowfoodusa.org.

synthetic fibers are . . . well, *synthetic*. They are cooked up in a lab from oil and chemicals with unpronounceable names. Their production requires high inputs of energy and water, and the process releases harmful volatile organic compounds (VOCs) into the air. VOCs are a major contributor to smog and a number of health issues. Synthetic fibers do have benefits; they are readily available and relatively inexpensive in today's markets, for starters. But they don't connect us to the earth. They aren't sustainable.

Natural fibers are part of our culture, our heritage. They have stories. They have a living, breathing animal (or a growing plant) behind them. They often have small-scale farmers or indigenous communities behind them, too — people and cultures whose livelihoods and historic identities can be supported by their continuing work with these fibers.

When the United Nations declared 2009 the International Year of Natural Fibres, that organization honored the role that natural fibers play in the lives of both producers and consumers. Inspired in part by this worldwide acknowledgment of the importance and vulnerability of the fibers that we love, we began writing this book with the aim of raising awareness among other fiber lovers. Perhaps knowing more about the source will spark a desire to seek out and experiment with natural fibers, which may in turn help keep the animals that produce these distinctive fiber resources alive and well. We hope that increasing awareness will also encourage (financially and culturally) the small-scale and indigenous farmers who raise these animals, as well as the mills and designers who bring their products to market. We think of this as sort of a Slow Fiber movement, similar to the Slow Food movement, and a way of bringing fiber lovers back to the roots of their craft and building on the heritage laid down by those who came before us.

Why Choose Natural Fibers?

According to the United Nations, there are many great reasons to choose natural fibers over synthetic substitutes. Here are some of them:

Natural fibers are a healthy choice. Textiles created from natural fibers absorb perspiration and release it into the air (in a process called wicking), creating natural ventilation. Because of their more compact molecular structure, synthetic fibers that are designed to wick can't "breathe" in the same way. In the case of animal fibers, the bends (or crimp) of the fiber effectively trap pockets of air within spun yarns and constructed fabrics, acting as insulation against both cold and heat. This is why the desert-living Bedouins wear lightweight wool clothing to keep themselves cool.

Natural fibers are a responsible choice. The production, processing, and export of natural fibers are vital to the economies of many developing countries, providing livelihoods for millions of small-scale farmers and workers in almost every corner of the planet. Natural fibers also play a key role in the emerging green economy, due to their potentially energy-efficient production. They provide not only traditional textiles but also a truly renewable raw material in the production of items such as car panels or furniture stuffing. There are lower carbon emissions in the production of natural fibers than in the production of their synthetic substitutes, as well as lower use of water. And at the end of their life cycle, natural fibers are 100 percent recyclable and biodegradable.

Natural fibers have industrial value. They have intrinsic properties — mechanical strength, low weight, and green credentials — that make them attractive to industry. For consumers, animal- and plant-fiber composites in automobiles and other applications provide better thermal and acoustic insulation than fiberglass, while reducing irritation of the skin and respiratory system. Their low density reduces weight, which cuts energy use in both production and transport. This saves wear and tear on machinery, thus cutting production costs by up to 30 percent. Worldwide, the construction industry is using natural fibers in light structural walls, insulation materials, floor and wall coverings, and roofing.

Natural fibers are chic again! They are deemed sustainable, green, ethical, and ecofriendly. Young designers are offering "100 percent carbon-neutral" collections that strive for sustainability at every stage of their garments' life cycle — from production, processing, and packaging to transportation, retailing, and ultimate disposal. Fashion collections are highlighting organic wool, produced by sheep that have not been exposed to pesticide dips, and we see increasingly creative ways for yarn and fiber aficionados to obtain their materials direct from the growers, including CSAs (Community Supported Agriculture programs), Etsy shops, and other marketing channels.

The Value of Diversity

IT IS ALMOST IMPOSSIBLE to say how many sheep breeds exist around the world, but an educated guess puts the number somewhere around 1,400. For goats, the number is closer to 100. That may seem a more-than-abundant number, but according to the United Nations' Food and Agriculture Organization (FAO), over one-third of these breeds are in danger of extinction — and that's not good news. As we'll show in this book, the fibers produced by the most endangered breeds have unique, irreplaceable qualities, and the animals themselves fit into valuable ecological niches.

Many of the animals included in this book are recognized as endangered by expert sources, such as the FAO; the American Livestock Breeds Conservancy, which tracks livestock breeds within the United States; Rare Breeds Canada; the Rare Breeds Survival Trust (United Kingdom); The Sheep Trust (United Kingdom); North SheD (focusing in Nordic countries); Heritage Sheep (European Union); Rare Breeds Trust of Australia; and Rare

"Breeds Breed True" and Other Truisms

A breed is a group of domestic animals with identifiable characteristics — visual, performance, geographic, and/or cultural — that allow it to be distinguished from other groups within the same species. When purebred members of a specific breed are mated to each other, those traits that make that breed unique will pass down through the generations in a consistent manner. Thus the saying, "Breeds breed true."

We use the term *landrace* from time to time. A landrace is essentially an old breed that developed in a very limited geographic area over many centuries. These old breeds were developed by farmers with little emphasis on modern breeding techniques, or sometimes strictly by natural selection due to environmental conditions.

Because a landrace has not undergone intensive selection, it generally has greater genetic variability (more variation in coat color or other discernible traits) than more highly developed modern breeds. A landrace usually lacks documentation through a breed registry, but it is always well adapted to the place where it developed. The Gulf Coast Native sheep (see page 124) is a landrace still found in North America. Some landraces are now being recognized and protected, although some of these breeds still fly under the radar of our awareness.

Clockwise from upper left: **Guanaco,** a South American wild camelid whose numbers decreased radically after the Spanish arrived in the Americas; **American bison**, representing an animal whose natural grazing territory has succumbed to human population pressures; **Santa Cruz sheep,** which are critically endangered; **Jacob sheep** and **Myotonic goats** (also known as Tennessee Fainting goats), which are also conservation breeds.

Breeds Conservation Society of New Zealand. In the case of some wild species, such as the guanaco (see page 379), the recognition comes from the Convention on International Trade in Endangered Species.

People can understand the need to protect wild species from extinction, but you may wonder, *why worry about conserving breeds of domestic livestock?* There are many reasons, but the first has to do with genetic diversity that could be crucial to us in the future. The industrialization of agriculture has consolidated domesticated animals into standardized systems of production. These systems rely upon climate-controlled confinement housing, sophisticated husbandry and veterinary support, chemical additives, and heavy grain feeding. Only a handful of breeds have acclimated to these systems, and though those breeds are highly productive, they are unlikely to adapt quickly enough to climate or environmental change, or to serious disease outbreaks. Monocultures of genetically similar animals may look good on spreadsheets, where short-term bottom-line economics is the main consideration, but they are like a house of cards that could collapse completely under pressure.

As our friend Don Bixby (a veterinarian and former executive director of the American Livestock Breeds Conservancy) says, "Genetic uniformity and genetic diversity are mutually exclusive." More diversity in the gene pool provides a much better shot at adapting to changing conditions. There are several other reasons to protect breed diversity:

- **Traditional breeds** work well in sustainable and small-scale agriculture and thrive in more natural farming systems. They can be used to improve the quality of the environment by reducing the negative impact of intensive agricultural practices.
- **Maintaining traditional breeds** helps support rural and regional communities by enhancing the profitability of small farmers and indigenous people.
- **Specific breeds,** well managed and placed in appropriate environments, can help reclaim damaged landscapes.

Some breeds are vulnerable because their population numbers have become very low, but even breeds that seem to have healthier populations can swiftly veer toward extinction. In 2001, the United Kingdom had a severe epidemic of foot-and-mouth disease (also called hoof-and-mouth disease); millions of sheep — whether they were sick or not — were destroyed to control the spread of the disease. When a breed is geographically isolated, an event like this can wipe out an important population even when the overall numbers seemed strong. For instance, Herdwick sheep (see page 266) are found mainly in northwest England's Lake District — an area that covers just 885 square miles (2,292 sq km). With an entire breeding population living in close proximity, it is difficult to protect them from the spread of disease. During the epidemic, at least 35 percent of the Herdwicks were destroyed, and their numbers are still significantly reduced. The survival chances of other breeds in similar situations were also dramatically affected.

As you read through the descriptions of breeds in this book, you will see that some are designated Conservation Breed or Critical Conservation Breed. For the record, these are not terms specifically used by the biodiversity groups listed above. Each organization has its own slightly different terms and reporting systems, and the breeds' classifications can change from year to year. So we came up with our own consolidated listing, based on our understanding of what these various groups say about the breed's overall picture for survival. The Critical Conservation Breeds are globally rare and stand on the precipice of extinction. All will truly benefit from your informed support. By seeking out the fibers produced by rare and endangered animals, you will help maintain them!

The Language of Our Lives

Thanks to the prominent roles fibers and textiles have played in human development, it isn't surprising that they have also played a role in our language and culture. Think of all the sayings we take from the world of fiber:

*He **knit** his brow in consternation.*

*She **spun** a great **yarn** about giants and elves.*

*He took a **cotton** to her.*

*The social **fabric** seems to be **unraveling**.*

*Their story is **interwoven** with the events.*

*Sleep that **knits up** the ravell'd sleeve of care (from* Macbeth, *Act 2, Scene 2).*

*I'm feeling fair to **middling**. (The best grade of cotton is called middling.)*

From the Shrouds of Antiquity

Developing the skills needed to turn fibers into string — and to turn that string into netting and cloth — is one of humanity's oldest accomplishments. Until quite recently, anthropologists and archaeologists focused on stone tools, and the story of human prehistory was told from the paradigm of the often-exalted male hunter. This painted an image of women as poor, defenseless beings who did little (other than provide offspring) to help humanity's forward momentum. But over the last 20 to 30 years, the story has begun to change, as researchers have taken up the study of "perishable technologies," including fiber. The latest work of people like Olga Soffer, professor of anthropology at the University of Illinois, and Jim Adovasio, director of Mercyhurst Archaeological Institute, shows that "women's work" is evident in the archaeological record when researchers look for it.

Before people like Soffer and Adovasio began looking at fibers, it was believed that humans began developing fiber arts about seven to eight thousand years ago. However, in our interviews with Soffer, we learned about strong evidence (eyed needles and the imprints of fine textiles on early clay statues, for example) of advanced usage of plant fiber at least 26,000 years ago in several archaeological sites spread around the globe. Soffer notes that it appears that "they already had some knowledge of retting, or extracting the fibers from the stems of plants; spinning to produce cordage; and some form of weaving or fabric production, including the use of non-heddle looms." (Heddles permit the weaver to raise, and sometimes also lower, groups of warp threads simultaneously, making what is called a shed and significantly speeding up the weft-insertion process. There are lots of different types of heddles, although most of them are made of either string or steel.)

"What we know for sure about the 26,000-year-old samples," she continues, "is that we are certainly not looking at the origins of human fiber production. The evidence is too sophisticated and complex to be newly invented. It shows a good long tradition."

How far back were people learning to use fiber? No one knows for sure, but Soffer referred to a 40,000-year-old example of Neanderthals using plant fibers, now petrified, as binding material on their stone tools. The one thing that Soffer is sure of: Plant fibers were used well before animal fibers. "We have no evidence for animal fibers that far back. Certainly they are using hides in the Paleolithic period, but the weaving we have from that period is not of hair. It is weaving of plant fibers."

So when did animal fibers come into use? That's still a mystery. The oldest known samples of woolen textiles date to 3,500 years ago, and samples of silk (produced by the larvae of certain moths) in China have been dated to 5,000 years ago. But as scientists look for the perishable arts in the archaeological record, and as DNA analysis helps us understand the history of domestication, these timelines are likely to change. For now, we know that alpacas were domesticated at least 7,000 years ago, and sheep were probably domesticated at least 9,000 years ago. It is a safe bet that humans were using the fiber of these animals in some way by the time they entered into a domesticated relationship with them.

The Biology of Fiber

NATURAL FIBERS are the result of biological adaptation. The organisms that produce them do so not to give us something to do with our hands, but for their own needs. Fibers provide the creatures who grow them with a host of services, including protection from the elements or predators, structural support, or a way to ensure seed dispersal. We are just lucky enough to be able to take advantage of their solutions.

From a chemical standpoint, all natural fibers fall into two major classes based on the predominant compounds that make up the fiber. Animal fibers of hair and wool are classified as protein fibers; plant fibers are classified as cellulose fibers. The chart below compares the pros and cons of the two fiber types in detail.

Rule-of-Thumb Fiber Comparisons

CHARACTERISTIC	PROTEIN FIBERS	CELLULOSE FIBERS
Susceptibility to moths	Greater	Less
Susceptibility to fungi and mildew	Less	Greater
Susceptibility to chemicals	More likely to be damaged by strong alkali chemicals and chlorine, but less vulnerable to acidic chemicals	Less likely to be damaged by strong alkalis and chlorine, but more vulnerable to acidic chemicals
Susceptibility to hot water shrinkage	Greater	Less
Susceptibility to wrinkling	Less	Greater
Insulation value and comfort	High insulation value; generally more comfortable in cool weather, yet fine wools can be very comfortable in warm weather	Low insulation factor; more comfortable in warm weather
Fire resistance	Greater	Less
Resistance to wear and abrasion	Greater	Less
Moisture absorbency and wicking ability	Greater	Less
Fiber elasticity	Much greater	Much less
Fiber strength	Less	Greater

A Closer Look at Animal Fibers

Most of us probably think of organs as things like a heart or the lungs. But biologists consider hair or wool to be an organ, too. This particular organ has its roots — or the live, growing portion — hidden in skin, and what we see outside of the skin is nothing but dead tissue.

Keratin is the main protein in wool and hair. But in addition to keratin, there are also carbon, hydrogen, oxygen, and sulfur molecules. The structure of these molecules varies a bit between hair and wool. Basically the molecules in wool are quite porous, which helps explain why wool can absorb up to 18 percent of its weight in water without feeling the least bit damp. It won't become completely saturated until it has absorbed at least 50 percent of its weight. Interestingly, as the wool takes on moisture a chemical reaction also occurs that produces heat.

Wool fiber

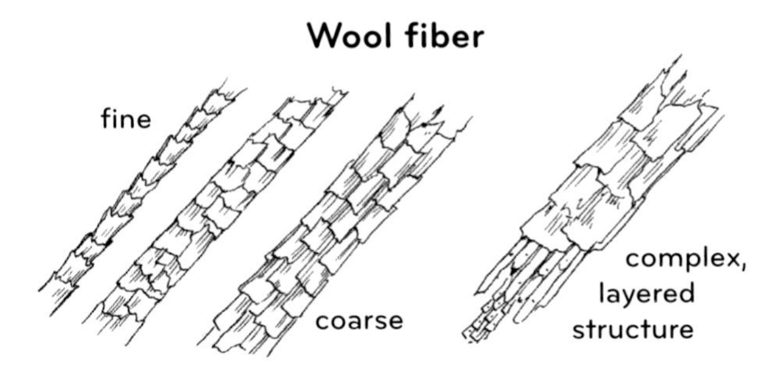

Fibers vary by breed and individual. Some are very fine, while others are much coarser. They are also all complex in structure, as suggested by the drawing on the far right. Some fibers, most often the coarser ones, have a central core, called a *medulla*, that is most often absent in fine fibers.

The term *wool* can mean the fibrous covering of a sheep in general, but it also refers to a specific type of fiber found in a sheep's fleece. Wool fibers are relatively fine and known for their crimp and elasticity. (Even wool fibers identified as coarse have diameters that are about half the size of the diameter of human hair.) In addition to wool fibers, sheep can have hair fibers, which are straight, smooth, and inelastic. These hairs are not only coarser than wool fibers, but also stronger. Some fibers are also called medullated (or med fibers), because they have an additional central structure called a *medulla*. The word *kemp* refers to the heaviest and coarsest of all hair fibers. Kemp fibers are often much shorter than the other wool and hair fibers in a fleece, although they may also grow to be quite long (see the South Wales Mountain sample on page 221), and they have a particular fault that can ruin certain projects: They don't display dyed colors well. They are also prickly and best avoided in projects that will be worn near the skin. Kemp isn't all bad, though. Its very nature of odd dye absorption is sometimes useful, as in the production of true tweeds.

Shepherds who are serious about fiber production for handspinning usually don't have kempy fleeces. They choose breeds without much kemp to begin with, and they make breeding selections to minimize it. But sometimes inexperienced producers, or those who make breeding choices based on other factors, may shear fleeces that contain quite a lot of kemp. Also, some of the coarser-fibered and older breeds that we really love, such as the Karakuls or the Scottish Blackfaces, may naturally have lots of kemp. Kemp has a short growth cycle and does tend to shed, especially in the spring, so by waiting a little later to shear, a shepherd can get a fleece with less kemp in it.

Sheep are not the only animals to have both hair and wool-type fibers (often found as a soft undercoat or down on other critters), but the term *wool* is more commonly applied to sheep than to other animals. Whether the fiber is called wool, down, or hair, it grows from follicles. Follicles are located in the epidermis, or the thin surface layer of the animal's skin, but as the fibers grow, they force their way both up out of the skin and down into the deeper dermis layer.

There are two types of follicles: primary and secondary. Primary follicles can produce all three types of fibers, while secondary follicles produce only true wool in sheep, or down fibers in species such as alpacas and yaks. Secondary follicles don't start producing until a couple weeks after birth, so babies typically have a higher percentage of med and kemp fibers at birth.

All animal fibers have an outer layer called the cuticle, which has scales like a fish's, and an inner layer called the cortex. As we've mentioned, in some fibers there is also a central core inside the cortex called the medulla. Kemp fibers, which can be as coarse as human hairs, can have as much as 60 percent of their diameter taken up by the medulla. The cortex is what gives wool its strength, elasticity, and durability. The medulla, when present, is spongy.

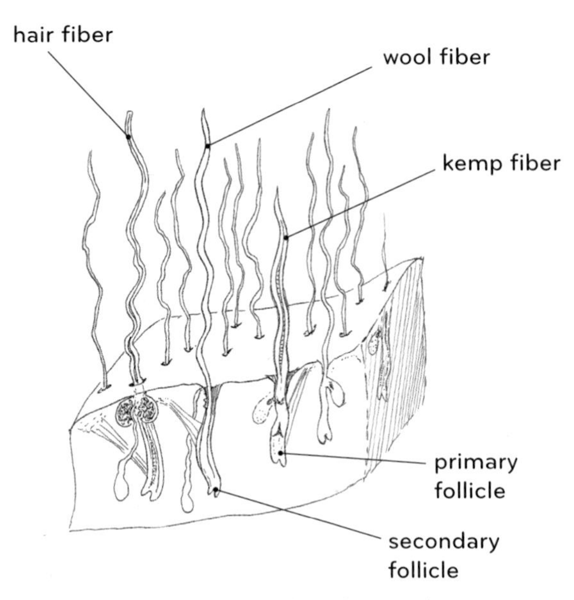

A cross section of sheepskin, showing the three types of fiber: wool, hair, and kemp. All fiber growth originates in the skin.

Examples of differences in *crimp*, or the wavy structure in the fibers. Crimp can be fine, broad, or anywhere in between and can be evident in individual fibers, in the locks, or in both.

Systems for Measuring Fiber Diameter

One of the confusing elements about the fiber world is that there are a variety of systems used for describing the fineness (or coarseness) of fibers: The Bradford count (or spinning count), blood count, micron count, and USDA grade systems are all approaches used to describe the average diameters of fibers in a selection of wool, whether that's a portion of a single fleece or a bale comprised of a whole flock's output.

Bradford Count. The Bradford count originally meant that 1 pound (0.5 kg) of a particular type of wool could be spun into that many hanks of wool (a hank is 560 yards or 512 m). So 70s would yield 70 hanks, and 60s would yield 60 hanks. The *s* stands for *singles*, so Bradford counts reflect the finest strand of yarn that could theoretically be spun from the fiber. The Bradford count, which is always expressed in even numbers, usually went only as fine as 80s, yet many Merino-family fibers have been known to grade 90s.

For many years, wool classers could order sample kits from the U.S. Department of Agriculture that contained locks representing the different Bradford grades. They would use these to compare to wools they were classing, and to train their eyes and fingers to determine what grade a particular fleece would fall into. Bradford counts were subjectively determined; of course, no fleece is completely consistent, so the wisdom and tactile sensitivity of the classers were critical to the effective operation of the system.

Although the Bradford system is gradually being replaced, as technology offers increasingly portable and practical alternate ways to obtain more objective measurements, it's still commonly used and referred to.

Blood Count. This system of grading came into use in the United States during the early 1800s. During this period, Merinos were being imported and crossbred with common sheep. The blood system indicated what fraction of the sheep's blood was from the Merino breed. Results were reported as fine, ½ blood, ⅜ blood, ¼ blood, low ¼ blood, common, or braid (for very coarse). Today, the blood count is the least commonly used system, and the designations are no longer restricted to Merino or part-Merino bloodlines. Instead, the terms represent the relative placement of fiber diameter on an arbitrary scale that refers to the Merino-influenced standards.

Historic Bradford

For centuries, the primary descriptor of fiber fineness was the Bradford count, named for the city of Bradford in northern England. Bradford has a long and important history in wool production. The city was founded in an area with a peat-type soil that resulted in very soft water — ideal for washing wool. In 1311, Bradford boasted a fulling mill, or processing center, that washed (and finished by lightly felting) woolen cloth. By the mid-nineteenth century, Bradford had become the center of woolen production in England, and was dubbed the Wool City. Today Bradford is a cultural center with a thriving tourist industry, and though wool is no longer processed there, it hasn't forgotten its past. The Wool Exchange building is an architectural gem of Victorian grandiosity, and an 1875 mill houses the Bradford Industrial Museum.

Micron Count. The micron system uses a laboratory test to measure the average diameter of the wool fibers: The larger the number, the coarser the wool. It is most often used by commercial buyers purchasing large quantities of wool. However, as the information is becoming easier and less expensive to obtain, fibers for handspinners are sometimes described with micron counts. Micron counts can be extremely useful — and extraordinarily misleading. Many people who refer to micron counts forget that wool is a natural substance, and that it varies, not only from one sheep to the next, but also for a single sheep, from one year to the next, and even along the length of a single fiber!

Micron count is not the be-all and end-all of fiber descriptors. For example, it's easy to think that two 28-micron fibers of a given length, say 4 inches (10 cm), would feel and behave the same. Yet they can be as different from each other as two pieces of paper that are cut to the same size and weigh the same amount — but one is a slick magazine-type paper, and the other is a natural rice paper. Your fingers would know the difference, even if the measurements used to describe them are identical.

In the old days, people didn't have the technology to actually measure fibers, so they depended on educated fingers. Then it became possible to measure by casting the enlarged shadow of a fiber on a wall and matching its width to a V-shaped reference point. Measurement technologies have become increasingly sophisticated since then, and we now have instruments that permit fast, thorough, computerized analysis of hundreds of fibers in a sample and allow us to average the results.

It's important to remember that there are no direct translation tables between Bradford counts and micron counts, while there *are* precise ways to translate from micron counts to the next classing system, the USDA grades.

USDA Grade. Once accurate lab measurement of fiber diameters became possible, the U.S. Department of Agriculture (USDA) assigned specific micron counts to the numbers previously associated with Bradford counts, to produce what are now called USDA wool grades. These USDA grades are still reported using the same types of numbers that were used for the Bradford system. The USDA was attempting to make the subjective Bradford count into an objective analytical system, which seems

Overview of Fiber Counts and USDA Grades

subjective	**Bradford counts**		Finer than 80s	80s	70s	64s	62s	60s
subjective	**General descriptions**			Fine				Medium
subjective	**Blood system**			Fine				½ blood
objective	**Microns**		less than 17	18 19	20	21 22	23	24 25
objective	**USDA wool grades (grease)**		Finer than 80s	80s	70s	64s	62s	60s
objective		average micron count	<17.69	17.70–19.14	19.15–20.59	20.60–22.04	22.05–23.49	23.50–24.94
objective		standard deviation	3.59	4.09	4.59	5.19	5.89	6.49

like a good idea, but since they used the same numbering system, you don't necessarily know which system was used. Is that 56s a Bradford count or a USDA grade? Unfortunately, it is hard to know, unless the source you're looking at specifically says Bradford or USDA. Where we give numbers in the style of Bradford or USDA grades, we generally call them spinning counts. This acknowledges the fact that people still find these designations useful, although our information sources were frequently unclear about which of these systems they were using. In a few cases, when a source did specify, we have kept the original indicator. Unfortunately, the confusion gets worse:

- **Decimal fragmentation.** The micron counts with the Bradford-appearing numbers are oddly decimal-fragmented; so a USDA 56s has an average fiber diameter ranging from 26.40 microns to 27.84 microns. It would have been nice to have even numbers, or even a single decimal place! (Grades assigned to mohair are much more logical in their breakpoints, although the way USDA numbers are assigned to that fiber create their own layers of confusion, described on page 338.)
- **Standard deviation.** On a USDA 56s, the fiber mass (individual fleece, bale, or whatever) can have a standard deviation or consistency (throughout the pile of wool) of plus or minus 7.59 microns. If the standard deviation or consistency of fiber diameter exceeds 7.59, the fleece is classed down a grade — in this case to 54s. The standard deviation is different for each grade; less variability is allowed for the finer grades.
- **Grease vs. clean.** If that's not befuddling enough, micron counts are different for grease fleece and clean wool — not a lot, but some. Bradford was always based on grease fleece. USDA grades may be based either on grease wool or on scoured (cleaned) wool.

Someday it will all be micron counts, and life will be easier — if less romantically complicated by lore — because the new technology for micron-count evaluations is fast! Accurate! Individualized! But remember that micron counts aren't all you need to know about a fiber, and your best source of information is still your own fingertips and your own judgment, as someone who loves and understands fiber. Feel it. Use it. Evaluate your results. Go feel some more.

58s	56s	54s	50s	48s	46s	44s	40s	36s	Coarser than 36s
	Medium				Coarse			Very coarse	
3/8 blood		1/4 blood			Low 1/4 or Common			Braid	
26	27 28	29	30 31	32	33 34	35 36	37 38	39 40	41+
58s	56s	54s	50s	48s	46s	44s	40s	36s	Coarser than 36s
24.95–26.39	26.40–27.84	27.85–29.29	29.30–30.99	31.00–32.69	32.70–34.39	34.40–36.19	36.20–38.09	38.10–40.20	40.21+
7.09	7.59	8.19	8.69	9.09	9.59	10.09	10.69	11.19	

Variability

Unlike synthetic fibers, which are thoroughly consistent unless purposefully inconsistent (as when they are used to make some types of novelty yarns), animal fibers show tremendous variability by their nature. The beauty, luster, elasticity, and strength of the fiber are affected by environmental conditions during its growth period, and by how it is handled after it has been harvested. For example, if a sheep has suffered from poor nutrition or been sick during the year, its fiber may have weak spots where it breaks easily. This variability is often surprising to people who don't work with the raw products regularly.

We have discussed the ways fiber is measured, but this variability creates some limitations that even the best technology can't overcome. While a full computerized analysis of a fiber sample contains abundant, fascinating, and useful information, we need to remember that labeling a fleece with a specific micron count is ultimately no more precise than subjectively assigning it a Bradford count. Wool is a natural, not a manufactured, product. The diameter of the fibers can vary (a little or a lot) throughout the fleece and even along the length of a single fiber. The true evaluation of a given sample will always need to be made by a set of human fingers, determining whether the fiber that's available will meet the perceived need.

In fact sheets throughout the book, we give standard numbers for staple length and fiber diameters for each breed and species. But please remember that within each group, every animal is truly an individual. Sometimes animals can produce significantly finer or coarser (or longer or shorter) fibers than their kind is known for. Some breeders select for these differences. Sometimes the quality of the feed or the climate where the animal is raised is a factor; other times the differences are unexplainable.

Wool Allergies and the Prickle Factor

Have you ever put on a wool sweater and felt itchy? It happens to all of us at times, though some people are more susceptible to the "prickle factor" than others. Folks will often say they have a wool allergy, but in fact very few people are truly allergic to wool. Think of it this way: Wool fibers are essentially the same as human-hair fibers; how many people do you know who are allergic to hair? More often, a negative reaction is caused by another substance that comes along with the fiber — possibly the lanolin. Or sometimes there are traces of grasses or seeds in the wool, or perhaps the chemicals used in its processing. In angora from rabbits, the bunnies' saliva may also be an allergen; once that has been thoroughly removed through washing, people often find they are just fine with the fiber. Others may have difficulties from inhaling fine particles, which may be more of a problem with commercially processed, rather than hand-processed, fibers.

When it comes to lanolin, some breeds produce lots of it, while others produce very little. A bit greasy, it protects the wool and keeps the sheep's skin smooth and supple (which is why it's often used as a natural base in cosmetic creams and lotions). A true lanolin allergy doesn't just cause itching; there is a severe rash (usually on the face, hands, and arms) accompanied by swelling and redness. This reaction can occur within a few hours or days of exposure and can last for several days.

Most wool from large-scale producers is chemically cleaned before spinning, then it is often treated with chemicals to repel moths and resist fire, and finally it may be dyed with other chemicals. The cleaning process includes several steps, involving alkaline baths to remove lanolin and acid baths to remove vegetable matter, such as burrs and hay chaff. These chemical baths are carefully balanced to minimize damage to the wool. Yet all these chemicals are harsh, and they may pummel the life out of the wool, or they may cause true allergic reactions in some chemically sensitive individuals.

So if only a few people have allergies, why do the rest of us sometimes get that itchy feeling? The answer is the "prickle factor," which is related to the diameter of the individual fibers used to produce the yarn. The micron count for this varies, depending on who's looking; anything over 28 microns in diameter is considered prickly by most people, though the official cutoff point is usually 30 microns. Larger fibers don't bend easily, and when the tiny ends of individual fibers poke out of the yarn or fabric, they produce a prickling sensation. When more than 5 percent of the fibers in a yarn have a micron count greater than 30, the itchy effect becomes readily noticeable to most people, especially if there is direct contact with sensitive skin, such as around your neckline or wrists. Some people get prickled by much-finer fibers, and some can stand hair shirts.

Chaffy (left) and clean wool from the same Oxford fleece. See the tiny specks in the chaffy sample? Broken-up vegetable matter affects the quality of the finished yarn.

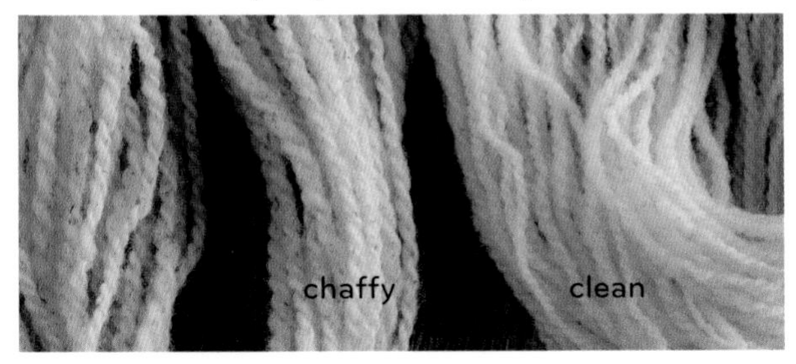

The way the fiber is spun and made into cloth can have a major effect on the perceived prickliness as well. Yarns that are spun with all of the fibers lying parallel to each other (worsted style) will have fewer ends sticking out and will not feel as prickly. Yarns where the fibers lie in a helter-skelter arrangement (woolen style) have both ends of each fiber potentially right at their surface, which could feel pricklier.

Any decision about what fiber is appropriate for use next to the skin depends on the type of fiber selected, the way it is prepared, and the sensitivity of the individual who will be using the fabric, as well as the way in which it's used. For a camisole, the finest fibers will be appropriate. For a cardigan, which may directly contact the body at only the back of the neck and the wrists, moderately fine fibers will probably be fine.

On the Fiber and Yarn Samples

For many animals, we obtained raw fiber that I processed and spun in an effort to discern subtle differences between fibers that on paper seem interchangeable. Later we decided to photograph the results. When we show multiple locks, they suggest variations in lock shapes, crimp patterns, and colors. Both raw and clean locks are often included, to demonstrate shifts in color, as well as in apparent length and crimp patterns. All samples not in boxes are shown at just over two-thirds of their actual size. (That's how we got the Rough Fell to fit on the page.)

These are "first take" yarns. I used a Lendrum single-treadle wheel, peasant combs, Louet mini-combs, hand carders, dog-grooming tools, a nostepinne, and a small niddy noddy. I spun just a few yards and took notes. I relied heavily on the two-row peasant or mini-combs, because they were fast to use. One fiber was spun on a takli (vicuña) and one on a high-whorl spindle (the finest Shetland).

The samples hint at possibilities and are only a beginning.

– Deb

Fiber in the Marketplace

Two traditional phrases recur in descriptions of wools: *carpet wools* and *hosiery wools*. These terms often contrast the fibers to the fine wools of Merinos and other finer-wool breeds, and they imply (unjustly!) that the so-called carpet and hosiery wools are inferior to the fine wools. Not so at all! In fact, there are many jobs that carpet and hosiery wools perform much better than the fine wools.

Carpet wools. Coarse, strong, and often lustrous, carpet wools are usually spun for a smooth surface, or worsted style. They may make sturdy outerwear, but you are unlikely to want to wear them next to your skin. They're fantastic for making hard-wearing rugs but are superb for a lot of other creative uses. Think of them when you want a tote bag that will be hard to wear out, a strap that won't stretch out of shape, upholstery or pillows that will still look great after the kids have grown up and quit throwing them at each other, or a well-constructed tapestry or other decorative textile.

Hosiery wool. Fibers that end up tagged as hosiery wools often come from Down breeds (see page 67) or other sheep that produce springy, elastic wools that are often matte rather than shiny. For *hosiery*, you could substitute the word *socks* and expand the application for this type of wool into sweaters, blankets, hats, and other general knitting — not the luxury items, but things you use every day to stay warm. These fibers are often spun woolen style, with the fibers evenly separated but going in all directions. This makes for a lot of insulating air space in the yarn and increases their coziness factor. For babies or lingerie, you'll probably want the finer wools, but many fiber applications need the extra durability provided by these workhorses of the textile world.

What's in a Name?

Many fiber names imply one thing but mean something else. Let's look at a few:

The term *Falkland* refers to wool grown on the Falkland Islands off the coast of Argentina, but there isn't a Falkland breed of sheep. The Falkland Islands are home to a significant number of Polwarths (an Australian breed developed from a mix of 75 percent Merino and 25 percent Lincoln; see page 289) and a fair number of purebred Merinos. There's also a strong history of Corriedales in the islands, as well as some Romney in the background. Falkland shepherds manage their flocks to produce increasingly heavy clips of ever-finer, high-quality fibers with low amounts of vegetable matter. Even though there is no specific breed name applied to their results, their work, and the overall quality of the wool, is exceptional.

There are no known sheep diseases on the islands, so the living animals don't go through the chemical dipping that occurs in other areas of the world to control pests. This, and the islands' climate, which minimizes bacteria and other factors that can shift wool color, may explain why the fleece from the Falkland Islands tends to be particularly white. Also, due to the cost of importing chemical fertilizers and herbicides, Falkland farmers never turned to the use of these additives; thus their wool meets organic standards. Fineness ranges from 18 microns to 33 microns. Staple lengths are most often 3 to 4 inches (7.5–10 cm), with the fleece showing good bulkiness and soft handle.

The word *guernsey*, while used to describe some yarns, also does not refer to a breed or a specific type of wool. Guernsey is the name of an island in the British channel. This island is part of the Bailiwick of Guernsey, which is an autonomous British Crown Dependency, and while it is known for its cattle and goats, it has no sheep. Nonetheless, you may hear of Guernsey yarns. Why? The island is also the home of many working fishermen, and "guernsey" refers to a five-ply yarn often used to make traditional fishermen's sweaters. The word also applies to the style in which those sweaters are knitted. The first usage is similar to when the term *worsted* refers to a weight of yarn.

Here's another interesting area for potential confusion. The well-known company Rowan produces a lovely series of British Sheep Breed yarns in its Purelife line. Although these yarns are 100 percent British wool, as far as we can determine (through inquiries and our own experience with wools), the finished yarns may not be individually breed specific. As the company's marketing materials state, the yarns are spun from wool that is "shorn and blended from four classic British sheep breeds," namely Jacob, Black Welsh, Bluefaced Leicester, and Suffolk. These wools are ultimately packaged as five yarns that have been named after the contributing breeds: the off-white yarn is called Bluefaced Leicester; the very dark brown is called Black Welsh; the dark gray is called Dark Grey Welsh; the light gray is called Steel Grey Suffolk; and the gray-brown is called Mid Brown Jacob.

Our only caution is that when you pick up this yarn, don't take its amalgam of qualities as representative of what you will discover if you obtain samples of the individual breeds. The Bluefaced Leicester yarn displays that breed's suppleness and spring and contains more luster than the other colors. The other yarns have a notably different texture from the Bluefaced Leicester, and they are much more similar to each other than handspun yarns from the breeds will be. The Black Welsh yarn is a very dark brown, whereas the breed is known for its unusually clear black fiber (see page 207). Gray Suffolks are quite rare; almost every sheep of the breed grows white, springy wool (see page 80). Because of the way industrial quantities of wool are graded and bundled in the marketplace, we (and Rowan) can be sure that these yarns have been spun from an assortment of classic British breeds. At the same time, we can't be certain that each ball of yarn consists of wool from the breed whose name it bears.

These yarns are wonderful wools produced in natural, undyed colors, and by designing and manufacturing them, Rowan gives us great material to work with while doing a fine job of supporting the British wool industry. If you want to understand breed-specific wools, use these great yarns as a jumping-off point and not a source of your final opinions.

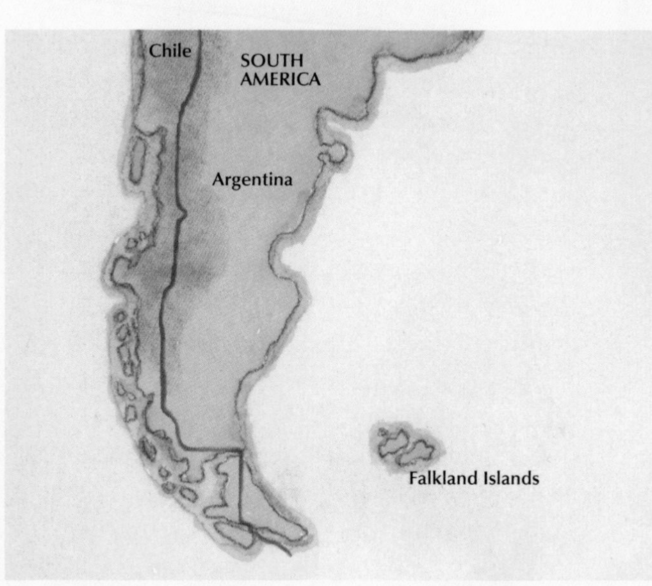

Shearing day. Fiber is most often removed from the animal by shearing. Here, the shearer's goal is to gently immobilize the sheep while removing the fleece in a single piece. The harvested fiber is evaluated, then placed on a rack (shown at right, next page), where the dirtiest parts are removed, or skirted off.

From Animal to Yarn

Most animals are sheared to harvest their fiber. In some cases, such as with some of the primitive sheep breeds, the fiber is plucked — much like you might pull clumps of loose hair from a shedding dog. In other cases, such as harvesting the exquisite qiviut from Alaskan musk oxen (see page 400), the fibers are combed out.

Small producers who intend to spin their own fleece may simply roll up the shorn mass of fiber, which often comes off the sheep in a single unit that holds together if treated gently, until they are ready to use it. If it is nice and clean, they may spin in the grease, or they may opt to wash the wool first. Some will prepare the wool into batts, rolags, or top, while others will do just a bit of light combing or teasing.

On the commercial scale, things are a bit different. Once the fiber is sheared, the farmer and the shearing crew bundle it into bales for shipping to the factory. A bale can weigh up to about 450 pounds (204 kg) and can contain the skirted fleeces of as many as 60 animals. The mass of wool is compressed into a large bag for transport to the industrial site. What that bag is made of is important, as is the type of fiber it contains. For instance, any stray foreign fibers (say, from the bag itself) are considered contaminants, as are any off-color fibers that may have been grown by the sheep just shorn or may have drifted in from a nearby pen. The entire batch of wool can only be considered as fine as the coarsest wool that it contains.

When the bale arrives at the factory, the wool will be scoured to remove dirt and may also be put through a precisely monitored chemical process called carbonization. This involves the application and then neutralization of acid solutions and helps eliminate especially stubborn bits of vegetation. After carbonization, which turns the scattered vestiges of plant matter into crisp and brittle flecks, the fiber is run through rollers that crush that treated vegetation into small pieces that just fall out (see the sample of Oxford wool, page 77, for an idea of the value of carbonization). Next the fiber goes through a carding process that disentangles the individual fibers and removes more vegetable matter that may have contaminated the wool, whether or not it has been carbonized, such as tiny bits of hay or seeds. The fiber emerges from this process as an untwisted length called sliver (pronounced SLĪ-ver, with a long I in the first syllable).

The next step varies a bit depending on the type of yarn to be produced:

- **Worsted yarn** is smooth, strong, and long-wearing, and used for the manufacture of wool cloth. To make a worsted yarn, the carded fiber goes next to mechanical combs. The combing process gets all the fibers fully parallel to each other and removes any *neps* or short fibers that survived scouring and slivering. It also removes the tiniest bits of vegetable matter that may have survived carding.
- **Woolen yarn** is the softer and loftier yarn typically used for knitting and crocheting. Woolen yarn may skip the combing process and go directly to the spinning frame, or it may be lightly combed.

Depending on the type of raw wool, how clean it is, and what the desired product is, the fiber may go through some additional steps:

- **Picking.** A picker is used before carding to remove large pieces of vegetation and open up the fiber locks so that they feed more readily through the carding machinery.
- **Dehairing.** If there are guard hairs, the raw fiber may go through a dehairing unit.
- **Dyeing.** After scouring and before spinning, the fiber may be dyed, though the exact point in the process where color is applied varies depending on the factory's design and the desired outcome for the dye.

Finding the Real Thing

Spinners can usually acquire breed-specific fleece, roving, or top and create their own yarn, but many fiber enthusiasts don't have the time, the knowledge, the tools, or the urge to spin their own yarn. Luckily, many breed- or species-specific yarns are available for sale. We've purchased a selection of yarns for illustration, but we found far more than we had time to gather or show you. Some of the yarns available to you may be commercial in the common sense of the word — produced in some quantity, by machine, and not too hard to find. Others may be available as handspun skeins or as skeins produced for individual farmers at small-scale custom mills. These may take a bit of dedication to locate, but they are worth the effort to seek out.

Of the purchased skeins we added to our stashes during the research on this book, a

Fiber shopping spree. We purchased the breed-specific yarns below in one afternoon at a fiber festival. They represent a broad variety of types of wools, from left to right (and finest to strongest): Merino; Cormo; a Suffolk/Dorset blend; Bluefaced Leicester; Romney; Wensleydale; Coopworth; and Leicester Longwool. In the following pages, you will find plenty of breed-specific yarns that we purchased while researching this project. You don't have to spin your own to experience breed-specific fibers, and the more we support the market the more choices we will all have.

number came from international sources. Globalization takes on a different feel when you are buying from farmers, small-scale co-ops, and artisans who are dedicated to their trade. This is the best of fair trade, and it is really exciting that the Internet opens the door for us to purchase from people across the state or around the world. In one Sunday-afternoon experiment, we searched the Internet and found we could purchase skeins of finished yarn for about 70 percent of the breeds listed in the book! If you add yarn from breed-specific crosses (animals) or breed-specific blends (fibers from individual breeds, blended during processing to make yarn), the number jumps to about 80 percent.

If you are a spinner looking for raw fleece rather than yarn, you may be intrigued by some of the breeds that we mention that are not grown in your region — or even in your country or on your continent! Occasionally you'll find wools from outside your area being imported by vendors who have booths at fiber festivals, or offered by mail-order and Internet suppliers who market primarily to your part of the world. However, fleece, like yarn, can also be purchased internationally. The British Wool Marketing Board offers hand-selected fleeces for sale to handspinners during the shearing season. We also purchased raw fibers and yarns from eBay, Etsy, and Local Harvest, as well as from the websites of individual shepherds and custom mills. When it comes to eBay, check out the eBay sites in other countries (listed on the bottom of the eBay home page) as well as those from the country you live in.

Overall, our Internet shopping experience was successful and satisfying: We found some absolutely marvelous fleeces, yarns, and products through these sites. But we also had a couple of disappointments. In those cases, the person selling the fleece wasn't the shepherd or even a real fiber enthusiast, and we hope the person who sent us what was, in at least one case, just trash, didn't know enough to realize that the fleece was poor quality. We corresponded with this seller, who was quite apologetic and offered to refund our money. If you do get a poor-quality fleece, take the time to let the supplier know.

We especially want to encourage you to look at some of the products that are available in the global marketplace that are constructed by people in other parts of the world from breed-specific or other identifiable natural fibers. We had great fun shopping for some of these specialty items, which adorn the pages that follow.

Experiencing (and Saving) the Real Thing

We like to think that as our readers seek out breed-specific fibers, yarns, and finished products, these will become even more readily available. As we all learn more and take time to obtain these special textile materials and fabrics, an important side effect is that the people who are raising the animals will benefit from the additional support and will feel our encouragement of their invaluable work in caring for their critters. In order to continue to have these irreplaceable resources available for our pleasure and delight, we need to support the living infrastructure of animals and people that makes their existence possible.

Enhance your life and other folks' lives, too. Adventure with us into the amazing world of the natural fibers grown by animals large and small, wild and tame. There's so much to learn here. We're still questing: The editors made us stop researching and print the book so you could share in some of our discoveries. If you learn something we need to know (we had to settle for a version of excellence here, rather than holding out for perfection), get in touch. We'd love to learn even more.

Terms to Know

If you come to this book from the yarn-user universe, there may be some terminology that isn't as familiar to you as it is to the shepherd or handspinner. Here's the skinny on the words that might be new to you. A note on yarn types: As is usual with everything in fiber, there are many hybrid types of yarns and a nearly infinite number of styles of preparation and spinning. Woolen and worsted are two of the most common types, but they are not the only game in town!

Badger-faced. This term refers to a pattern of coloring on the body of an animal. Although it mentions the face, the whole body is involved. The animal is primarily light-colored, with dark markings not only on the face (around the eyes, on the muzzle, and stripes down the cheeks) but also on the underbelly, legs, chest, and under the tail. There is also a reverse badger-faced pattern, in which a dark-colored animal has light markings in the same locations.

Batt. A form of prepared fiber, ready to spin or as an intermediate step in getting ready for spinning. Fibers are arranged with relatively even density and somewhat helter-skelter directions. Batts are airy. An old-fashioned quilt batt is an example. Batts are most often prepared on carding equipment, for example a drum carder or hand carders.

Bulky. Wool that is described as bulky feels full in the hand, incorporating a lot of air for its mass. Bulky wools can be spun into yarns that are lightweight for their size. The term can be confusing, because the word is also used to describe a specific size of yarn (the Craft Yarn Council of America's yarn weight category 6). We have substituted the description *high-bulk* in our discussions of wool qualities.

Card. Carding is a means of separating locks and clumps of fiber into thin, even layers prior to spinning. The process is used somewhat differently in industry than in handspinning, but in both cases the goal is to end up with a maximum amount of air between fibers. In handspinning, the fibers are arranged in different directions, so they are not parallel to each other, and they are spun directly from that preparation. In industry, the fibers end up headed mostly in the same direction, just because of the way the equipment works. Commercially carded fiber may go through an additional combing process, which enforces the parallel arrangement and removes shorter fibers, resulting in a more homogeneous mass of wool. Traditionally, carded fibers are spun woolen style (see Woolen) to produce yarns that emphasize the fibers' softness and warmth (enhanced by the insulating air spaces).

Card clothing. Also called carding cloth, card clothing is a rubberized fabric with stainless-steel teeth embedded in it. It is applied to the wooden "paddles" of hand carders or to the cylindrical drums of drum carders. This part of either tool set makes the carding action take place, much as the teeth of a saw do the cutting, whether a hand saw or a more industrial saw.

Chalky. Wool that is chalky has a chalklike white surface, noticeably without luster.

Clip. This word has a couple of meanings. Sometimes it's almost synonymous with *shearing*; when people talk about first-clip wool, for example, they mean the animal's first shearing. Most often, however, a clip is all of the fleeces shorn at a single time. If breeders are looking for a heavy clip, they want a greater mass of wool than they have previously obtained, and their goal can be an increased yield from either the individual sheep or the flock as a whole.

Rolag
Peasant combs
Hand carders
Lock
Peasant combs
Flicker
Mini-combs
Top, commercial
Top, hand-combed

Combing. In industry, fibers are sometimes carded and then put through a second (combing) stage that removes short fibers and arranges the remaining fibers (of similar length) parallel to each other. In hand processing, locks of clean wool are taken directly from the fleece and combed, without the intermediate carding step. In both cases, the end result of this orderly arrangement of fibers is smooth-feeling yarn that is dense (because it contains a minimum of air). Traditionally, combed fibers are spun worsted style (see Worsted) to produce yarns that emphasize the fibers' durability and smoothness. Hand combing of fibers can be carried out with a variety of different tools, most often used in pairs. These include: dog combs (inexpensive and readily available but not speedy or thorough); mini combs (small, handheld combs especially useful for fine fibers); Viking or peasant combs (also handheld, although one comb may be fastened to a table with a mounting device, and versatile for a variety of fibers); English wool combs (heavy-duty and heavy, equipment for thoroughly processing large quantities of fiber).

Crimp. Refers to the natural kinks, waves, or bends in individual fibers. Crimp in natural fibers occurs as part of the growth process. Sometimes the fiber waves line up across bunches of fibers, so there is crimp in the lock as well. Sometimes the waves are small and close together, and sometimes they're larger and more spread out. Sometimes they are very organized, and sometimes they occur quite randomly. In addition, crimp can be two-dimensional or three-dimensional. (See photo on page 100.)

Draft. Between fiber preparation and the formation of a strand of yarn comes a stage called drafting, in which the mass of combed or carded fibers is drawn out, drafted, or attenuated. The ends of the fibers nearest the spinning device (spindle or wheel) are pulled gently forward so they slip past each other, forming a long, slender mass of fiber. As soon as twist enters this mass, the fibers hold together and become a stable length of yarn. What happens in drafting determines the weight of the finished strand and its regularity. More fiber in the drafting area (sometimes called a drafting zone) means thicker yarn; less means finer yarn.

Elasticity. The amount of stretching an individual fiber can do without breaking and still return to its original shape and length when released.

Felt. A textile in which fibers have joined together so they can't be pulled apart. This usually happens through the application of heat, moisture, agitation, and a bit of soap, and it can be intentional or accidental. The fibers actually lock onto each other, so felting is a permanent condition. Some fibers felt readily, and wools are one of the best-known felting fibers. However, not all wools will felt. See also Full.

Fleece. The coat of wool from a wool-bearing animal, obtained by shearing.

Flick. Flicking is a method of opening out locks, so the fibers don't stick together and yet remain in the same relative positions in which they grew. This can be carried out in different ways. For one, hold each end of the lock in turn and flick its other end against a hand carder (or a similar but differently proportioned tool called a flicker). Or place the lock on your knee (with a protective layer of leather or cloth between leg and fiber) and flick the tool down on the lock. The interaction between the tool's teeth and the lock "bounces" the fiber into a looser configuration. This latter kind of flicking doesn't work well with a regular hand carder and is best managed with the specialized tool.

Full. To full a fabric is to wash it under conditions resembling those used to produce felt but to stop the process before felt is produced, at the point where the distinct separations between the component yarns become less obvious and the fabric feels more like a unified whole but has not undergone the full transformation to felt.

Grease. A combination of lanolin and suint (see Suint) in animal fiber. Some spinners prefer to spin "in the grease" and then wash the yarn. Others prefer working with clean fiber. Wool of some breeds lends itself more readily to spinning in the grease. Wools spun with heavy grease in place are sticky and reluctant to draft (see Draft), and they may end up looking grubby even after they've been washed.

Grist. A yarn's grist is its thickness, which comes from a combination of how much fiber the yarn contains and how tightly it's been spun.

Guard hairs. Coarse, water-repellent fibers that overlay and protect an animal's soft downy undercoat. Found in some sheep breeds, as well as in camels, musk oxen, some types of Angora rabbit, and some other fiber animals.

Heterotypic hair. A type of hair that changes consistency with the season, becoming more wool-like for warmth in winter and more hair-like for shedding rain in the summer. Because of heterotypic hairs, among other reasons, fleeces from some breeds vary in character depending on when they are sheared.

Kemp. A coarse, hollow fiber found in the fleeces of some sheep breeds. Kemp is brittle and scratchy, has less elasticity than other fibers, and doesn't take dye the same way. It either appears not to have taken the dye at all — although the dye molecules may be inside the fiber and obscured by the cells' opacity — or it displays a lighter version of the dyed color than the surrounding fibers do.

Lanolin. An oily or waxy substance that sheep exude from sebaceous glands in their skin. Yes, it's the same stuff that shows up in cosmetics, and playing with unwashed grease wool can nicely soften the skin of your hands.

Locks. Natural divisions in a sheep's fleece, composed of small clumps of fiber that hold together. When you grasp the tip of a defined grouping of wool in a fleece and gently pull, you can separate out a lock. Some breeds' fleeces have well-defined locks and others do not (see Staple). To spin a fleece from the locks means spinning it directly, without an intervening step of carding or combing. However, it's common to use the fingers to slightly separate the fibers of the lock so they move past each other more freely when they are being drafted (see Draft).

Loft. Refers to the airiness of a yarn, lock, or fleece. Higher loft means more airiness and generally warmer finished products.

Longwool. A category of wools without distinct boundaries. Although this is not a technical definition, longwools generally have staple lengths greater than 4 inches (10 cm). They also display clearly defined crimp with less frequency (that is, number of crimp waves per inch or centimeter) than the shorter wools, like the Down breeds and the fine wools, like Merino. Many, but not all, longwools also display luster, or shine, due to the comparatively large, light-reflecting scales that cover their surfaces. Wools with smaller scales are not as shiny because the edges of the scales fragment the light reflections. English Longwools are the standard to which all other longwools are compared, and are known for their especially dramatic luster.

Luster. The shininess of a fiber. Some fibers are lustrous (reflect light), some are not, and some are in the middle (semilustrous). The ends of this continuum — the very shiny and the obviously matte — are easy to discern.

Medullated fiber. Animal fibers always consist of a main body, called a cortex, and a surface layer of scales, called, as a whole, the cuticle. Some fibers also have a core portion called a medulla, which makes these fibers stiffer than nonmedullated fibers.

Molt. See Shed.

Mouflon coloring. Body coloring that resembles that of the mouflon, a mixture of brown and other colors, with a dark area running down the spine and lighter portions on the muzzle, underbody, and legs.

Nep or noil. A tangled knot of short, broken, or disarranged fibers. Bunched wads of neps can form slubs, or thicker and less twisted areas in the yarn.

Overdye. To dye one color over another, whether the first is a natural color or one that has been produced by a previous dyeing session.

Pick or tease. Picking or teasing fiber means plucking the locks apart to fluff them up and separate any stuck spots.

Pointed tips. The tips of locks have different and frequently characteristic configurations. Some are pointed, some blunt or blocky.

Rolag. This form of fiber is prepared for spinning with the use of hand carders. The carded fiber is rolled into a sausagelike form.

Roo. To pluck wool from a sheep after the fiber has loosened and is naturally shedding. Wool can be collected by rooing only from sheep that retain the primitive characteristic of seasonally molting or shedding their fleeces.

Roving. Washed and carded fiber that is ready to spin (see Top).

Scour. Washing wool can result in greater or lesser degrees of cleanliness. Generally, you want to remove anything that will get in the way of making the yarn you want, whether you are processing the wool by hand or with the help of equipment. Some people wash wool so it retains a portion of the natural lanolin. Scouring is the most thorough stage on the continuum of washing possibilities. It refers to the complete removal of all lanolin and suint. Scouring done well results in very clean fiber that can be easily processed by mechanical equipment. If you are a shepherd or spinner sending your wool to a small mill to be turned into roving or top, you will almost certainly need to scour it first, or the mill will do this job for you. Insufficiently scoured, gummy fiber can wreck machinery and is frustrating to work with, but overzealous scouring can damage fiber, causing it to become brittle or lose its luster. (Mohair is particularly vulnerable to damage in scouring; see page 340.)

Second cuts. When a shearer takes two passes with the shears or clippers and the cuts do not land in exactly the same location, small bits of wool may be snipped from the bases of the locks. These cling to the fleece and if not removed will make it difficult to spin a smooth yarn from the fiber. However, they can often simply be brushed away from the cut surface of a raw fleece, or they can be removed by combing.

Sett. Weaving term that refers to how closely or openly the warp threads (which are threaded on a loom to form the foundation for weaving) are spaced.

Shed. Just as some dogs shed a lot and others nary a bit, some breeds of sheep naturally shed or molt their fiber. At a particular season, the fiber loosens and lifts away from the skin on its own and doesn't need to be shorn. Shedding is a characteristic of breeds closer to the ancient types of sheep, rather than those developed through intensive modern genetic selection.

Skirt. Skirting is the act of removing undesirable portions of a fleece, such as the belly wool, the edges of the fleece (where the sheep lies down and gets the wool dirty), bits of manure, or any felted or matted areas.

Staple. Most often refers to the length of the individual fibers; the word is sometimes used in the same sense as *lock.*

Strong wool. A wool that has a relatively coarse fiber diameter and is comparatively durable. Strength in wools is a relative, not absolute, concept. For example, the English Longwools are strong when compared to Merinos, while at the same time some Merinos are stronger than others.

Suint. Salt that is given off in an animal's sweat. It mixes with the lanolin to make the grease that coats and protects the fibers.

Teased lock. A lock that has been lightly worked with the fingers to open the fibers, releasing any stuck portions.

Top. Ideally, clean fibers that are combed and aligned, with all vegetable matter, debris, and short or broken fibers removed. Classically used for producing worsted yarns. Fiber prepared for spinning as top can be processed by hand or by machine; hand-combed top is delightful and easy to draft (see Draft). We say that top ideally corresponds to this definition because the terms *top* and *roving* (all of which refer to fibers prepared in slightly different ways for spinning) are often applied incorrectly.

Twist. Holds fibers together to make yarn. Twist can go in one of two directions, producing either S-twist or Z-twist yarn. The direction of twist is determined by the rotation of the spindle or the wheel during spinning.

Woolen. In the context of Western European-based yarn construction, woolen refers to methods of preparation and spinning that produce yarns that incorporate a lot of air, are lightweight, and have high insulation qualities. The fibers are arranged in helter-skelter fashion within the yarn to maximize the air-to-fiber ratio. Carding is the fiber preparation that encourages this. Spinning techniques can be selected to preserve the yarn's loft.

Worsted. In Western European-based traditions, worsted yarns are prepared and spun to enhance the sleekness and draping qualities of the fibers. Combing is the fiber-preparation method that precedes worsted-style spinning techniques to produce these yarns. The term *worsted* can also refer to a particular weight of yarn, regardless of how it is processed or spun.

Yield. The yield of a fleece is the amount of clean wool left after the vegetable matter, grease, and other contaminants have been washed out or mechanically removed. A fleece with a 90 percent yield has almost nothing other than wool in it when freshly shorn; a fleece with a 50 percent yield came off the animal with a lot of lanolin, suint, and/or contaminants.

PART 1 Sheep: Oodles and Boodles of Wool

I think I could sing
and shear a few sheep at the same time.

— Robert Plant, Led Zeppelin singer/songwriter

If you were to travel the globe, judiciously noting what kinds of livestock people keep, you would see far more sheep than any other type of farm animal. China leads the flock with over 143 million head of sheep; Australia follows with more than 98 million.

By comparison, the United States has a paltry 7.7 million — a number that has been steadily declining as agriculture and food production have moved from small-scale farms to industrial production. Yet if you look for them, you'll find sheep in almost every corner of the country. A handful of large flocks in the western states and provinces accounts for the majority of the North American sheep flock, but small-scale farmers and backyard shepherds help maintain an amazing diversity of breeds, and lots of these folks got into shepherding in the first place thanks to their love of fiber. They may have only a handful of animals, yet these individuals and their families often concentrate on producing gorgeous fiber from happy sheep.

Britain may not have the most sheep, but the residents of these islands can reasonably lay claim to living in the sheep capital of the world. With a landmass equivalent to about 1½ percent of that of the United States (an area just slightly larger than the state of Minnesota), the British Isles have more than three times as many sheep as the United States (around 25 million) and more recognized breeds than any other country in the world. Thus it shouldn't come as a surprise just how great an influence British sheep have had on fiber development and production, and just how many of the breeds we write about here are of British descent.

Some blackfaced sheep, such as the Scottish Blackface (at left), belong to the Blackfaced family, while others, like the Suffolk (at right), are not part of this family, even though they have dark faces.

Blackfaced, Blackface, and Black Face

Blackfaced sheep have black faces because they are blackface breeds, but not all sheep with black faces are Blackfaced Mountain breeds. The usage of some sheep-related words presents conundrums to anyone who loves consistency, but breed names of sheep just aren't consistent. Nor are the spellings of the names for tools or fiber-shaping processes. We've carefully considered the spelling options, and there is method behind our apparent madness. If we call a breed *blackfaced,* we mean that it belongs to what we have defined as the Blackfaced Mountain family; if we say it has a black face, the color of its face is dark but it is not a member of that family.

Swaledale Sheep, Yorkshire, England

Sheep must be marvelously brave or stupid,
or carriers of some instinctual flaw,
to seek comfort from the cold night
in wintry wet walls of Yorkshire stone,
outcroppings in a lush hill,
mossed gray walls far above the lights of the town
where children play on the village green
and old men sip ale at the local pub.

While silent smoke puffs from scattered chimneys,
and warm barns, stacks of hay, privet hedges,
massed heather and clumps of blackberries
temper the sweetness of lower ground,
the sheep gravitate towards dark caves in the hills beyond,
frozen shoulders of rocks dripping icy shards,
and draw solace from bitterness or peaceful ignorance,
teaching us better than any Stoic philosopher
how to embrace the inhospitable.

— Donna Pucciani, reprinted by permission
First published in *MacGuffin* magazine (Fall 2004)

U.K. Means What, Exactly?

We Americans tend to think that *Britain*, *England*, the *British Isles*, and the *U.K.*, or *United Kingdom*, all mean the same thing. With so many islands and subdivisions, not to mention name changes throughout history, it's easy for outsiders to get confused. For the record, *Great Britain* refers to the largest island in the group, which includes the countries of England, Scotland, and Wales. Each of these three countries also includes many smaller islands. The *United Kingdom of Great Britain and Northern Ireland*, or the U.K. for short, is the preferred term for all territory under the same government. Ireland, also known as the Republic of Ireland or Éire, is a separate country. To make matters more confusing, it's officially okay to use the term *Britain* (not *Great Britain*) instead of the U.K., and some international codes still use G.B. instead of U.K. The term *British* refers to all citizens of the United Kingdom, including the Welsh, Scottish, English, and Northern Irish.

Changelings

SURPRISINGLY, there may be more variability between individuals within a breed than there is between breeds, and there are usually regional differences in flocks around the country or around the globe. We knew this before we started this project, but through our research we have come to appreciate just how dramatic these differences can be. Unlike many other books on fiber, which usually report information based on other published sources (including those that are generally the most reliable: the industry associations), we actually got workable samples for almost all of the breeds that are written up in this book. So the data you see reflect not only conventional wisdom but also our personal analysis of samples for the breed. We often received fibers that challenged our assumptions or contradicted published resource materials. Some were so outside the realm of our expectations for the breed that we sent samples to a laboratory to get a bit of scientific analysis, and yes, they were different indeed!

Since our samples consisted of measurements taken from a very small number of locks, the results cannot be used to make sweeping generalizations about the breeds. However, they gave us a way to objectively compare the wools we were holding to readily available information on the Internet or in other publications describing the associated breeds' qualities. For some very rare breeds, there is a dearth of published information, so our fiber and the analyses gave us at least one point from which to begin developing an objective understanding of that breed's fiber. As to the geographic differences, we can use Border Leicesters (see page 88) as a good example of the kinds of things we discovered. In North America, Border Leicester fiber is reported to typically range from 30 to

38 microns (spinning counts 50s–40s), while sheep of the same breed in the British Isles ease off the fine end of that span, at 29 to 32 microns (spinning counts 54s–48s). Those from Australia overlap the coarser portion, at 34 to 38 microns (spinning counts 44s–40s), and Border Leicesters in New Zealand go even farther in the coarse direction, at 37 to 40 microns (spinning counts 40s–36s). Yet these are all still Border Leicesters!

We frequently report a much wider range of stats than you might see in other books or articles. Our goal was to give a good estimate of what you're likely to find in the field. We gathered information from a variety of sources, plotted the ranges reported for each breed or species on graph paper, and looked for patterns. We compared this to our samples, and then we drew tentative conclusions. This is all a work in progress! And because we are studying a naturally produced resource grown by ever-changing and evolving populations of animals, the fibers shift even as we study them. You may find, as we did, examples that far exceed the range of specifications found in the current literature, or even found here. Natural fibers are not manufactured — they do not have to fall within precise tolerances — although we'll note when breeds are especially consistent, or inconsistent, in the fiber they produce, and why.

Geography is just one factor affecting fiber variability. The animal's age, nutritional plane, general health, and stage of production (such as nursing ewe, growing lamb, or ram during the nonbreeding season) all have an impact on fiber characteristics. Climate, soils, and cultural approaches to husbandry also play large roles in how animals and their fiber grow.

Hair Sheep

Sheep are usually thought of as being woolly, but there is a class of sheep whose members are known for *not* producing wool, or producing only a small amount of wool that sheds out (shedding being common among wild sheep). These are known as hair sheep. The breeds of hair sheep raised in North America include Barbados Blackbelly, Dorper, Katahdin, Painted Desert, Royal White, St. Croix, and Wiltshire Horn. A dedicated and talented spinner may be able to spin some of the wool that these sheep shed in the spring and produce some interesting yarns from their fiber, but for the most part, these animals are of limited interest to wool enthusiasts.

Katahdin

This mix of crimpy wool and stiff hairs was balky, but quite possible, to spin.

shown at 40% of actual size

A Starter Guide to Breed-Specific Wools

SOFT

For supersoft infants' garments, luxury items (camisoles, ultrasoft shawls, special socks), and people who think they can't wear wool

Bond*
Booroola Merino
California Variegated Mutant
Cormo
Corriedale lamb
Debouillet
Delaine Merino
Finnsheep*
Gotland lamb
Merinos (finer grades)
Polwarth*
Polypay*
Rambouillet
Romeldale
Santa Cruz*
Saxon Merino*
Sharlea
Shetland*
Targhee*

The perfect underlayer. Merino does some jobs exquisitely well. This Merino shirt, found in a Rocky Mountain ski shop, is perfect as an underlayer for sports activities. Yet Merino isn't an all-purpose wool, and other breeds' fleeces work far better in other applications.

Merino shirt

RELIABLY VERSATILE

For everyday sweaters, mittens, hats, knockabout socks, blankets — usually soft enough for most people to have in contact with skin, and the softer wools within most of these breeds are also suitable for many types of baby garments

American Tunis
Black Welsh Mountain*
Bleu de Maine
Bluefaced Leicester
Bond
Booroola Merino (stronger grades)
California Red
California Variegated Mutant
Clun Forest*
Columbia
Coopworth*
Corriedale
Debouillet (stronger grades)
Delaine Merino (stronger grades)
Dorset Down
Dorset Horn*
Dorset Poll*
Finnsheep
Gotland*
Hampshire*
Hill Radnor*
Île-de-France
Karakul*
Kerry Hill
Llanwenog
Lleyn
Manx Loaghtan*
Merinos (stronger grades)
Montadale
Navajo Churro
Norfolk Horn
North Ronaldsay*
Oxford*
Panama
Perendale*
Polwarth
Polypay
Portland
Rambouillet (stronger grades)
Romeldale
Romney*
Rouge de l'Ouest
Ryeland
Santa Cruz
Saxon Merino (stronger grades)
Sennybridge Welsh Mountain
Shetland*
Shropshire*
Southdown
Suffolk*
Targhee
Texel*
Welsh Hill Speckled Face*
Welsh Mountain Badger-Faced*
Whitefaced Woodland*
Zwartbles*

STURDY

For hard-wearing sweaters, blankets, pillows, bags, and other uses that benefit from increased durability

Black Welsh Mountain
Border Leicester
Clun Forest
Coopworth
Cotswold*
Dorset Horn
Dorset Poll
Gotland
Hampshire
Hill Radnor
Karakul
Leicester Longwool*
Lincoln Longwool*
Lleyn
Navajo Churro
North Ronaldsay
Oxford
Perendale
Romney
Shetland
Shropshire
Suffolk
Teeswater
Texel
Welsh Hill Speckled Face
Welsh Mountain
Welsh Mountain Badger-Faced
Wensleydale
Whitefaced Woodland*
Zwartbles

TO SPUR YOUR CREATIVITY

A lot of fun, and they'll push you out of your creative ruts, if you let them

Cotswold
Herdwick
Leicester Longwool
Lincoln Longwool
Manx Loaghtan
Navajo Churro
North Ronaldsay
Rough Fell
Scottish Blackface
Shetland
Soay
Teeswater
Wensleydale

Reading the yarn. The texture and luster of a breed's wool can be seen easily even in finished yarn. You can discern the effects of crimp patterns, fiber diameter, and other qualities more readily by touch than by sight. From left to right, these commercially spun skeins are Cormo, Bluefaced Leicester, and Wensleydale.

*Some grades of this breed's wool may be fine enough for the indicated use; others will not. Trust your fingers. This list is garment biased. If you want to make rugs, begin your quest in the "sturdy" and "creativity" columns.

Rough Fell
Scottish Blackface
Swaledale

Blackfaced Mountain FAMILY

The breeds of this clan, emanating from the hill country of northern England, Scotland, and Wales, are referred to as blackfaced mountain, blackfaced hill, blackfaced heath, blackfaced moor, or just blackface sheep. Today there are six recognized breeds in the family: Dalesbred, Derbyshire Gritstone, Lonk, Rough Fell, Scottish Blackface, and Swaledale. There are other sheep with black faces, such as the Suffolks and Hampshires (which belong to the Down family), but they are not considered part of the Blackfaced Mountain family.

Until the early years of the twentieth century, when flock books were started for the different Blackfaced Mountain breeds, these were considered regional landraces of similar heritage, influenced more by their environment than by human selection. Although the exact origin of these breeds is unclear, it is known that monks kept sheep of similar description more than 800 years ago. Both written documents and textiles from the period support the assertion that sheep in the Blackfaced Mountain family have been around a long time in the hills and dales of the United Kingdom. Most agricultural historians consider the Lonk and Swaledale breeds to be the oldest breeds, with Dalesbred, Derbyshire Gritstone, Rough Fell, and Scottish Blackface as their progeny.

As the name implies, these breeds have dark faces, but they all sport distinct white facial markings as well. They also share other similarities, such as rounded, protruding snouts (known in sheep as "Roman noses"). They're quite hardy, in response to the harsh environments where they developed. In fact, it was their very hardiness that kept shepherds from creating flock books and practicing pedigree breeding; the shepherds believed that pedigree breeding would lead to selection for appearance over productivity, thus making them less robust.

Scientists recently demonstrated the mechanism that helps these sheep withstand the often cold, wet, and bleak conditions of the hill country: In their bodies, the blood vessels near the surface constrict when exposed to cold, thus helping to retain their inner body heat.

The only breed from the Blackfaced Mountain family that is readily found in North America is the Scottish Blackface.

Marley, Me & Sheep

If you saw the movie *Marley & Me*, you may recall a scene where the characters played by Jennifer Aniston and Owen Wilson take a much-needed trip to Ireland, leaving the irascible Marley with a doggy sitter. During their Irish trip, the travelers are stopped by a flock of Blackfaced Mountain sheep blocking the road.

Dalesbred

Conservation Breed

THE DALESBRED HAILS from the Yorkshire Dales of northern England and a bit of Wales (an area not much larger than Rhode Island). Ninety-five percent of all Dalesbred sheep are confined to that small area. The animals sport distinctive white patches, suggestive of large teardrops and known as *smits*, on either side of their nostrils. The muzzle around each sheep's mouth is gray, and its legs bear black and white markings.

The Yorkshire Dales are known for 70-plus inches (178 cm) of rain each year, and that rain is often bitterly cold. Because of this climate, the breed's wool needs to be rugged enough to protect the sturdy sheep that grows it. Many Dalesbreds are *hefted*, or very strongly attached to the particular piece of land on which they were born. They also play a role in maintaining the ecological balance of the landscape in which they traditionally live.

The animal and the wool are both similar to the Swaledale (see page 49), another producer of carpet-quality wools, which can also be used to make tweedy or tough outerwear fabrics. Our Dalesbred sample spun up with more of a fine-wiry texture than we found in the Swaledale sample. If you made yourself a tightly knitted or woven jacket from one of the finer Dalesbred fleeces, you'd want to wear a softer layer underneath, but you'd probably feel well protected in rainy or snowy weather.

I See You

"Faces are highly emotive," says Keith Kendrick, a cognitive and behavioral neuroscientist at the Babraham Institute, Cambridge, England. "Even as very young babies, we find smiling or familiar faces attractive and comforting."

Having grown up around sheep, Kendrick decided they would make ideal candidates for research to learn if nonhuman animals have similar associations with faces. He said, "I wondered if sheep have sophisticated face-recognition skills, and whether they would respond emotionally to certain faces."

Using Dalesbred and Clun Forest (see page 237) sheep in several different experiments, Kendrick has demonstrated that, indeed, sheep do recognize individuals (humans and other sheep) by their faces, sometimes long after they last saw the face (shown to them in a photo). They can have either a positive or a negative response, depending on the association with the particular face.

Dalesbred Facts

▶ **FLEECE WEIGHT**
4½–7 pounds (2–3.2 kg)

▶ **STAPLE LENGTH**
5–8 inches (12.5–20.5 cm)

▶ **FIBER DIAMETERS**
36–40 (or more) microns (spinning counts coarser than 36s–40s)

▶ **LOCK CHARACTERISTICS**
Long staples with curly, pointed tips; thick at the bases. Loose, wavy crimp.

▶ **NATURAL COLORS**
White, and mostly (but not completely) free of black and gray fibers.

Using Dalesbred Fiber

Dyeing. Should be receptive to dye, except kemp.

Fiber preparation and spinning tips. Combing will separate the longer, coarser, hairy fibers (which form the curly tips of the locks) from the shorter, finer fibers of the predominant undercoat, but perhaps not cleanly; you're likely to end up with some hair in the undercoat, and some undercoat accompanying the hair. Carding is an option for those who are comfortable working long fibers that way. Picking the locks thoroughly and then spinning from the mass is one possibility; loosening the locks and spinning them individually, one after the other, is also an option. When spinning, it's difficult to get the varied fibers to feed in at the same rate. Hold the fibers lightly, and don't get caught up in perfection.

Knitting, crocheting, and weaving. Best used commercially for making carpets, Dalesbred is a natural for rugs and other textiles that need to hold up well to wear. It can offer interesting textured effects. Finer examples of the breed can be suitable for garments that won't be worn next to the skin.

Best known for. Providing the ewes that bring about Mashams (see page 328), as well as good, strong (yet hairy) wool of its own.

2-ply

Undercoat plus hair plus fine kemp. Resisted combing and carding; spun from plucked locks.

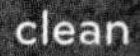

Dalesbred

raw

Derbyshire Gritstone

THE DERBYSHIRE GRITSTONE originated in the 1770s in a valley called the Dale of Goyt, so the breed was once known as Dale o'Goyt sheep. *Goyt* is an old Celtic word for "water," and the valley of the River Goyt (located on the border of Derbyshire and Cheshire counties in the Peak District National Park in England) is better known today for its two giant reservoirs and recreation than for its sheep.

Like other Blackfaced Mountain breeds, Derbyshire Gritstones are hardy sheep, developed to endure harsh conditions and marginal feed. Their black faces bear white markings, yet there is no particular pattern to the white. Derbyshire Gritstones are polled (or hornless) in both sexes.

Grown primarily for meat, Derbyshire Gritstones nonetheless produce a versatile fleece that warrants, and rewards, exploration by people who work with fiber by hand (including felters, who find it to have the best feltability of the Blackfaced Mountain family). Consistent, dense, and with enough crimp to give yarns good loft and resilience, Gritstone wool is easy to spin and feels pleasant on the needles or loom. It is typically finer than the wool from the other Blackfaced Mountain breeds. In fact, Gritstone fleeces have won top honors at major shows in Britain, beating out entries from finer-wooled and long-wooled breeds.

Derbyshire Gritstone Facts

▶ **FLEECE WEIGHT**
5–6½ pounds (2.3–3 kg)

▶ **STAPLE LENGTH**
4–8 inches (10–20.5 cm), most likely around 6 inches (15 cm)

▶ **FIBER DIAMETER**
27–31 microns (spinning counts 50s–56s)

▶ **LOCK CHARACTERISTICS**
Blocky staples with very short, pointed tips; the locks are not particularly distinct from each other and tend to disengage from the fleece in long strips. Crimp well developed but disorganized. May contain some black fibers or kemp (although our sample had none).

▶ **NATURAL COLORS**
White.

Using Derbyshire Gritstone Fiber

Dyeing. Our samples had a very subtle bit of luster, which suggests clear colors with more light reflection than in the other breeds in this group.

Fiber preparation and spinning tips. The length suggests picking and spinning from the locks (which need to be loosened up) or combing. Carding will work well with a shorter fleece or a long-staple selection if the staples are cut in half. Easy to draft. The longer-stapled fleeces can be low-twist bulky singles, and other lengths will make nice yarns in weights ranging from lace to bulky plied yarns.

Knitting, crocheting, and weaving. Unusually fine for this group of breeds, Derbyshire Gritstone is one of the workhorse wools, like a classic knitting worsted; it is a versatile choice for projects that call for mid-range wools, like sweaters, blankets, and weft-faced or balanced weave structures.

Best known for. Exceptionally good quality and fine wool for a breed that is part of this family.

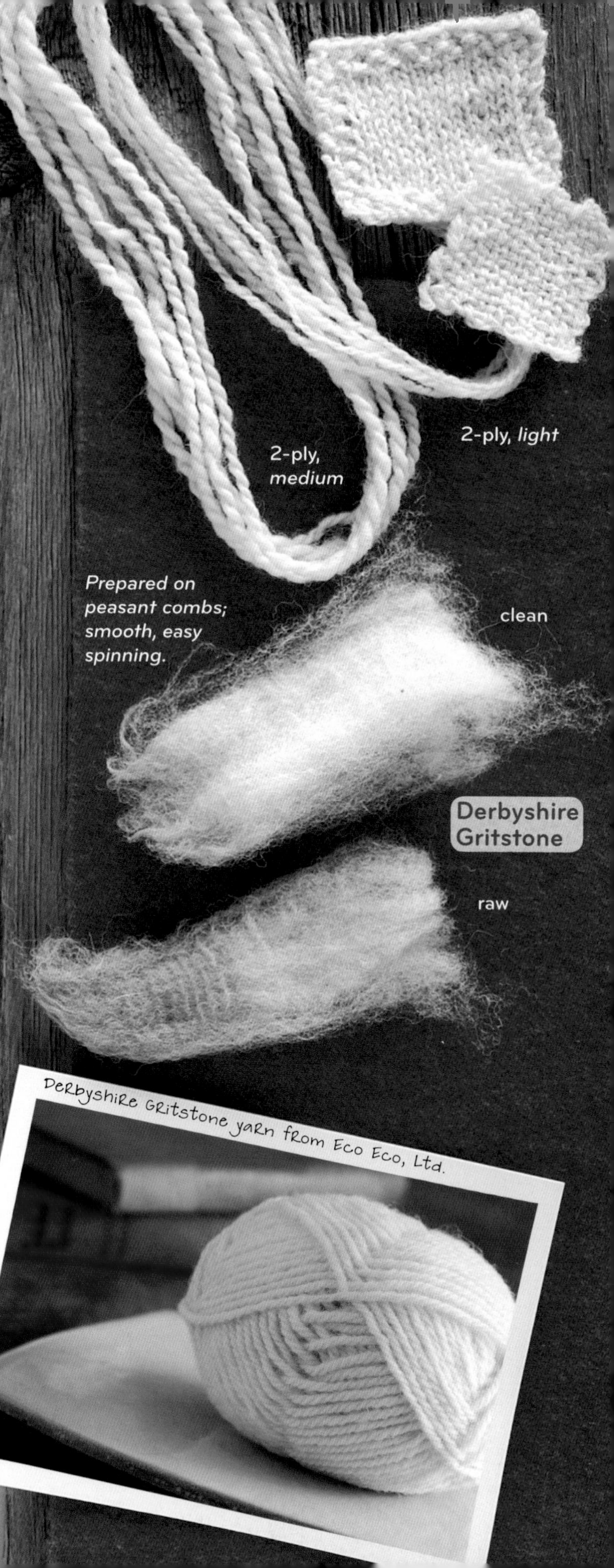

Derbyshire Gritstone yarn from Eco Eco, Ltd.

Lonk

Conservation Breed

THE ORIGIN of the Lonk's name is a bit of a mystery. Some agricultural historians link it to Middle English (the period from one thousand to six hundred years ago), when the word *lonk* was used as we would use *long* or *lanky* today. These sheep are fairly tall, but they are stout, too, so other historians speculate that the breed's appellation has been truncated from the name Lancashire, the county where the breed has been most concentrated for centuries. This makes sense, as many British breeds of livestock derive their names from the county, city, or district where they developed.

However Lonk sheep came to be named, this very old breed has been documented in the county of Lancashire since the early eighteenth century. Today the Lonks' range is slightly larger, extending into Yorkshire and Derbyshire. These sheep make their living on the moors, or uplands, of the Pennines, a low-slung range of mountains that runs up the center of England and into southern Scotland.

Lonks look very similar to Derbyshire Gritsones, except they're larger and have horns in both sexes. Although Lonk fleeces are described as being like those of Derbyshire

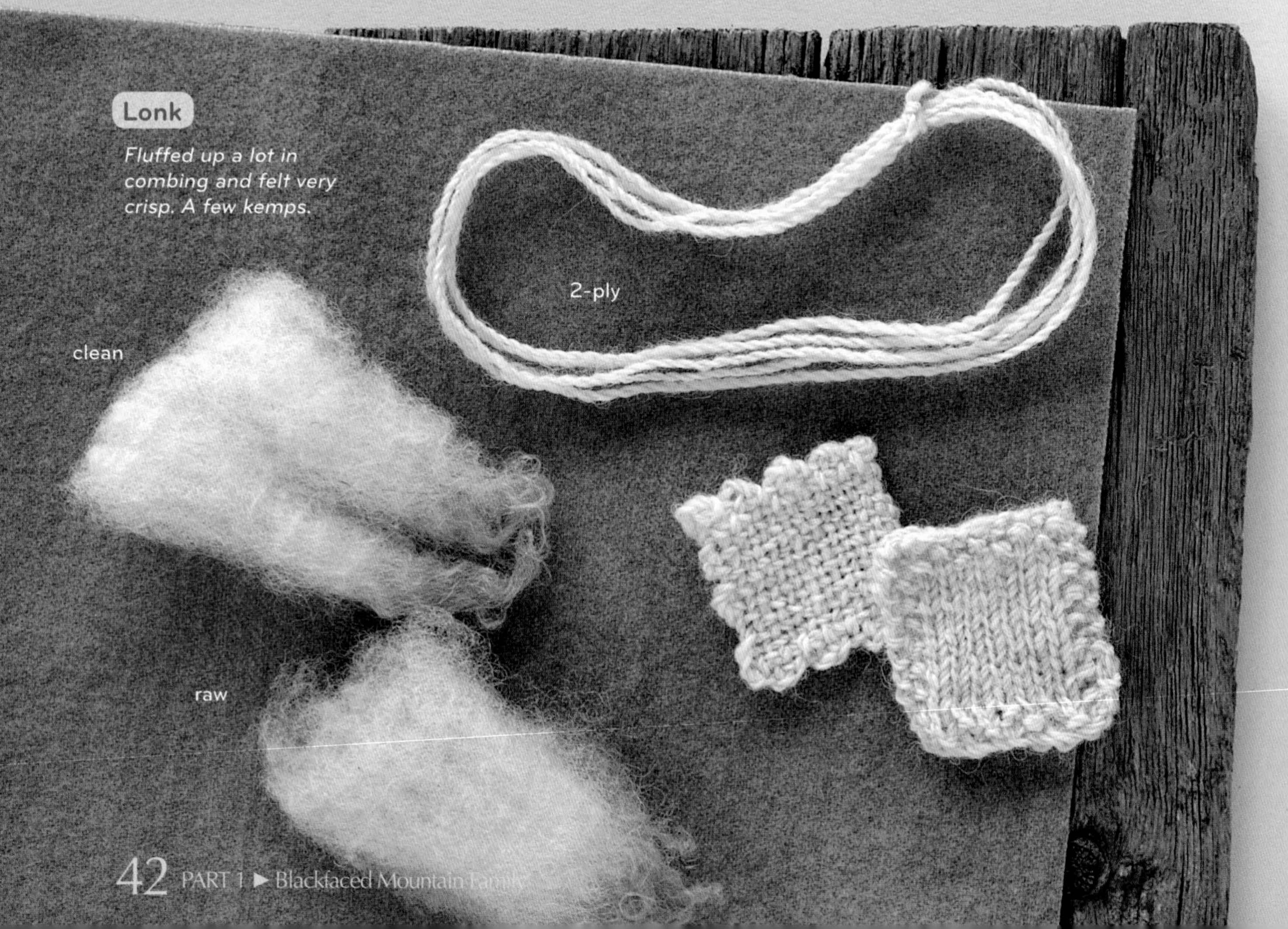

Lonk

Fluffed up a lot in combing and felt very crisp. A few kemps.

Gritstones, the Lonk's fleece tends more toward the carpet-wool end of the spectrum than the Gritstone's does. Some Lonk fleeces may be soft enough for next-to-the-skin contact for less-sensitive folk, but most will be on the rougher side. Nonetheless, the wool feels more robust than harsh.

Lonk Facts

▶ **FLEECE WEIGHT**
5–6½ pounds (2.3–3 kg)

▶ **STAPLE LENGTH**
4–6 inches (10–15 cm)

▶ **FIBER DIAMETER**
Approximately 28–36 microns (spinning counts 44s–54s), generally 28–34 microns (spinning counts 46s–54s)

▶ **LOCK CHARACTERISTICS**
Blocky staples with tiny, pointed tips; crimp well developed and generally disorganized, although there may be some wave coordination visible in the staples. There may be some black fibers, and although the breed in general aims to be kemp-free, a small amount may be present.

▶ **NATURAL COLORS**
White.

Using Lonk Fiber

Dyeing. A matte finish on locks (perhaps a touch of luster on the tips) means clear colors but a soft effect.

Fiber preparation and spinning tips. The fleece is amenable to any preparation technique you're comfortable with. Longer fleeces will want to be flicked or combed; more moderate lengths can be carded. Combing will minimize the protrusion of fiber ends and thus tame the prickle factor as much as possible. Easy to spin.

Knitting, crocheting, and weaving. Depending on the fibers' strength, it is suitable for a range of durable textiles, including outerwear, bags, pillows, upholstery, rugs, and tapestries.

Best known for. Sturdy, white wool of good quality and durability.

Tastes Good, Too

Great British Menu on BBC features competitions between some of Britain's top chefs. In 2009, Nigel Haworth, chef at the award-winning Northcote Manor in Lancashire, won the top place with Lonk Lamb Lancashire Hotpot, a dish of lamb, onions, and sliced potatoes. Haworth has also helped promote other traditional breeds of livestock from Lancashire, and the farmers who produce them.

Rough Fell

Conservation Breed

In England, the term *fell* refers to uncultivated high grounds that were traditionally used for common grazing by multiple farmers. The Rough Fell breed occupies a small area; about three-quarters of the population is found within a 7- to 10-mile radius of the fells in Cumbria. This northwestern-most county of England is bordered by the Irish Sea on its western edge and Scotland on its northern edge.

This small geographic range presents particular concern for the Rough Fell's future. During the 2001 foot-and-mouth disease outbreak in the United Kingdom, 35 percent of the Rough Fell flock was put down in an effort to contain the spread of the disease. With only 18,000 breeding ewes at the start of the outbreak, this was a devastating blow for the breed, whose numbers have not recovered much since the culling.

Cumbria is home to England's highest peak, Scafell Pike, and its largest lake, Windermere, but what is significant for these sheep is the area's harsh climate. It is extremely wet, receiving 60 to 100 inches (152.5-254 cm) of rain per year, and is blanketed by heavy snowfalls during winter. To survive in this climate, the Rough Fell has developed an exceptionally strong and heavy fleece. Because there is not much of a contemporary market for wool of this type, the breed is primarily raised for meat production. As with the other Blackfaced Mountain breeds, Rough Fells have white markings on their faces, but the markings tend to be large, dominating the snout and making the black base coloring look like a mask. The breed is known for its size (it's one of the largest of the mountain sheep) and its docile nature.

Even in the wide and varied world of fleeces, Rough Fell wool is an aptly named and peculiar inhabitant. Take a look at this breed in conjunction with the Herdwick and the Swaledale, fellow denizens of the English Lake District. All these hardy mountain sheep grow fleeces that mix wool, hair, and kemp in varying proportions. Although the statistics for the three sound similar on paper, Rough Fell wool is by far the coarsest overall.

While people often remark on the connection between the Rough Fell's name and the animal's robust nature, *rough* also applies resoundingly to the wool. Felting is not its strong point, and

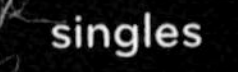

Rough Fell

Wool plus stiff hair and kemp, spun from loosened locks.

the fiber is sometimes used to stuff mattresses. That connection makes perfect sense as soon as you hold a sample: The outercoat resembles horsehair. Resilience and durability are built in. The locks could be woven into a base fabric as accents or added to baskets during their construction.

These fibers are also independent cusses, so plan to vacuum your surroundings when you're through working with them. Most of what you'll pick up is kemp, as well as some hair. Rough Fell fiber is amazing. It'll help you enlarge your preconceptions about what wool is.

Rough Fell Facts

▶ **FLEECE WEIGHT**
4½–7¾ pounds (2–3.5 kg)

▶ **STAPLE LENGTH**
6–12 inches (15–30.5 cm)

▶ **FIBER DIAMETER**
36–40 (or more) microns (spinning counts coarser than 36s–40s)

▶ **LOCK CHARACTERISTICS**
Long, triangular locks, thicker at their bases where the wool layer resides and narrower at their tips, where the hair extends; remarkably hairy and kempy.

▶ **NATURAL COLORS**
Predominantly white; the tips may be darker than the main parts of the locks.

Using Rough Fell Fiber

Dyeing. The kemp won't display dyed colors, and the hair and wool components are likely to take color in different shades.

Fiber preparation and spinning tips. Almost anything you do to this fiber will separate the coats, to a greater or lesser degree. You might want to separate the two primary fiber types and make one hairy (ropelike) yarn and one more woolly yarn, both with kemp accents (there will be more kemp in the woolly yarn). Separate by holding the base ends and pulling the hair loose. You'll get a lot more hair than wool, and the hair, though bristly and slick, will be the easier component to spin. Or try spinning from the whole locks, teasing them out a little bit to separate the fibers so they will draft past each other. The length and wiriness make spinning a challenge. Keep your grip very light. Breathe. Enjoy the strangeness of the experience. The ends of the yarn won't want to stay twisted when you let go of them.

Knitting, crocheting, and weaving. These locks raise the temptation to highlight them, without any interference, as fascinating objects in themselves. They could be used, unspun, as accents in weaving or basketmaking. Rough Fell is an obvious candidate for making woven rugs. Spin some into a ropelike yarn and then use your imagination to discover applications for the strand; with sturdy hands and tools, Rough Fell yarns could be knitted or crocheted into items similar to those that can be more easily (but perhaps less interestingly) constructed with weaving, like doormats. The density will result in heavy fabrics, and the coarseness will be challenging to manage on a large scale, so keep your knitting or crocheting ambitions small until you're sure you want to go big with this fiber.

Best known for. Hairiness, texture (from kemp), and durability.

Scottish Blackface

THE SCOTTISH BLACKFACE breed is, economically, the most important type of sheep in the United Kingdom. It accounts for about 30 percent of all sheep across England, Scotland, Wales, Ireland, and Northern Ireland, as well as for 50 percent of the wool harvested in Scotland. Thanks to its prominence and range across the British Isles, the Scottish Blackface has several recognized strains:

- **The Lanark,** the dominant strain in Scotland, has the heaviest fleece, though its wool is shorter and finer than that of the other strains.
- **The Northumberland** (found in northern England) is the largest of the strains and is known for especially soft wool from among the Scottish Blackface alternatives (it's not a contender for softness among non-Blackfaced Mountain breeds, of course).
- **The Perth strain** is the most widely distributed and is found across northeast Scotland, as well as in parts of England and Northern Ireland. It is also large framed, with a medium to heavy fleece.
- **The Newton Stewart strain** is found in the wettest coastal areas of Scotland, Ireland, Northern Ireland, and the Hebrides islands off the coast of Scotland. It has a bit shorter fleece that is more resistant to rain than the other strains.

Though Scottish Blackface is the most common name for the breed, these sheep are also called Blackfaced Highland, Kerry (in Ireland), Linton (after the village that was once the main sale center for the breed), Scottish Mountain, Scottish Highland, Scotch Blackface, or Scotch Horn. The ewes and rams both have horns, like those of wild sheep, with the ram's horns especially sturdy and curling.

Did You Know?

Scottish Blackface wool is used extensively in producing Scotland's famous tweeds. Harris Tweeds, perhaps the most famous tweeds of all, are produced on the islands of the Outer Hebrides from a mixture of North Country Cheviot (see page 58) and Scottish Blackface fiber. The colors of Harris Tweeds were traditionally obtained from plant sources, including lichens.

Bag, blended Scottish Blackface and Cheviot, from the Harris Tweed Shop

Scottish Blackface Facts

▶ **FLEECE WEIGHT**
3–6½ pounds (1.4–3 kg)

▶ **STAPLE LENGTH**
6–14 inches (15–35.5 cm)

▶ **FIBER DIAMETER**
28–40 (or more) microns (spinning counts 28s–54s)

▶ **LOCK CHARACTERISTICS**
Long, hairy, and pointed staples, with an undercoat of fine wool. There may be some kemp.

▶ **NATURAL COLORS**
White; black fibers are not supposed to be in the fleece but may occasionally be present.

Using Scottish Blackface Fiber

Dyeing. This breed's wool is known for its clear white quality, and the fibers take and display dyes well. Any kemp that is present will not display the dye with the same intensity as the other fibers.

Fiber preparation and spinning tips. This is a great candidate for spinning from the lock or for combing, preferably with Viking-style combs, which have widely spaced teeth. Scottish Blackface needs a very light touch in the preparation and spinning, because the fibers stray apart until they are twisted together. It is prone to static electricity in a dry climate and may benefit from misting with water or a light mix of spinning oil (try adding a touch of olive oil to water in a spray bottle).

Knitting, crocheting, and weaving. The outstanding factor in working with Scottish Blackface is its stability, or lack of elasticity. It feels more like working with linen than with other wools. Use it when you want crispness, body, and exceptional strength.

Best known for. Long, strong, hair-predominant fleece, with effectively no points for softness, compensated for by outstanding durability and resilience. It has a beautiful recovery from compression.

Swaledale

THE SWALEDALE SHEEP is easily distinguished from the other Blackfaced Mountain breeds thanks to white to grayish stripes over the eyes or circles around the eyes, as well as a white muzzle. The Swaledale is named for the valley in which the breed originated, located in Yorkshire Dales National Park (in fact, a Swaledale ram's head is used as the park's logo). But today the breed's influence has expanded. There are millions of Swaledale ewes spread across much of the northern English hill country, often referred to as the English Lake District.

Although wool was once the main product for Swaledale shepherds, today the breed's primary use is for crossbreeding Mashams and mules (see page 328), usually with Swaledale ewes being bred to Bluefaced Leicester rams to produce what is called the North of England Mule. Like other Lake District breeds — the Herdwicks and Rough Fells — Swaledales have remarkably sturdy, rough fleeces, though Swaledale fleece is (and this is entirely relative) the finest of the group. It's interesting to take a look at the wools from these breeds together, and to think about the relationship between these sheep and the unique and storied landscape of the English Lake District. Sheep husbandry has been practiced in the area since Roman times; individual sheep tend to be hefted, or *heafed*, to a specific piece of the mountains, or fells, and remain where they belong (and where they were born) without fencing.

The wools of these three breeds also demonstrate how the numbers we depend on to understand a breed's fleece — primarily fiber diameter and staple length — fail to capture the true qualities of the wools. On paper, Swaledale, Rough Fell, and Herdwick are all in roughly the same range for fiber diameter (36–40 microns or more) and have overlapping ranges for staple length: 4 to 8 inches (10–20.5 cm), 6 to 12 inches (15–30.5 cm), and 3 to 10 inches (7.5–25.5 cm), respectively. All three types of fleeces also contain hair and kemp, fibers that help protect the animals from wet weather, along with finer wool to keep them warm. If you acquired a 6- to 8-inch (15–20.5 cm) lock of wool of the appropriate micron count, you might think, based on these criteria, that it could be any one of the three — except that each is entirely distinct from the other when you have them in hand. They are far from interchangeable.

Among the three, Swaledale feels the finest. It has the largest proportion of finer fibers. It's still a rough, tweedy wool, with the possibility of dark fibers showing up randomly and kemp that doesn't color well with dye. Consider it for projects that need to stand up to weather and knocking about: possibly outerwear, definitely rugs, perhaps a duffle bag, or as part of a basketry project.

Swaledale Facts

▶ FLEECE WEIGHT
3½–6½ pounds (1.6–3 kg)

▶ STAPLE LENGTH
4–8 inches (10–20.5 cm)

▶ FIBER DIAMETER
36–40 (or more) microns (spinning counts coarser than 36s–40s)

▶ LOCK CHARACTERISTICS
Locks have the wide bases and tapered tips typical of breeds with short, insulating undercoats and longer, rain- and snow-shedding outercoats; there may be kemp and some black hairs. Swaledale and Dalesbred are said to be similar, but our Swaledale samples felt more rough and sturdy, while the Dalesbred felt more wiry.

▶ NATURAL COLORS
Generally described as white or off-white, although likely to contain kemp and black fibers; some yarns and prepared fiber marketed for spinners are gray.

Using Swaledale Fiber

Dyeing. Coloring Swaledale, with its dark fibers and kemp, will emphasize its heterogenous nature in a nice way.

Fiber preparation and spinning tips. Spin from the teased locks, card, or prepare on Viking combs. During combing, the fibers will want to pull off the final comb in sequence: longest, medium, then kemp. Decide if you like the effect or recombine the lengths of fiber before spinning. Carding will keep them mixed. The fiber types will also separate out during spinning unless you keep your grip nice and light. Use the shortest fiber length in the bunch as one of your guides to the amount of twist to use.

Knitting, crocheting, and weaving. Make yourself a hair shirt, or something more appropriate, like a bag or lined jacket or rug. Some people may be able to handle Swaledale in a hat, but don't plan on it for a sweater.

Best known for. Off-white to white fleece with a medley of fiber types.

Say Cheese

Ewe's milk from Swaledale sheep is used to make an artisanal cheese, based on a tradition dating back to the eleventh century. At one time, the cheese was made as a farmstead product. Today, the Swaledale Cheese Company (www.swaledalecheese.co.uk) carries on the practice. The company also makes goat's milk and cow's milk cheeses.

Swaledale yarns in two weights from eBay/UK

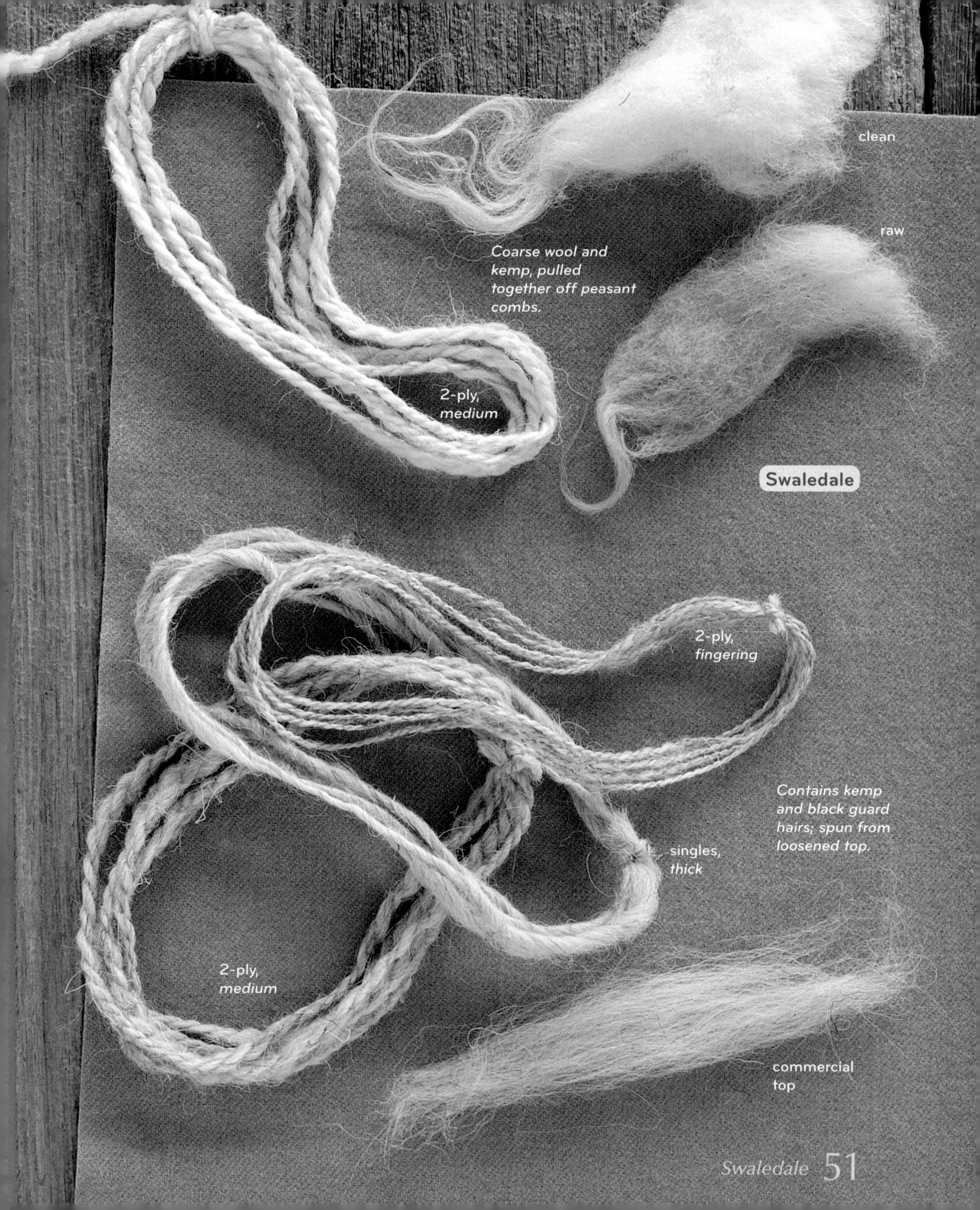
clean
raw
Coarse wool and kemp, pulled together off peasant combs.
2-ply, medium
Swaledale
2-ply, fingering
Contains kemp and black guard hairs; spun from loosened top.
singles, thick
2-ply, medium
commercial top

North Country Cheviot
Cheviot

Cheviot FAMILY

The Cheviot family is in a bit of a state of flux on our side of the Atlantic, with new names, and possibly new breeds, popping up. Traditionally there were just three breeds in this family, all with origins in the British Isles and with names derived from the Cheviot Hills, which run along the border of England and Scotland. The first was historically called the Border Cheviot, or sometimes the South Country Cheviot. Today you may also simply hear it referred to as the Cheviot, or the American Classic Cheviot. The other two traditional British breeds are the Brecknock Hill Cheviot, which is Welsh, and the North Country Cheviot, which has played a large role in the Scottish Highlands. There's also a relatively new breed in North America called the American Miniature Cheviot, which was originally called the Brecknock Hill Cheviot in North America, although its origins are slightly murky. Fiber and body traits suggest this breed is related to the traditional Brecknock Hill sheep, yet it is a much smaller sheep and many breeders believe that it has been developed on this side of the Atlantic from Border Cheviots. Without DNA analysis, this question remains up in the air.

Whatever you call them, the members of this family share distinctive looks: They are white-faced sheep with "clean faces," meaning they have no wool on their faces. Their ears are perky and upright, and they have Roman noses. Their eyes are ringed dark, looking almost as though someone dabbed them with kohl eyeliner, and their noses are dark, too.

The wool is suitable for soft-enough, sturdy textiles with nice body. Most sheep of these breeds grow only white fleece, though the Brecknock Hill and American Miniature Cheviots also have some strains that produce natural colors. Cheviot wool feels friendly, perfect for knitting garments or weaving fabrics with both body and bounce. The fibers have a unique, three-dimensional crimp that gives them resilience. Although Cheviots aren't among the core Down breeds (see page 67), their wool falls into the same general range, though it's usually longer and just a bit stiffer than the characteristic Down springiness. While you may be able to produce felt with Cheviot wools, look to other breeds for quick, reliable felting results.

The Cheviot Hills

The hills of the Cheviot massif were formed by a volcanic flow of lava 400 million years ago. The hills were once taller than Mount Everest, but they have been ground down over the eons by erosive forces. Today the tallest point, known as the Cheviot, is at 2,673 feet (815 meters) above sea level.

Cheviot

THESE CHEVIOTS are the dominant sheep in the southern portion of the Cheviot Hills. The breed has been recognized for centuries in this area. No one can say for sure what their early heritage is, but there's a story that these sheep came ashore when a Spanish ship went aground off the English coast in the fourteenth century. At that time, the Spanish were developing the forebears of the Merinos (see page 135), and Spanish sheep were already renowned for their high-quality, fine fleece. The escaped sheep, so the story goes, quickly established a large population in the rugged southern hill country.

There were also quite possibly additions of some Merino and Lincoln (see page 106) bloodlines to these local Cheviot Hill sheep during the fifteenth to seventeenth centuries, but mostly these Cheviots were developed through selection and adaptation to the local environment. By the eighteenth century, the breed was well dispersed and easily recognized throughout the Northumberland region of England. Cheviots were exported to other parts of Britain during the nineteenth century, providing the foundation for development of the Brecknock Hill Cheviots and North Country Cheviots.

These very hardy sheep were first imported to New York in 1838 from Scotland, followed by numerous additional importations to the United States and Canada. They are very active sheep; because of this, many people who raise and train Border collies and other herding dogs prefer to use Cheviots for training their canine helpers.

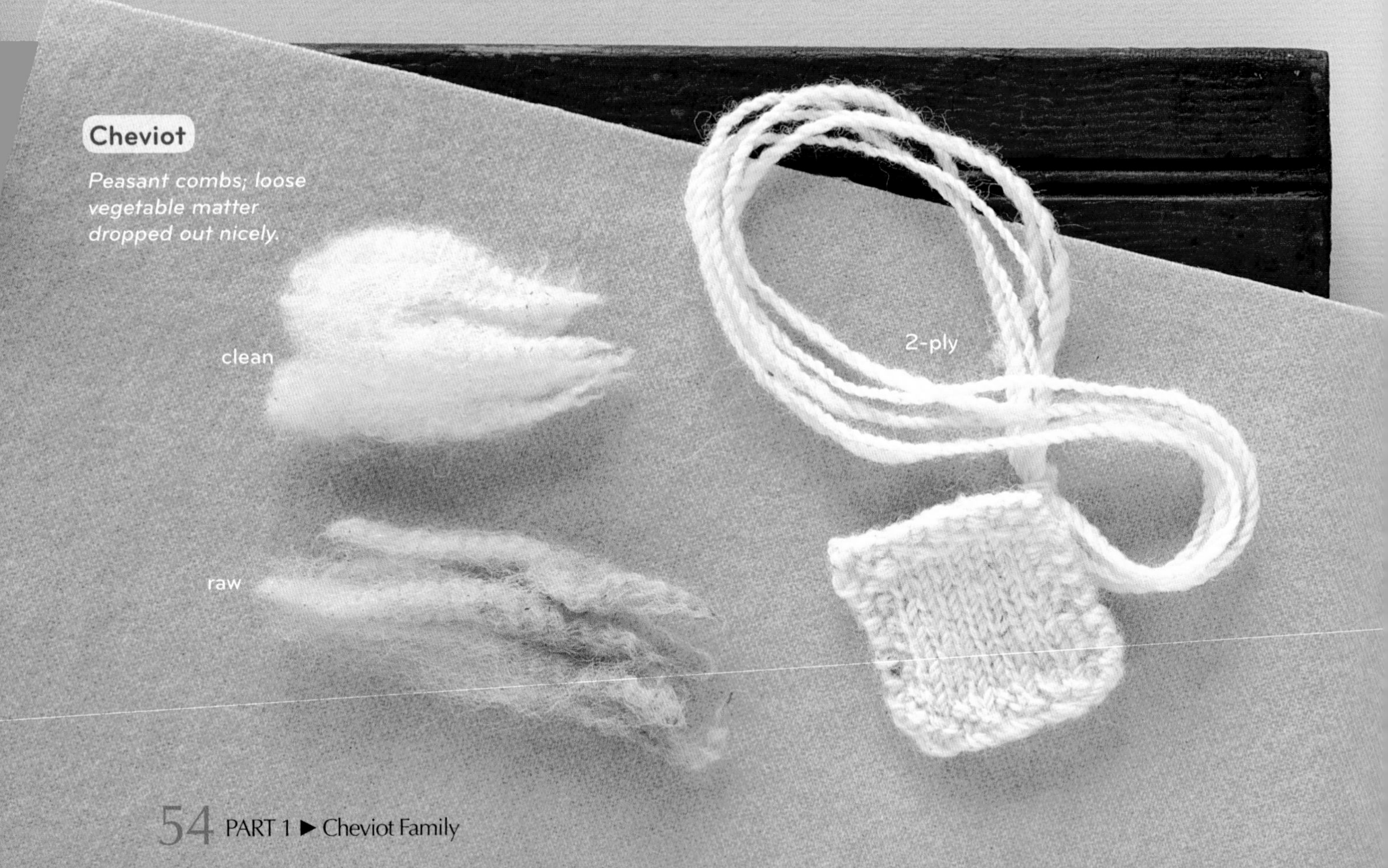

Cheviot

Peasant combs; loose vegetable matter dropped out nicely.

Cheviot Facts

Cheviots are grown primarily for meat these days, although they produce high-bulk, resilient, good-quality wool. Detailed information on the wool is difficult to locate, so we've made educated guesses.

▶ **FLEECE WEIGHT**
5–10 pounds (2.25–4.5 kg); up to 14 pounds (6.4 kg) for a ram; Australian Cheviots may be on the lighter side (with finer, longer wool); yield 50–65 percent

▶ **STAPLE LENGTH**
4–5 inches (10–12.5 cm); Australian Cheviot wool may run a bit longer, up to 6 inches (15 cm)

▶ **FIBER DIAMETERS**
27–33 microns (spinning counts 46s–58s); Australia's Cheviots may run as fine as 24 microns (spinning count 60s)

▶ **LOCK CHARACTERISTICS**
Rectangular staples with slightly pointed tips. The Cheviots all have a unique, three-dimensional crimp. The crimp is bold and uniform, with consistent quality from butt to tip. The breed associations specify no hair, kemp, or colored fibers in the ideal fleece.

▶ **NATURAL COLORS**
White.

Using Cheviot Fiber

Dyeing. Called "chalky," the wool dyes well and clearly, but without the brilliance of the longwools.

Fiber preparation and spinning tips. The shorter staples will hand-card well, but there's usually enough length for flicking or combing. Good Cheviot will be a pleasure to spin; it's a neighborly fiber (one that works well with others and is versatile).

Knitting, crocheting, and weaving. Cheviot yarns will make great sweaters, socks, and other everyday garments, as well as blankets, pillows, and the like. The coarser fleeces will be a bit much for next-to-the-skin wear, while the finer ones will be pleasant in that role, although not of luxury texture.

Best known for. Being formerly known as the Border Cheviot or South Country Cheviot, the core breed from which the other Cheviots branched. Smaller than the North Country Cheviot and larger than the Miniature Cheviot, a reliable producer of dense, firm wool that is durable without being harsh.

Cheviot (white) and Cheviot/Black Welsh Mountain (brown) yarns from Bearlin Acres, Etsy

Brecknock Hill Cheviot

THIS BREED was developed during the 1800s in the Brecon Beacons hills of south Wales and was named for the historic county of Brecknockshire. The sheep originated from a cross of the original Scottish Cheviots (the ones that are also known as Border or South Country Cheviots) with native landraces, with some minor additions of Leicester Longwool bloodlines.

The Brecknock Hill Cheviot is very closely associated with the landscape within which it thrives. John Williams-Davies says, in *Welsh Sheep and Their Wool,* "The history of no other breed demonstrates so clearly the peculiarly close and complex relationship which exists between the sheep and its environment."

Brecknock Hill Cheviot Facts

▶ **FLEECE WEIGHT**
3½–6 pounds (1.6–2.7 kg)

▶ **STAPLE LENGTH**
2½–4 inches (6.5–10 cm)

▶ **FIBER DIAMETERS**
25–32 microns (spinning counts 48s–58s)

▶ **LOCK CHARACTERISTICS**
Fleece is crisp, dense, even, with a slight luster. The Cheviots all have a unique, three-dimensional crimp. Along with their overall finer fiber profile, the Brecknock Hill Cheviots may have some kemp, although it tends to be relatively unobtrusive, and excessive kemp is considered a fault.

▶ **NATURAL COLORS**
Full range of natural colors.

American Miniature Cheviot

Using Brecknock Hill Cheviot Fiber

Dyeing. Like the other Cheviots, these are described as "chalky." The wool dyes well and clearly, but without the brilliance of the longwools.

Fiber preparation and spinning tips. The shorter staples will hand-card well, but there's usually enough length for flicking or combing.

Knitting, crocheting, and weaving. As with the other Cheviots yarns, think of great sweaters, socks, and other everyday garments, as well as blankets, pillows, and other household textiles. Some of the finer fleeces in these breeds may be compatible with next-to-the-skin wear.

Best known for. The Brecknock Hill Cheviots grow slightly finer wool on average than the other Cheviots (except for the American Miniature Cheviot), and also may offer a range of natural colors.

American Miniature Cheviots

American Miniature Cheviots are raised primarily as pets, and can provide nice wool as well. Because of a historic naming problem, they may be confused with (and may or may not be related to) the Welsh (Brecknock Hill) Cheviots.

These sheep are docile, hardy, and quite cute! Like the other Cheviots, the wool they grow is high-bulk, resilient, and of good quality. Although American Miniature Cheviots are about half the size of the Brecknock Hill Cheviots, they produce large fleeces (up to 8 pounds, or 3.6 kg), with greater length (3–7 inches, or 7.5–17.8 cm). Micron counts are identical to those of the Brecknock Hill Cheviots, as is the possibility of natural colors in the wool. American Miniatures can grow fleeces that are white, black, tan, or dilute (mixed with white). In 2008, the breed's registry also approved a "painted" category, with two or more colors in clearly defined color patches.

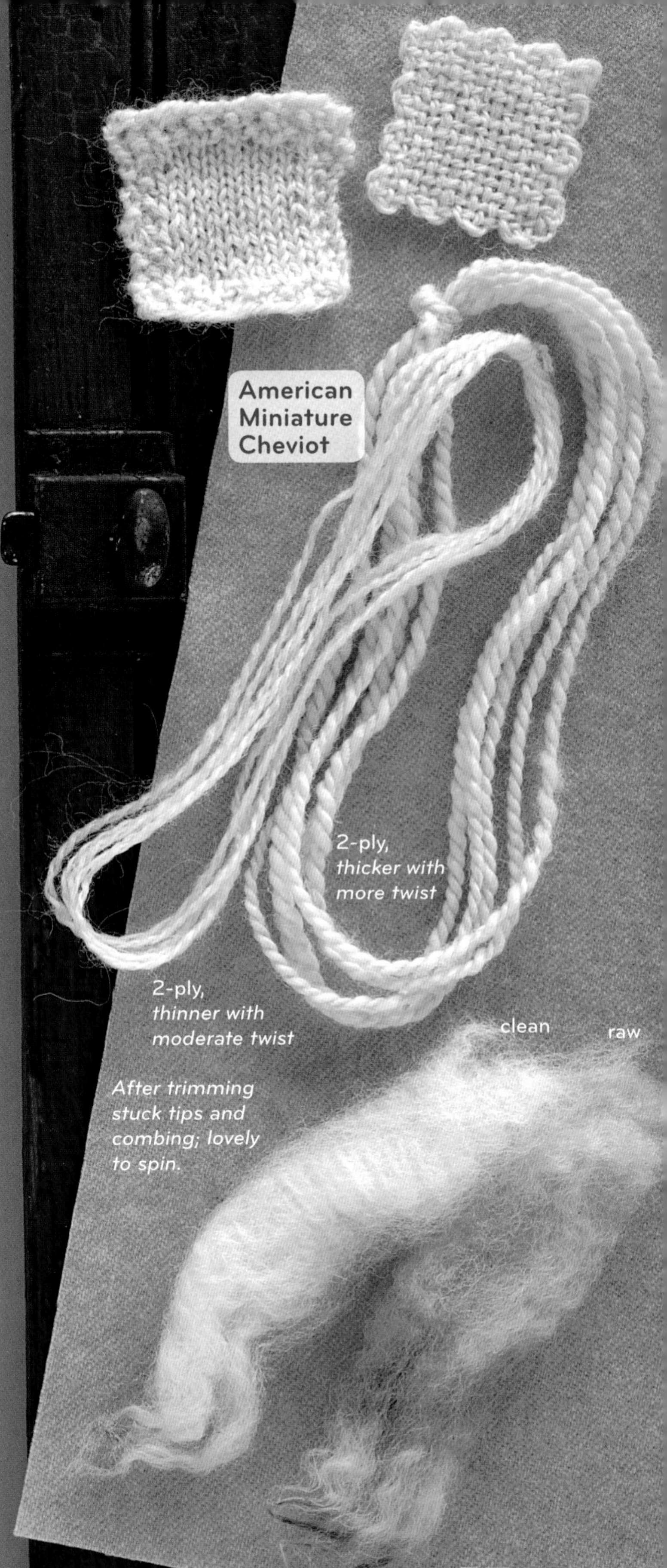

North Country Cheviot

NORTH COUNTRY CHEVIOTS hail from northern Scotland, where Sir John Sinclair — a fellow better remembered for coining the word *statistics* than for his sheep husbandry — started developing the breed in 1791. Sinclair brought five hundred Border Cheviots up from the southern part of Scotland and crossed them with Leicester bloodlines (both Leicester Longwools and Border Leicesters).

North Country Cheviots first came to Canada in 1944, with the importation of 10 ewes and 2 rams to MacDonald College, Quebec. The lambs from this small flock were so impressive that in 1949 the Canadian Department of Agriculture brought in 51 ewes and 5 rams. In 1953, several more shipments, totaling 120 head, arrived. North Country Cheviots quickly became popular with Canadian shepherds and made their way across the border to the United States in the early 1950s. They are hardy, independent, and especially long-lived, with many animals continuing to produce lambs (and fleeces) into their teens, a ripe old age, considering that the average life expectancy for a sheep is about 12 years.

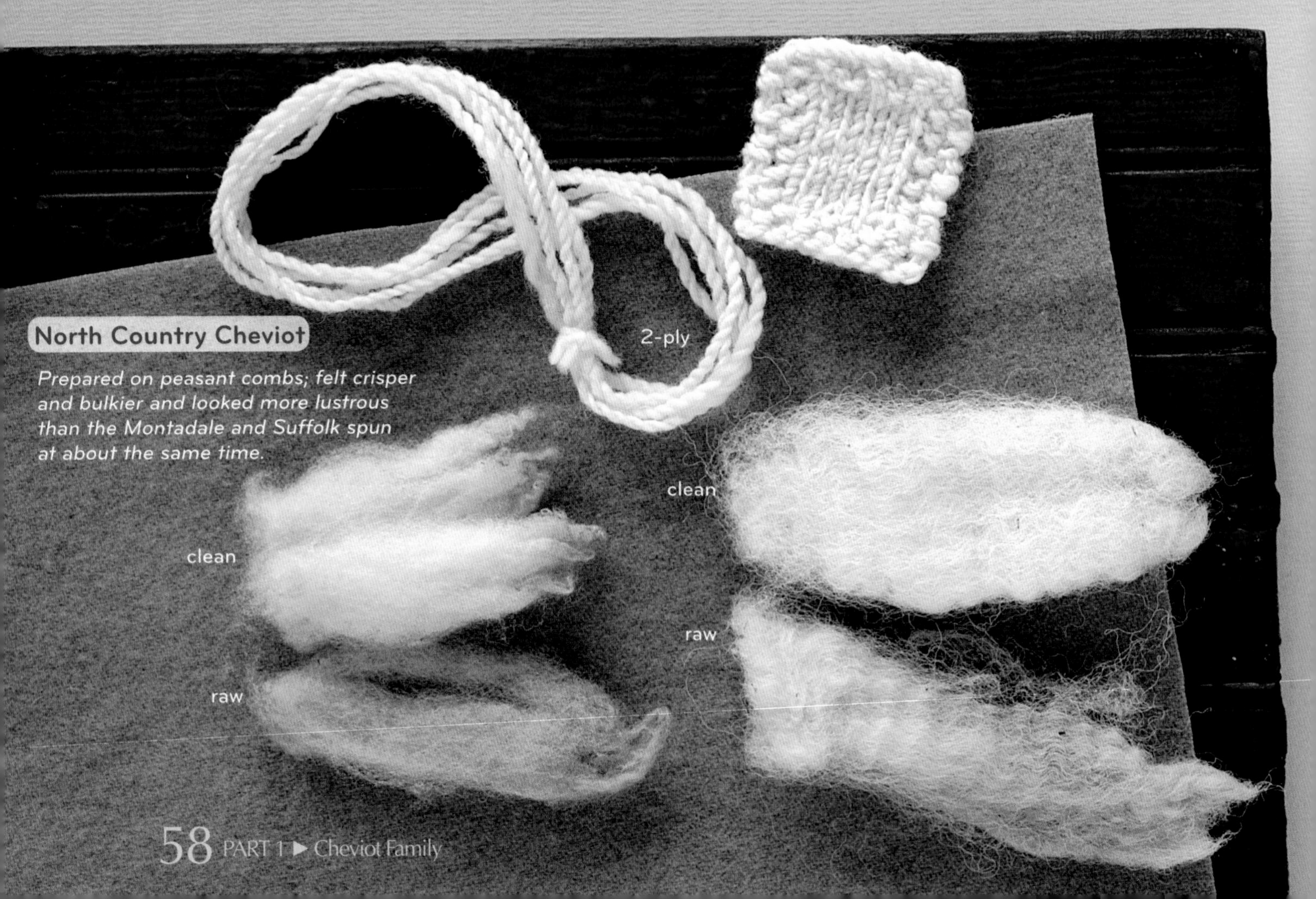

North Country Cheviot

Prepared on peasant combs; felt crisper and bulkier and looked more lustrous than the Montadale and Suffolk spun at about the same time.

North Country Cheviot Facts

Cheviots are grown primarily for meat these days, although they produce high-bulk, resilient, good-quality wool. Detailed information on the wool is difficult to locate, so we've made educated guesses.

▶ **FLEECE WEIGHT**
5–10 pounds (2.3–4.5 kg); more for a ram; yield 50–65 percent

▶ **STAPLE LENGTH**
3½–6 inches (9–15 cm)

▶ **FIBER DIAMETERS**
In general, 27–33 microns (spinning counts 46s–56s), apparently most often between 27 and 30 microns (spinning counts 50s–56s)

▶ **LOCK CHARACTERISTICS**
Rectangular staples with slightly pointed tips. The Cheviots all have a unique, three-dimensional crimp. The crimp is bold and uniform, with consistent quality from butt to tip. The breed associations specify no hair, kemp, or colored fibers in the ideal fleece.

▶ **NATURAL COLORS**
White.

Using North Country Cheviot Fiber

Dyeing. Called "chalky," the wool dyes well and clearly, but without the brilliance of the longwools.

Fiber preparation and spinning tips. The shorter staples will hand-card well, but there's usually enough length for flicking or combing. Good Cheviot will be a pleasure to spin; it's a neighborly fiber, easy to work with and versatile.

Knitting, crocheting, and weaving. Cheviot yarns will make great sweaters, socks, and other everyday garments, as well as blankets, pillows, and the like. The coarser fleeces will be a bit much for next-to-the-skin wear, while the finer ones will be pleasant in that role, although not of luxury texture.

Best known for. White, bouncy wool. Larger sheep than the southern Cheviots; there are three distinct strains within the breed.

Dorset Horn

Dorper, a hair sheep in the Dorset group

Dorset GROUP

Notice that we don't call this a family! It's a group of like-named sheep, but they aren't all kissin' cousins. There are essentially two breeds of sheep with *Dorset* in their names, but to complicate things, these sheep have multiple names in common use. The two breeds share no close relationship, but they acquired similar names because both were developed in Dorset, a county in southwest England with a coastline on the English Channel.

Dorset Down is one of the classic English Down breeds, and we include our primary coverage of it in that family's section. (Family coverage begins on page 67, with additional information on the Dorset Down on page 72.) Down breeds share a trait of having some facial color, ranging from tan or light gray with smudging on the skin of the nose to dark brown or black heads with black noses. The Dorset Down's facial color is gray-brown.

The other Dorsets, which are white-faced sheep, are the Polled Dorset and the Dorset Horn. They're essentially two types of the same breed — one without horns, the other with. The naturally hornless (polled) type is often referred to simply as a Dorset (a matter causing significant confusion), while the horned type is called either a Dorset Horn or Horned Dorset.

These polled and horned Dorsets have an interesting relative we won't cover in detail: the Dorpers. Classed as a hair-sheep breed, Dorpers were developed in the 1930s in South Africa by crossing Dorset Horn rams with Blackhead Persian ewes. The Blackhead Persian, a hair sheep, was misnamed when it found its way to South Africa in the 1860s; the breed originated in Somalia on the east coast of Africa, but Somalia was never part of Persia. Those early South African farmers may have confused their geography, but they did recognize that, in the heat and humidity of their new environment, the native hair sheep could outperform sheep they had brought from Europe. So they began producing the Blackhead Persian (whose solid-black-caped head resembles an executioner's hood), a new breed that worked well in their climate.

By the 1930s, though, they wanted to export lamb to Europe, and the meat of their Blackhead Persians, which had a strong taste, didn't fare well in that market. They began a breeding program with various European breeds to improve export marketability, settling on the Dorset as the best sire for the new breed they were developing. Like the other hair breeds, Dorpers are not used for fiber production, although thanks to their Dorset heritage they do have a fairly abundant and spinnable undercoat that sheds out in the summer. Spinners with access to the breed could find it fun to play with. Dorpers can have either pure white faces, like their original sires, or solid black heads, like their first dams.

Dorset Horn and Polled Dorset

Conservation Breed (Dorset Horn)

THE DORSET HORN came into existence hundreds of years ago, though its origins are a mystery. Many historians think it was developed through crossbreeding Spanish Merinos (page 135) with native, horn-bedecked Welsh sheep. Others speculate that it came strictly from centuries of selection within the native sheep population of southern England. Whichever way these Dorsets came into being, they spread across a wide swath of England and Wales, and then around the globe.

One of the things that drove Dorsets' popularity was their prolificacy. They were able to mate throughout the year; ewes regularly produced twins, and triplets were common. For many sheep breeds, estrus comes seasonally, in the winter months, so ewes can only give birth once each year in the spring. A sheep that can breed throughout the year can produce lambs for the Christmas meat market, or it can lamb three times in two years — both advantages from a nonfiber productivity standpoint. The Dorset Horn was the only major British breed that exhibited this nonseasonal breeding trait.

Polled Dorset

Interestingly, the polled version of the Dorset was developed in two independent locations. The Aussies started on the project in the 1940s. They bred horns out of the Dorsets with minor infusions of hornless-breed bloodlines such as Ryeland (see page 304) and Corriedale (see page 250).

Polled Dorsets in North America came from a breeding project at North Carolina State University that began in 1956. Dorset Horns were first imported to Oregon in the 1860s, and the scientists in North Carolina selected for the horn-free trait after it showed up as a mutation in a flock they kept for research purposes.

Polled Dorsets are now far more common than their horned siblings, which are on conservation lists.

From a fiber standpoint, the fleeces of these Dorsets are quite similar. The wool is very white, with an organized, regular, and relatively fine crimp pattern in both fiber and lock. It feels somewhat crisp or firm, with good body that carries over to the yarn. Micron counts range between 26 and 33, which means that finer fleeces will be appropriate for next-to-the-skin wear, and coarser ones will need to be directed toward outerwear and household textiles that don't need to be supersoft. Polled Dorsets from the Australian and New Zealand sources (now found in the United Kingdom as well) are more likely to have wool with some Down-like qualities: denser, squarer staples and more bounciness. The Dorset's fiber is reluctant to felt.

Dorsets are frequently crossed with other breeds; the most common cross you're likely to encounter is the Dorset-Finn.

Dorset Horn

Bodybuilder Sheep

A new genetic mutation showed up in a Dorset ram in Oklahoma in 1983. This guy had Mr. Universe muscling, especially in the buttocks. Sheep with this mutation are called *callipyge*, from the Greek for "beautiful butt." Only lambs that inherit the callipyge mutation from their fathers develop the bodybuilder-style muscles.

Polled Dorset yarns, natural and plant dyed, from Renaissance Dyeing, eBay/UK

Dorset Horn and Polled Dorset Facts

▶ **FLEECE WEIGHT**
4½–9 pounds (2–4.1 kg)

▶ **STAPLE LENGTH**
2½–5 inches (6.5–12.5 cm)

▶ **FIBER DIAMETERS**
26–33 microns (spinning counts 48s–58s); U.S. breed standard calls for 26–32 microns

▶ **LOCK CHARACTERISTICS**
Dense locks, with strong but irregular crimp in the fibers that is also evident in the staple formation.

▶ **NATURAL COLORS**
Most of the wool is white, and colored fibers or kemp will disqualify animals from some of the breed societies; there are some colored Dorsets.

Using Dorset Horn and Polled Dorset Fiber

Dyeing. The whites are very white and so will take colors clearly.

Fiber preparation and spinning tips. This is a versatile, moderate wool, amenable to carding or combing, depending on length. It can also be spun from the staple. The longer fleeces with more open crimp patterns are quite easy to spin, and the more Down-like fleeces with shorter staples may require more experience to control.

Knitting, crocheting, and weaving. The fleece can be spun either for airiness (woolen) or compactness (worsted) and is appropriate for use with all techniques for making everyday garments, blankets, and the like.

Best known for. Versatile medium wool, cleanly white in most cases, with some black sheep as well.

The "Other" Dorsets and Dorset Down

For reasons of history and clarity, we're making a distinction in this book between the six classic British Down breeds and the many other breeds with wools of similar color, lengths, crimp patterns, and fiber diameters.

Many breeds around the world trace their ancestry at least in part to the Down breeds, and even more have wools that fill comparable niches. We think that the use of the term *Down wool* to describe fibers unrelated to the core Down breeds tends to hinder more than it helps our understanding of the world of wool.

As we've noted, the Dorset Horn's and Polled Dorset's origins may have little in common with those of the Dorset Down. Sample locks of Polled Dorset and Dorset Down are at left. While they are similar in many ways, we don't want to inhibit our ability to see their individual qualities by applying the same label to both.

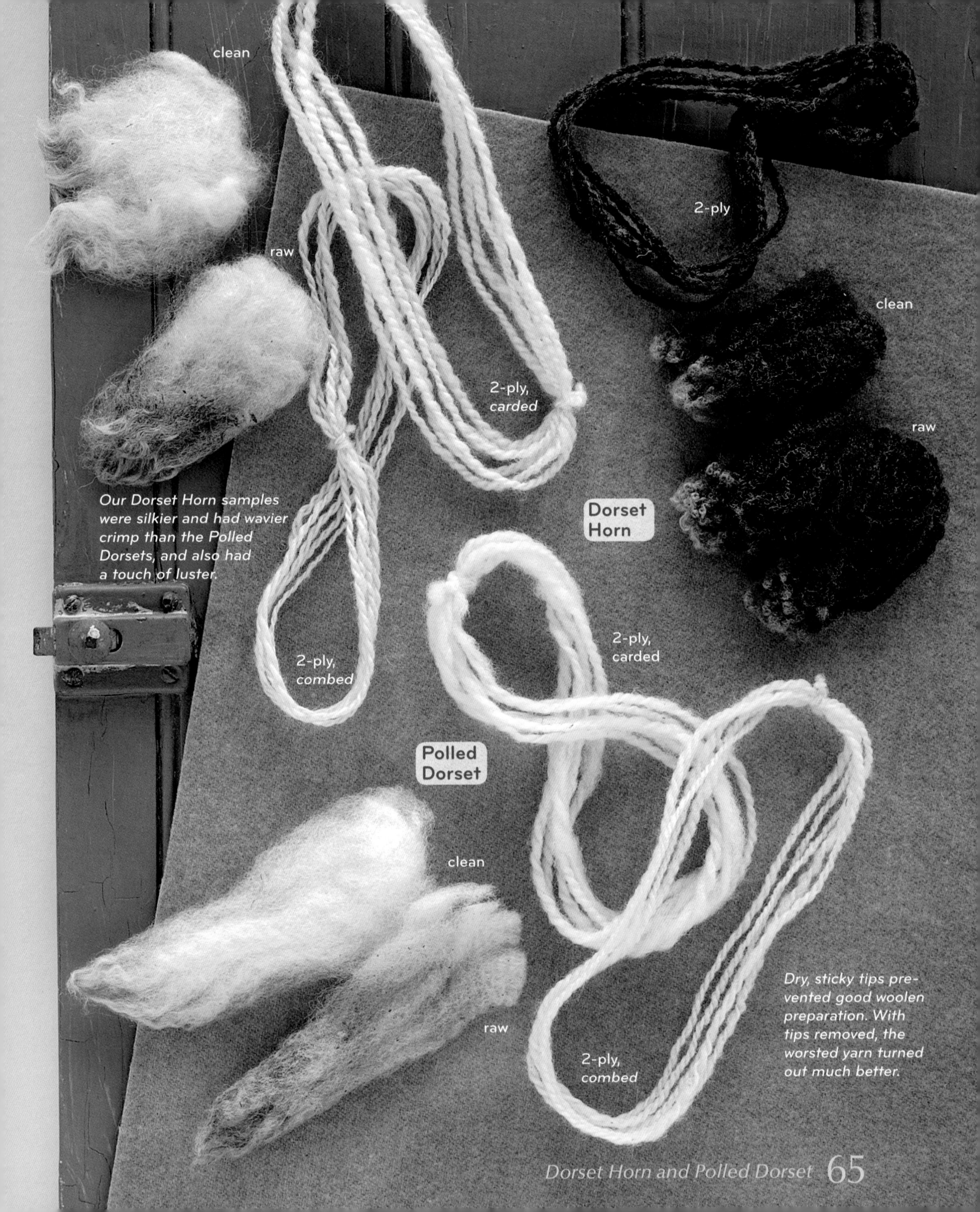

Our Dorset Horn samples were silkier and had wavier crimp than the Polled Dorsets, and also had a touch of luster.

Dry, sticky tips prevented good woolen preparation. With tips removed, the worsted yarn turned out much better.

Suffolk
Shropshire
Hampshire Down

Down FAMILY

The breeds included in the Down family are Dorset Downs, Hampshire Downs (simply called Hampshires in North America), Oxford Downs (or just Oxfords), Shropshires, Southdowns, and Suffolks. They all have colored faces, and grow "Down wools."

Down wools tend to be relatively short, with staple lengths in the 2- to 4-inch (5–10 cm) range (although there can be pleasant, longer surprises). As we mentioned in our discussion of the Dorset group, these sheep all have colored faces, ranging from tannish brown to dark brown, or charcoal to true black. They also have some color on their lower legs. Their fleeces should not, traditionally, have color in them, although dark-faced sheep are prone to having tiny amounts of scattered dark fibers here and there. A few breeders have located and cultivated dark-fiber genes that have produced full coats of colored wool.

Down breeds are primarily raised for meat, so their wool is generally sold to the

Getting Down to Basics

In the context of fiber, the word *down* can be downright bamboozling. The word is used in a lot of ways, and all of the usages are borrowed from other contexts. Our goal in these paragraphs isn't to totally clear up the muddle — a mission-impossible task if ever there were one — but instead to acknowledge the chaos of down and to help you chart a course when you hear it or read it.

One type of down refers to the soft undercoat grown by some animals. In this sense, the term comes from the softest insulating feathers on birds. Down serves the same function on other creatures as it does for ducks and geese, although in hair rather than feather form. A prime example is qiviut, the down that keeps musk oxen warm in their Arctic clime (see page 400). Down fibers are a type of wool, but they are the finest of fine-wool fibers. They grow as a separate layer, covered over (and protected by) a layer of hair-type fibers, a coat of longer and somewhat coarser wool, or a combination of the two. Down fibers tend to shed seasonally.

Then there are the Down breeds of sheep. The name, in this case, does not refer to the animals' fiber, which is definitely not downlike in the sense of the soft, fine, insulating fluff. Instead, *Down* refers to breeds that originated at a particular time (the nineteenth century) in a specific geographic region — the downs or downlands of southern England. Because they are used to characterize an ecosystem, these terms are generally not capitalized. When capitalized as a place name, Down refers to a county in northern Ireland — but let's not even go there. We capitalize Down when referring to these breeds to honor their geographic ties, and to distinguish their wools from the undercoat type of down fiber.

The Wool Pool

Down breeds dominate the sheep industry in the United States. The breeds with the largest populations of registered animals are the Suffolk and Hampshire. Southdowns and Shropshires are among the top 10 breeds numerically. Yet finding Down fiber for hand textile work is extremely challenging. What's behind such a paradox?

Down breeders focus primarily on meat production, and secondarily on the "club lamb" market. Club lambs are sold for the show ring and are so-called because they are often sold to kids participating in 4-H, Future Farmers of America, or other youth programs. The largest producers sell their wool clips directly to industrial woolen mills, and smaller producers sell through "wool pools."

A wool pool is similar to a cooperative. It acquires wool from numerous smaller producers across a region, then grades the wool (according to fiber length and diameter, crimp, cleanliness, color, and strength) and combines similar lots into bales. It takes somewhere around 50 to 60 fleeces to make a bale. Bales are sold by highest bid to industrial mills at the end of the shearing season. The pool manager pays the shepherds for their wool.

How can you find Down fleeces? Check the 4-H barn at your county fair, contact breeders in your area directly (they are often listed on the Internet), or place ads on craigslist. Top wool pool prices in 2009 were a mere 30 cents per pound for short white wool (that is, the Down breeds), so local breeders may welcome your interest. Remember that dirt washes out and almost all vegetable matter (not including the nearly pulverized type shown in the photo on page 77) will drop from the fiber as it is prepared for spinning.

commodity market, where, in almost all cases, dark fibers in the wool are considered detrimental; a single dark fiber can ruin a finished, industrially produced textile. (See The Wool Pool, at left, for how most of these fleeces are sold.) These are generally large sheep, and the ewes are prolific, frequently having twins and triplets. Their lambs grow to market weight quickly. Thus there are lots and lots of these sheep around, though only the most die-hard spinners may actually be familiar with their fiber. That's a shame, because Down wool is springy, with great elasticity and strength, making it an excellent choice for hard-wearing items (can you say, "Socks, socks, and more socks"?), as well as good everyday sweaters, hats, mittens, and blankets — and its springy crimp is evident in each and every fiber! Here are a few other points about Down wool:

- It's a bit chalky, which means it lacks luster.
- It has enough crimp to be nicely elastic.
- It doesn't felt especially well, so you may be able to machine wash it.
- It adds significant resiliency when blended with a longwool.

While the Down wools usually vanish into the unnamed mix of wools in standard knitting worsted-weight yarns, breed-identified Down wools are starting to become available as ready-to-use yarns and are worth scouting out.

But wait, says our die-hard spinner, the breeds you mentioned aren't the only breeds that grow Down-like wool! Absolutely correct. We also have Down-descendant and Down-like breeds, some known to be related genetically and others that are simply similar in the characteristics of their fiber. The Down descendants include the South Dorset Down, the South Hampshire, and the South Suffolk.

For other breeds whose wool has Down-like characteristics and makes similar types

Southdown

of yarns, there are myriad options. Some even have Down breeds in their backgrounds. One completely unrelated breed with many comparable qualities is the Ryeland (see page 304), which in its distant past grew a much finer fleece but has become more Down-like over the centuries. For other possibilities, almost everything we have described as midrange or "used to make standard knitting worsted-weight yarns" behaves to some extent like a Down wool. Some breeds that fit this description are American Tunis, California Red, some Cheviots, Clun Forest, Colbred, Columbia and Panama, Devon Closewool, Derbyshire Gritstone, the not-officially-Down types of Dorsets (Horn and Polled), Exmoor Horn, some Gulf Coast Natives, Hill Radnor, Kerry Hill, some Lonks, Montadale, Norfolk Horn, some Portlands, and Texel. Even Jacob and Shetland have been described as, and often can be handled as, Down-like wools. Yet each of these fibers has its own personality as well, and they're not among the six core Down breeds. So now you know why they're considered Down-like, and why we haven't put them here with the Down family.

Southdown

Conservation Breed

AMONG THE DOWN family members, the Southdown is the grand ancestor, the breed from which the other Down breeds were developed, and thus warrants our attention at the beginning of this section. From medieval times, there are records that short-wooled, black-faced, speckle-legged sheep were found on the South Downs (an area along the English Channel encompassing Hampshire and Sussex counties). Some historians believe that the first sheep to arrive in the Jamestown Colony in Virginia, in 1609, came from these flocks.

In the 1780s and 1790s, several breeders began improving these animals for better mutton production. They produced very high-quality meat. By the late 1790s, the improved Southdowns had become the most important breed in England. As Juliet Clutton-Brock says in *Two Hundred Years of British Farm Livestock*, "Throughout the nineteenth century, the Southdown was the central element on every Sussex hill Farm."

The Southdown's drawback was that it was a relatively small animal at maturity, so breeders quickly began using the improved Southdown for crossbreeding with the native stock in their downland counties. The improved Southdown's bloodlines therefore run through all the other Down breeds.

Today there are different types of Southdown sheep, with the distinctions between them based on size. One is a medium-sized animal still used in commercial agriculture; one is the much smaller Baby Doll Southdown; and a third is the Miniature or Toy Southdown. The latter is a recent type, developed by breeders in the 1990s, who used selection to get a sheep that is less than 24 inches (61 cm) tall at the withers. All types are docile and easy to handle, with rather affectionate dispositions. The Baby Doll and Miniature Southdowns are largely raised as pets and for fleece, but today's standard Southdown is first and foremost a meat breed, producing a lean yet tender carcass with good taste and doing so with high feed efficiency; in other words, they produce a lot of meat in relation to what they are given to eat.

The Southdown may have been the foundation of the Down family, but it gave way to a number of other breeds that are still important in the sheep world. In fact, Suffolks (see page 80) and Hampshires (see page 74) have become some of the most common sheep in North America.

Southdown Facts

▶ **FLEECE WEIGHT**
With a lot of size variety among the different strains of Southdowns, estimates range from 4–6 pounds (1.8–2.7 kg) to 7–12 pounds (3.2–5.4 kg); middle ground gives a working average of 5–8 pounds (2.3–3.6 kg); yield 40–55 percent

▶ **STAPLE LENGTH**
1½–4 inches (3.8–10 cm), mostly 2–3 inches (5–7.5 cm)

▶ **FIBER DIAMETERS**
23–29 microns (spinning counts 54s–60s) for white; 27–31 microns (spinning counts 50s–56s) for black

▶ **LOCK CHARACTERISTICS**
Dense, resilient, medium-grade fleeces, with blocky, rectangular staples that hold together and may be hard to distinguish from each other.

▶ **NATURAL COLORS**
White. There may be a few black fibers, because Down breeds have colored faces, but any off-color fibers lower the commercial value of the wool. There are some colored Southdowns.

Using Southdown Fiber

All of the Down breeds have similar usage guidelines. The individual fleece or yarn you are evaluating will be somewhere within this breed's range, and its individual qualities will guide your decisions about how to use it best.

Dyeing. These wools dye nicely. They aren't lustrous, but the colors won't be flat.

Fiber preparation and spinning tips. These are nice, versatile, medium-handling wools. Shorter fleeces can be carded; longer ones will want to be flicked or combed. Spin to maintain the loft and springy character, keeping the drafting on the light side and the twist at moderate levels.

Knitting, crocheting, and weaving. Great for socks, mittens, hats, and everyday sweaters. The finer fleeces will work next to the skin but still have good durability.

Best known for. Being unnecessarily overlooked as a fiber resource for handspinners.

Southdown yarn from Two Boy Farm

Dorset Down

Conservation Breed

THE DORSET DOWN was a relative latecomer to the field of Down breeds. In the 1840s, breeders in Dorset began breeding Southdowns to the local native ewes. After several generations, they incorporated Hampshire bloodlines, then continued to add more Southdown and Hampshire for several additional generations. Thus they stabilized a type of animal that closely reflected its Southdown heritage yet was well suited to Dorset — which has a slightly wetter climate and more arable farmlands than the South Downs. We don't know of any Dorset Downs in the United States, but they are still common in the United Kingdom. There are quite a few in New Zealand, and we've learned of an importation to Mexico a few years ago.

Dorset Down Facts

▶ **FLEECE WEIGHT**
4½–6½ pounds (2–3 kg)

▶ **STAPLE LENGTH**
2–4½ inches (5–11.5 cm), mostly 2–3 inches (5–7.5 cm)

▶ **FIBER DIAMETERS**
25–29 microns (extrapolated from Bradford grades 54s–58s)

▶ **LOCK CHARACTERISTICS**
Dense, resilient, medium-grade fleeces, with blocky, rectangular staples that hold together and may be hard to distinguish from each other.

▶ **NATURAL COLORS**
White. There may be a few black fibers, because these breeds have colored faces, but any off-color fibers lower the commercial value of the wool.

Using Dorset Down Fiber

All of the Down breeds have similar usage guidelines. The individual fleece or yarn you are evaluating will be somewhere within this breed's range, and its individual qualities will guide your decisions about how to use it best.

Dyeing. This wool takes dye well. It is neither luminous nor dull.

Fiber preparation and spinning tips. A nice, versatile fleece, Dorset Down is a medium-handling wool. You can card shorter fleeces and flick or comb longer ones. To maintain the loft and springy character, spin while keeping the drafting on the light side and the twist at moderate levels.

Knitting, crocheting, and weaving. Wonderful yarn for making for socks, mittens, hats, and everyday sweaters. The finer fleeces will be soft enough to wear against the skin but durable enough to withstand a lot of wear.

Best known for. Like all Down wools, being needlessly ignored as a fiber resource for handspinners.

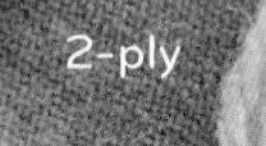

2-ply

clean

from
New Zealand

raw

Dorset Down

The pattern of dirt is typical of meat breeds. The part of the lock nearest the skin is cleanest. The tips on both samples were not dry and combed out beautifully.

2-ply

clean

from England

raw

Hampshire

SOUTHDOWNS HAD MUCH better conformation and carcass qualities than the old native landraces of Hampshire County (Berkshire Notts, Old Hampshires, and Wiltshire Horns), so in the first decades of the nineteenth century, shepherds began crossing their various breeds with Southdowns and later some Cotswold (see page 96) bloodlines. No one is sure of the exact combination of these old breeds that formed the modern Hampshire breed, but one thing is certain: When population numbers are used as the criterion for determining success, this breed scores very highly. Hampshires are found around the world, and they are one of the most dominant breeds.

Hampshire Facts

▶ **FLEECE WEIGHT**
4½–10 pounds (2–4.5 kg); yield 50–60 percent

▶ **STAPLE LENGTH**
2–4 inches (5–10 cm)

▶ **FIBER DIAMETERS**
Variable, ranging from 24–33 microns (spinning counts 46s–60s), most likely 25–29 microns (spinning counts 54s–58s)

▶ **LOCK CHARACTERISTICS**
Dense, resilient, medium-grade fleeces, with blocky, rectangular staples that hold together and may be hard to distinguish from each other.

▶ **NATURAL COLORS**
White. There may be a few black fibers, because these breeds have colored faces, but any off-color fibers lower the commercial value of the wool. There are some colored Hampshires.

Using Hampshire Fiber

All of the Down breeds have similar usage guidelines. The individual fleece or yarn you are evaluating will be somewhere within this breed's range, and its individual qualities will guide your decisions about how to use it best.

Dyeing. These wools are neither lustrous nor flat and take dye nicely.

Fiber preparation and spinning tips. These are nice, medium-handling wools that are versatile. Flick or comb longer fleeces and card shorter ones. To maintain loft and spring, spin while keeping the drafting light and the twist moderate.

Knitting, crocheting, and weaving. The yarn works well for making socks, mittens, hats, and everyday sweaters. The finer fleeces are soft without giving up their durability.

Best known for. Like all Down wools, being overlooked as a good fleece for handspinners.

Oxford

Conservation Breed

OXFORD COUNTY in England may be famous for its university, but it is also recognized as one of the most fertile agricultural areas in the United Kingdom. Because of the area's rich farmland, the shepherds who began developing the Oxford breed (circa 1830) were seeking a large animal that could gain exceptional muscle on the available forage and crops, while also producing a high-quality fleece. They combined native sheep with Hampshire, Southdown, and Cotswold genes to yield a giant sheep!

Oxfords weigh in at up to 325 pounds (147 kg) for a mature ram and 200 pounds (91 kg) for a mature ewe, making them the second largest sheep in the British Isles, right behind the Lincoln Longwool (see page 106). (For comparison purposes, sheep of the larger Blackfaced Mountain breeds are a little over half those sizes.) When well fed, Oxfords produce both excellent carcasses and superior fleeces. They are known for having gentle dispositions.

Oxford Facts

▶ FLEECE WEIGHT
At one time almost twice as large, the current fleece weights run 6½–12 pounds (3–5.4 kg), most likely 8–10 pounds (3.6–4.5 kg); yield 50–60 percent

▶ STAPLE LENGTH
Usually 3–5 inches (7.5–12.5 cm), although some may reach 6–7 inches (15–18 cm)

▶ FIBER DIAMETERS
The overall range spans 25–37 microns (spinning counts 44s–58s), though the wool is described most often as falling between 28 and 34 microns (spinning counts 46s–54s); in New Zealand it may be even stronger, at 33–37 microns (spinning counts 40s–46s)

▶ LOCK CHARACTERISTICS
Dense, resilient, medium-grade fleeces, with blocky, rectangular staples that hold together and may be hard to distinguish from each other.

▶ NATURAL COLORS
White. There may be a few black fibers, because these breeds have colored faces, but any off-color fibers lower the commercial value of the wool. Colored strains may exist or emerge.

Using Oxford Fiber

All of the Down breeds have similar usage guidelines. The individual fleece or yarn you are evaluating will be somewhere within this breed's range, and its individual qualities will guide your decisions about how to use it best.

Dyeing. These wools take dye nicely. The colors aren't radiant but won't be flat.

Fiber preparation and spinning tips. These medium-handling wools are nice and versatile. You can card shorter fleeces, but longer ones will want to be flicked or combed. It is best to keep the drafting light and twist moderate to preserve the loft and springy character of the fleece.

Knitting, crocheting, and weaving. These yarns work well for making socks, mittens, hats, and casual sweaters. The finer fleeces will be soft yet durable.

Best known for. Like all Down wools, being overlooked as a fiber resource for handspinners.

Get Smart

Say "Oxford" and most people think about the oldest English-speaking university in the world, found in the city of the same name, on the banks of the Thames River. Classes are documented at the university since at least 1096 CE, although some people think the institution actually started a couple of hundred years earlier than that. Oxford University is the home of the Rhodes Scholars, such as Bill Bradley (former basketball star and U.S. senator from New Jersey), Wesley Clark (former U.S. Supreme Allied Commander and candidate for president during the 2004 elections), Bill Clinton (42nd president of the United States), and Kris Kristofferson (actor and musician).

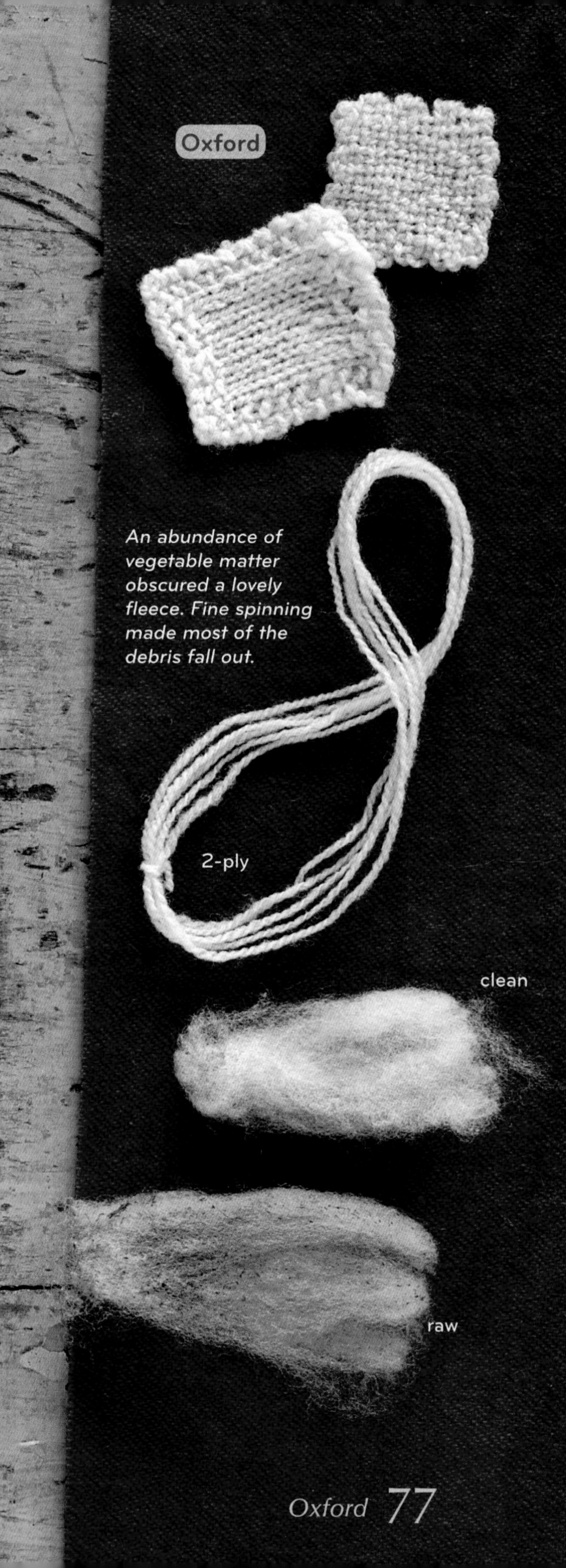

An abundance of vegetable matter obscured a lovely fleece. Fine spinning made most of the debris fall out.

Shropshire

Conservation Breed

SHROPSHIRE IS A DOWNS county in southwestern England, on the border with Wales. Southdown, and perhaps Leicester (see page 83) and Cotswold (see page 96), rams were introduced to the native sheep to improve their productivity. By 1853, the Shropshire was recognizable as a breed and began finding favor with shepherds throughout England. It was first imported to North America through Virginia in 1855 and quickly found great success beyond its point of introduction, becoming the most common breed in North America by 1930. Though not quite as common today as they once were, Shropshires are still quite popular. Shropshire ewes commonly have twins and triplets, and they are long lived, often producing lambs into their teens.

Functional shearing on meat-emphasis breeds often results in many second cuts. Preparation with mini combs removed them.

Shropshire Facts

▶ **FLEECE WEIGHT**
4½–10 pounds (2–4.5 kg) for ewes; up to 14 pounds (6.4 kg) for rams; yield 50–75 percent

▶ **STAPLE LENGTH**
2½–4 inches (6.5–10 cm)

▶ **FIBER DIAMETERS**
Variable, from 24.5–33 microns (spinning counts 46s–58s), most probably 26–29 microns (spinning counts 54s–58s)

▶ **LOCK CHARACTERISTICS**
Dense, resilient, medium-grade fleeces, with blocky, rectangular staples that hold together and may be hard to distinguish from each other.

▶ **NATURAL COLORS**
White. There may be a few black fibers, because Down breeds have colored faces, but any off-color fibers lower the commercial value of the wool. Colored strains may exist or emerge.

Using Shropshire Fiber

All of the Down breeds have similar usage guidelines. The individual fleece or yarn you are evaluating will be somewhere within this breed's range, and its individual qualities will guide your decisions about how to use it best.

Dyeing. These wools dye nicely.

Fiber preparation and spinning tips. These are nice wools that can be handled in a variety of ways. When working with shorter fleeces, it is best to card, but longer fleeces can be flicked or combed. Spin with moderate twist to maintain loft and elasticity.

Knitting, crocheting, and weaving. Excellent yarn for hats, mittens, socks, and casual sweaters. The finer fleeces are soft enough to be worn next to the skin but durable enough for everyday wear.

Best known for. Like all Down wools, being unnecessarily overlooked as a fiber resource for handspinners.

Shorter and crisper than the sample opposite, but prepared the same way for the same reasons.

Suffolk

Early in the development of the Down breeds, Suffolks came from crossing Southdown rams with old-style Norfolk Horn ewes. They were a recognized breed in 1810. The first documented import to North America was made in 1888 by G. B. Streeter, a New York farmer who acquired some prize breeding animals to begin a flock. Suffolks are now the most common breed in North America, thanks to their superior conversion of feed into meat.

Because Suffolks are grown mostly for meat, their wool tends to be overlooked. It's usually sold for commercial processing, and thus growers customarily don't make the effort needed during the year to produce handspinning-quality fleeces. Yet if you can locate a nice Suffolk fleece, it can be a delight to spin and will reward you with great yarn for winter woollies, because it's bulky and has good insulating properties.

Suffolk Facts

▶ **FLEECE WEIGHT**
4–8 pounds (1.8–3.6 kg)

▶ **STAPLE LENGTH**
2–3½ inches (5–9 cm)

▶ **FIBER DIAMETER**
25–33 microns (spinning counts 46s–58s)

▶ **LOCK CHARACTERISTICS**
Dense, resilient, medium-grade fleeces, with blocky, rectangular staples that hold together and may be hard to distinguish from each other.

▶ **NATURAL COLORS**
White. There may be a few black fibers, because these breeds have colored faces, but any off-color fibers lower the commercial value of the wool.

Note: At the time of our research, there is a processed fiber sold as "grey Suffolk" that demonstrates few of the qualities that characterize Suffolk wool, differing from breed expectations in length, crimp pattern, presence of kemp, and other regards, in addition to the expected color. We've traced this fiber back as far toward its sources as we can and have not been able to determine definitively what it is. It's enjoyable wool to use, but it is completely unlike any Suffolk we have ever seen or spun or read about. Spinners who use it should not make assumptions about Suffolk based on their experiences with this fiber.

Suffolk yarns from Fresh Isle Fibers on Manitoulin Island in Ontario

Using Suffolk Fiber

All of the Down breeds have similar usage guidelines. The individual fleece or yarn you are evaluating will be somewhere within this breed's range, and its individual qualities will guide your decisions about how to use it best.

Dyeing. This wool isn't luminous, but colors won't be dull. Takes dye nicely.

Fiber preparation and spinning tips. This is a medium-handling wool that is nice and versatile. Keep the drafting light and the twist moderate to preserve loft and spring. Card shorter fleeces and flick or comb the longer ones.

Knitting, crocheting, and weaving. Finer fleeces are soft yet durable. Yarn makes great socks, mittens, hats, and everyday sweaters.

Best known for. Like all Down wools, being unnecessarily overlooked by handspinners.

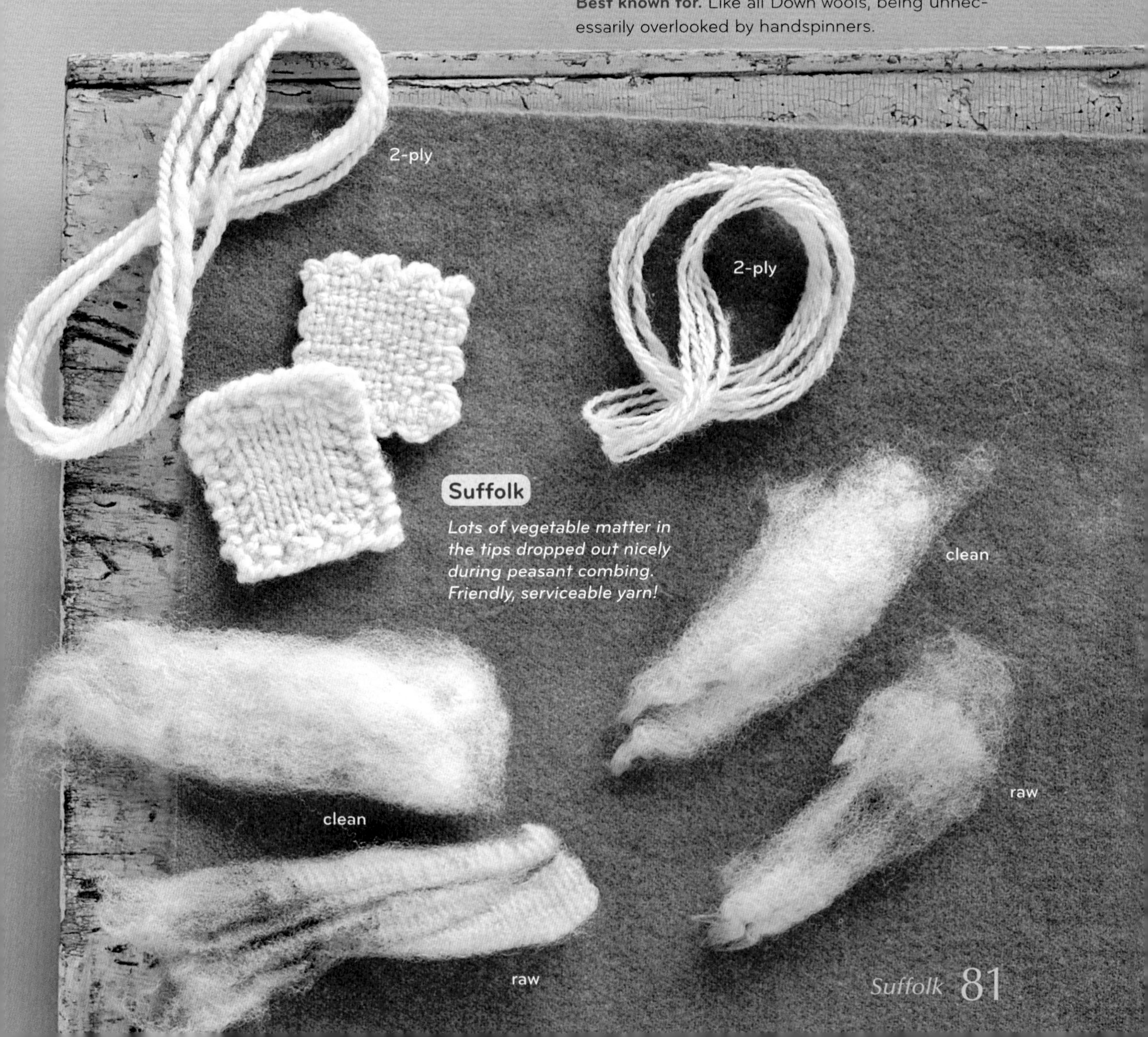
2-ply
2-ply
Suffolk
Lots of vegetable matter in the tips dropped out nicely during peasant combing. Friendly, serviceable yarn!
clean
clean
raw
raw

Cotswold
Leicester Longwool

English Longwool FAMILY

The English Longwool family of sheep is a large and important group from the fiber standpoint. Many of our favorite fiber breeds come from this clan. Some historians speculate that these breeds descended from a common ancestor brought to "Britannia" by the Romans during the reign of their empire (43–410 CE). The Romans were shepherds from their earliest history, and by the time Pliny the Elder wrote *Naturalis Historia (Natural History)*, around 79 CE, there were already distinct breeds known in Rome. These included both short-wool and longwool varieties, as well as types suited to lush grazing, sparse hilly areas, and rugged mountains.

The Leicesters

We start with the Leicesters (pronounced like the man's name "Lester"), because these really are the foundation of the modern English Longwool group. Dozens of breeds trace back in part to the Leicester Longwools, including the Miniature and North Country Cheviots, the Cotswolds and Charollais, the Texels and Wensleydales, and many more.

In this section we focus not on the many branches of the family tree but on the three pivotal breeds that carry Leicester in both their names and their genetics: the Leicester Longwool, Border Leicester, and Bluefaced Leicester. First in this group is the Leicester Longwool, which you may also hear referred to as the English Leicester or Improved Leicester outside the United Kingdom. This breed has been around since the 1790s. Its progenitors were the now-extinct Dishley Leicesters (see Robert Bakewell's Sheep, page 86), which were really the first "improved" breed of sheep. The Border Leicester and the Bluefaced Leicester came in direct descent from the Leicester Longwools.

Physically, these three types of sheep display strong family resemblances. They all have Roman noses, ranging from slight in the Leicester Longwool to more prominent in the Border Leicester and Bluefaced Leicester. Their ears are upright, from barely vertical in the Longwool to extremely so in the Border.

But how do their wools compare with each other? When it comes to fiber — and speaking, of course, in averages — Leicester Longwools grow the longest fiber, around 10 inches (25.5 cm) per year, though some may reach 14 inches (35.5 cm) in a year. Border Leicesters' wool grows around 7 inches (18 cm) per year. Both breeds can be shorn twice a year, so the staple length can be more manageable when taken in half-year batches — although the full length of a year's growth can be useful. Just think about the ways you could use a shiny, supple fiber 8 to 10 inches (20.5–25.5 cm) long!

Bluefaced Leicesters typically yield a shorter fleece, say 3 to 6 inches (7.5–15 cm), but their wool has a fineness often described as silky. Their fiber is the most predictable of the group, with diameters of 24 to 28 microns (spinning counts 60s–56s). This is definitely the Leicester you'd want to consider for fabrics that will be next to your skin, although the others make sturdy, smooth fabrics that wear extremely well.

Bluefaced Leicester

THE BLUE-FACED TRAIT — actually an illusion, created by a haze of short white hair over rich black skin — has shown up occasionally in Leicesters since the days of Bakewell. Yet the Bluefaced Leicester breed is a relatively new addition to the world, initially developed to play a role in the production of mules (see page 328). In the early twentieth century, breeders around Hexham in Northumberland (the northernmost county of England) began purposefully selecting for the blue faces. You may occasionally hear the breed referred to as Hexham Leicester, Bluefaced Maine, Blue-Headed Leicester, or Blue-Headed Maine. Thanks to its lovely, fine, longwool fleece, the breed has become a favorite among fiber lovers.

If the fleece in front of you looks like a pile of finely wound springs, there's a darn good chance it's Bluefaced Leicester (or possibly one of the related mules). The fiber is fine, silky, lustrous, and a longwool. Its springy appearance is called tightly purled, which brings to mind the purling movement of liquids flowing with a circular motion, as well as purled in the sense of a fine cord of twisted fibers. You can see both of these images in the locks. When this wool is spun and plied smoothly, the yarn can look like a string of glistening, white pearls as well.

Bluefaced Leicester is also one of the most predictable fleeces out there in many aspects: micron count or grade, fiber length, and fleece weight. A Bluefaced Leicester fleece should be uniform throughout, with no kemp or hair. Knitters and spinners use this wool frequently enough to refer to it in shorthand as BFL. It is an incredibly versatile wool: fine enough not to be scratchy; durable enough to wear well; shiny enough that dyes produce shimmering colors; and moderately feltable. BFL also blends well with other fibers. For example, it lends a touch of resilience to yarns when blended with silk or mohair (two fibers that aren't known for bounce), without muting those fibers' natural qualities of luster and drapability.

Bluefaced Leicester fleece is fairly easy to find, and it's recently become one of the most readily available breed-specific yarns for knitters, crocheters, and other yarn users to locate. It's showing up commercially in sock yarns, a great application for this fiber; the wool holds up well to intense wear and is also soft enough to be comfortable next to the skin. Bluefaced Leicester wool makes pleasant, hard-wearing sweaters, as well as woven fabrics for clothing and household use.

Bluefaced Leicester yarn from Spinning Flock Farm

Bluefaced Leicester yarn

Bluefaced Leicester scarf from Makepiece

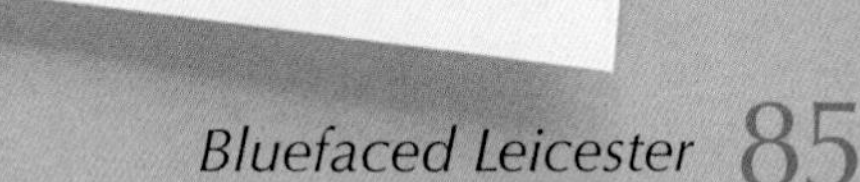

Robert Bakewell's Sheep

We have a lot to thank Robert Bakewell for. While still quite young, Bakewell (born in 1725) traveled throughout Europe studying the farming practices of the time. He took over his father's farm in 1755 and began applying and documenting new ideas in livestock breeding. His ideas later influenced the thinking of such renowned naturalists as Gregor Mendel (the father of genetics) and Charles Darwin.

Bakewell began his breeding program with the native (or landrace) sheep in his region of Leicestershire and neighboring Lincolnshire. At the time, both sexes of livestock animals were simply kept together in the fields, so breeding just happened naturally, but the first thing Bakewell did was separate the rams from the ewes. By controlling which rams were allowed to enter the flock and when they were allowed to breed, he found he could successfully "select" for specific traits. The rams he used were big, yet delicately boned, with good fleeces and fatty forequarters to appeal to the market of the day (which favored fatty mutton, the meat of mature sheep).

He named the distinct sheep he developed the New Leicesters, or Dishley Leicesters, after the farm's name, Dishley Grange. Not only did Bakewell develop the breeding method that soon gave rise to numerous improved breeds, he also originated the idea of the breed society or association. In 1793, he established the Dishley Society, whose members were required to follow a list of rules that maintained the purity of the breed. Bakewell's sheep reached North America in pre-Revolutionary days; George Washington was among the Colonial farmers who kept Dishley sheep.

Bakewell's Dishley Leicesters died out soon after his death in 1795, but the modern Leicester Longwool breed traces its lineage to Bakewell's animals.

Bluefaced Leicester Facts

▶ **FLEECE WEIGHT**
2¼–4½ pounds (1–2 kg); yield 75–80 percent

▶ **STAPLE LENGTH**
3–6 inches (7.5–15 cm)

▶ **FIBER DIAMETERS**
24–28 microns (spinning counts 56s–60s)

▶ **LOCK CHARACTERISTICS**
Extremely distinctive, spiraling, thin "locklets."

▶ **NATURAL COLORS**
Mostly white, with a recessive black gene that shows up to produce individuals with black and gray fleeces.

Using Bluefaced Leicester Fiber

Dyeing. The luster means it takes colors clearly and well.

Fiber preparation and spinning tips. The fleece is a little challenging for a newcomer to prepare for spinning, since the locks are so springy, slim, and slippery. Pick it, comb it (using combs with fine teeth), or flick it; if you get a fleece with one of the shorter staple lengths, you can card it. Once it's prepared, Bluefaced Leicester is easy enough to spin, so less experienced spinners might start with commercially prepared top to get the feel for the fiber before they move to fleece. Be sure to loosen the top before spinning, so the fibers flow smoothly instead of clumping.

Knitting, crocheting, and weaving. This is an exceptionally adaptable wool, well suited for garments that need to be soft but also must stand up to wear, such as socks, sweaters, mittens, and hats. In weaving, it will drape well without feeling heavy.

Best known for. Its unusual fineness for a longwool, its luster, and those bouncy locks.

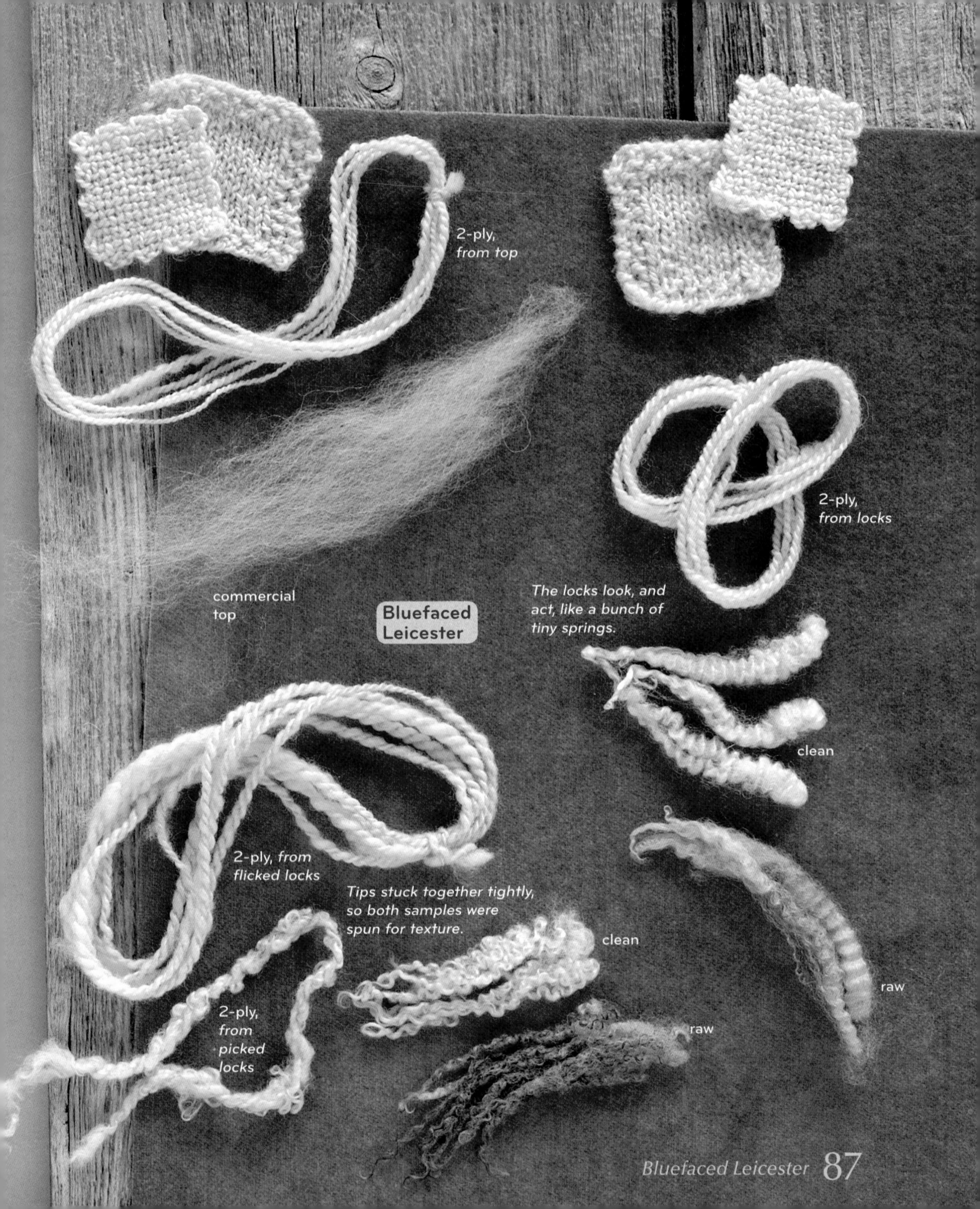

2-ply, from top

2-ply, from locks

commercial top

Bluefaced Leicester

The locks look, and act, like a bunch of tiny springs.

clean

2-ply, from flicked locks

Tips stuck together tightly, so both samples were spun for texture.

clean

raw

raw

2-ply, from picked locks

Border Leicester

The Border Leicester came into being in 1767. A pair of Scottish brothers, George and Matthew Cully, acquired some Dishley rams from their friend Robert Bakewell (see page 86). Crossing these with either Teeswater or Cheviot ewes, or perhaps a combo of the two (no one's quite sure), the siblings developed a breed that became popular throughout much of the United Kingdom by the middle of the nineteenth century. In addition to the mystery about their development, another mystery conceals the date when Border Leicesters first arrived in North America. A 1920 agricultural census documented 727 purebred animals, but no one knows when or how they crossed the Atlantic Ocean.

In many breeds, wool is an afterthought for breeders. For people who raise Border Leicesters, wool is important — so much so that the sheep are shown in competition with enough fiber growth for the judge to take the fleece into account. This versatile, easy-to-handle fiber makes a good alternative to explore if you like Romney (see page 110) and want to gently expand your horizons. It tends to have a bit more fullness in the hand and a less silky texture. It's an upright, reliable sort of wool.

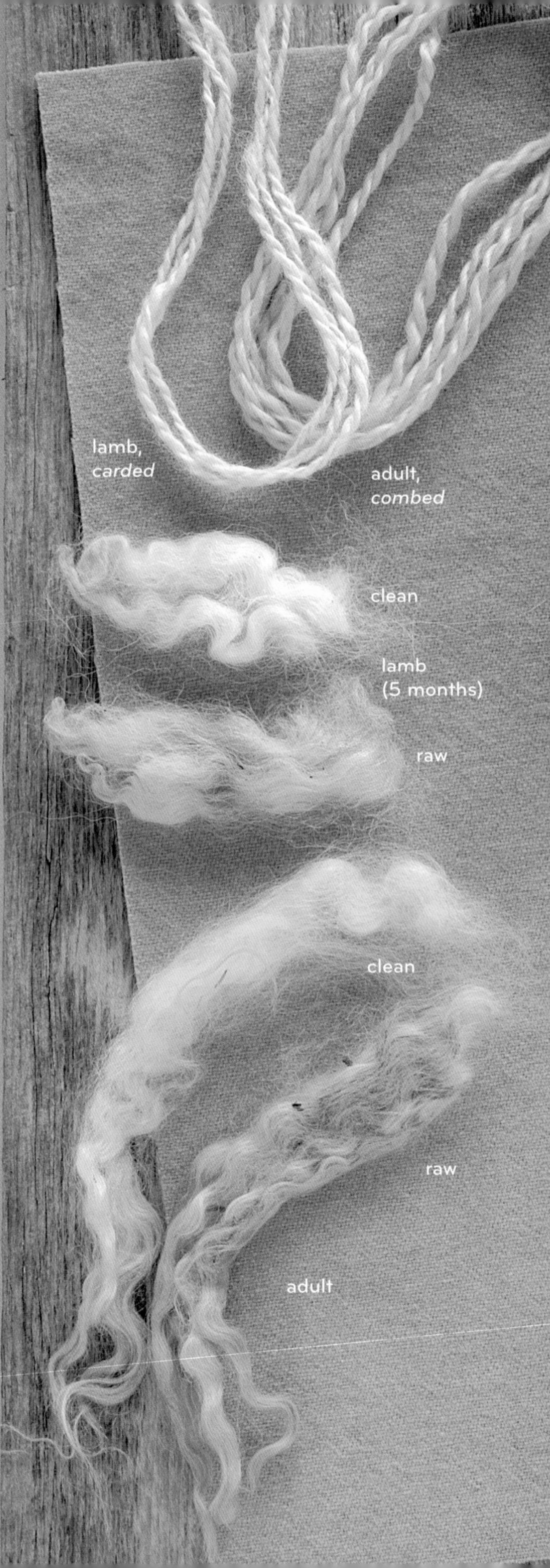

Border Leicester

2-ply,
from picked locks

2-ply,
from picked locks

2-ply,
combed

The crisp texture lends itself to novelty treatments, so two of these samples were spun for texture from picked locks. The third demonstrates fine smoothness, made from fiber prepared on mini combs.

clean

clean

clean

raw

2-ply,
carded

Despite hand carding because of its unusually short staple length, this wool produced yarn with characteristic Border Leicester body and luster.

Fineness of the fiber varies quite a bit, depending on where in the world it's grown. North American Border Leicesters have the widest range, while British animals have a narrow band at the finer end of that spectrum. New Zealand standards hang out on the coarse end. Australian Border Leicesters are right in the middle, with the colored animals finer than the white; colored fibers in many breeds are coarser than their white counterparts, but in this case that pattern reverses.

Some Border Leicester wool may feel a bit wiry, especially from the coarser end of the range or when commercially scoured with too much enthusiasm or chemicals. Border Leicester fiber diameters are coarser and the wool feels heavier than Bluefaced Leicester, but the fibers' fineness correlates pretty closely to that of the Leicester Longwool. However, the range of fiber lengths and the crimp patterns are different between the Border Leicester and Leicester Longwool, and it's easy enough to tell these wools apart.

Border Leicesters have a lot of genetic potential for producing wool in a range of colors. In the United Kingdom, the breed description specifies white wool without any colored fiber; by contrast, the American Border Leicester Association offers a coded registration to natural-colored animals in North America. The New Zealand and Australian breed associations don't specify, most likely because they take "white" for granted. Border Leicester wool felts reasonably well.

Border Leicester yarn sample card from Thirteen Mile Wool in Montana

Border Leicester yarns, also from Thirteen Mile Wool

Hand-dyed Border Leicester locks from eBay

Border Leicester Facts

▶ **FLEECE WEIGHT**
8–12 pounds (3.6–5.4 kg) average, but top fleece weights may reach 20 pounds (9.1 kg); yield 60–80 percent

▶ **STAPLE LENGTH**
4 (half-year)–10 (full-year) inches (10–25.5 cm); frequently shorn twice a year

▶ **FIBER DIAMETERS**
U.S. 30–38.5 microns (spinning counts 36s–50s)
U.K. 29.25–32 microns (spinning counts 48s–50s)
New Zealand 37–40 microns (spinning counts 36s–40s)
Australia White, 34–38 microns (spinning counts 40s–46s); colored, 32–34 microns (spinning counts 44s–46s)

▶ **LOCK CHARACTERISTICS**
Tight curls and luster, not as openly wavy as Leicester Longwool nor as slim as Bluefaced Leicester, although Border Leicesters also have the purling characteristic. The locks are individually distinct, and their tips end in charming curls.

▶ **NATURAL COLORS**
White; breed associations in North America permit registration of natural-colored animals, often blacks and grays.

Using Border Leicester Fiber

Dyeing. Because of the luster, Border Leicester will take dyes clearly and well.

Fiber preparation and spinning tips. Like other longwools, Border Leicester likes to be spun from the lock, picked and spun, combed, or flicked. It can be spun smooth and sleek, to produce a characteristic buffed-looking gleam in the finished yarn. At the same time, its crisp hand and the fiber length lend themselves to novelty-yarn treatments, where the shine is downplayed in favor of the fiber's other qualities.

Knitting, crocheting, and weaving. The length and fiber diameters make this a great wool for household textiles (like pillows or upholstery fabric), bags (which need to be durable and also comfortable to carry), and the like. Finer fleeces make hard-wearing and comfortable everyday garments like sweaters, mittens, hats, and socks.

Best known for. A friendly, sturdy wool, Border Leicester is easier to prepare and spin than both Bluefaced Leicester (which is finer and trickier) and Leicester Longwool (which is longer and a bit more challenging to manage). It can be a workhorse for constructing fabrics with any textile technique.

Leicester Longwool

Conservation Breed

AT ONE TIME the Leicester Longwool was the most common breed in England. Not so today — the breed is critically endangered — and it's a treasure we certainly want to keep alive. In the British Isles, there are only around four hundred breeding ewes. Leicester Longwools became extinct in the United States during the 1930s and '40s, and there were only a handful left at that time in Canada. In the 1980s, the Colonial Williamsburg Foundation began slowly working to reestablish the breed in North America and in 1990 imported a foundation flock from Australia to put more muscle behind its efforts. Today several dozen conservation breeders keep flocks around the United States and Canada, helping to keep this historic breed alive.

Leicester Longwool is an unsung wonder among the luster wools. Spin a bit, and you'll know why; in length, shine, and strength, it's a star. It's a pity the breed is so rare, because no other fleece can match its staunch presence, intense shine, and potential for long staples. The fleece is very dense, with great heft in the hand that translates to drape in the finished fabric. There's so much luster that the shine doesn't need to be emphasized, but it will gleam even more if you do pay some attention to featuring it. For example, spin the wool worsted

Commercial Preference vs. Potential for Handcrafters

It is heartbreaking. The Longwool breeds have not fared well in the industrialized world, and most are considered conservation breeds by the international groups working to preserve breed diversity. The Longwools include some of the largest sheep we have, but they are slow growing, which isn't a trait industrial farmers are interested in. Also, most wools used in commercial milling operations are medium-length fibers, and the large mills prefer to buy the most consistent product they can — no colored fibers or longwools need apply. But the good news is that handspinners, knitters, crocheters, and weavers can really help to keep these animals and the farmers who raise them viable! Yes, you can find fleeces. And yes, you can find ready-made yarns eager to meet your needles, hooks, or loom.

Leicester Longwool yarns from Row House Farm

Leicester Longwool yarn from Hill Farm

style, which will be the easiest approach due to its length, although that is not the only way to go about making yarn from it.

The amount of body in the yarn will impress you. Leicester Longwool can be spun fine and smooth, as well as heavy and textured (or heavy and smooth, or fine and textured), and will perform well in a variety of crafts. The crisp hand can highlight texture or lace patterns in knitting; the robust character will hold texture in art textiles and novelty yarns; and the durability will make a well-woven rug or art weaving into an heirloom that can be passed down through generations. Because of its length, wavy crimp, shine, affinity for dyes, and body, Leicester Longwool is also used for dolls' hair.

Leicester Longwool Facts

▶ FLEECE WEIGHT
5 (half-year)–18 (full-year) pounds (2.3–8.2 kg)

▶ STAPLE LENGTH
5 (half-year)–14 (full-year) inches (12.5–35.5 cm), averaging 6–10 inches (15–25.5 cm); frequently shorn twice a year

▶ FIBER DIAMETERS
U.S. White and colored, 32–38 microns (spinning counts 40s–46s)
U.K. White, 32–38 microns (spinning counts 40s–46s); colored, 32–46 microns (spinning counts 40s or coarser)
New Zealand White and colored, 37–40 microns (spinning counts 36s–40s)
Australia 32–38 microns (spinning counts 40s–48s); colored, 32–35 microns (spinning counts 44s–48s)

▶ LOCK CHARACTERISTICS
Beautiful, long, distinct locks with crimp that is well defined from pointed tips to flat bases.

▶ NATURAL COLORS
White, black, and a varied, shimmering gray (called English blue).

Using Leicester Longwool Fiber

Dyeing. The white tends to be clear, and the wool's luster means dyes take cleanly and to good effect. The grays can be overdyed.

Fiber preparation and spinning tips. Usually too long for comfortable carding, the wool can be spun from the lock, picked, flicked, or combed. For spinning, whenever you are moving fibers past each other, remember the staple length and keep your hands well separated. When drafting, hold lightly, as if you were spinning mist, and rearrange the fiber mass as needed to keep it flowing evenly and smoothly. Spin for texture or for fine, even yarns — Leicester Longwool welcomes either treatment. Because of the length, you can work with low twist, although you can use more twist for reasons of aesthetics or durability.

Knitting, crocheting, and weaving. Considered a strong wool, Leicester Longwool is extremely sturdy. Its entire fiber diameter range is above the 30-micron threshold for next-to-the-skin wear, yet when spun for a smooth surface, the yarn can feel sleek enough to use in something like a knitted lace or woven shawl; you may surprise yourself in the number of applications you find for it. Exceptional durability means it makes a fantastic warp for weaving and terrific rugs.

Best known for. Length, strength, luster, clear white, beautiful grays and blacks. This wool is a class act.

Leicester Longwool

A nice length for many types of processing. Prepared on peasant combs and spun with a little texture, with a vision of making rug weft.

So much luster, it gleams even if you don't spin to maximize the shine. Singles from hand-picked locks spun for texture; 2-ply prepared on peasant combs.

2-ply

singles

singles

clean

raw

clean

raw

Cotswold

Critical Conservation Breed

THE ROMANS originally introduced sheep to the Cotswold Hills, a wide-open, uplifted limestone landscape in the south-central area of England. By the Middle Ages, this region was well known for wool production. Sheep generated tremendous wealth and respect for the area, funding the construction of numerous churches and historic buildings that still stand today. But the breed we know today as the Cotswold was developed in the late eighteenth and early nineteenth centuries, when shepherds introduced Leicester Longwool genetics to their native sheep.

Cotswolds are fairly large sheep, each of which has a shock of curly bangs hanging over its forehead. They have dark noses and lips, as well as dark skin inside their ears. On the animal, the fleece can easily become matted and dirty, which will prove a challenge for a spinner and may create a negative impression about what can be one of the wool world's greatest delights.

Yarns produced from Cotswold fleece have good body and can demonstrate an almost pearl-like texture. Because of their length, the fibers don't need much twist to hold them together, although more twist may be appropriate for the yarn design or intended application. The fiber length also means that a spinner (or mill) can construct novelty yarns with both character and integrity. The curls of Cotswold can contribute great texture or can be smoothed over, depending on what's desired. Cotswold can be used without spinning, in lock form, for the weaving of fleece rugs and other textured applications, as well as for dolls' wigs. The Cotswold's fleece is a willing felter. Cotswold is sometimes called "the poor man's mohair," a comparison we don't think does justice to either of these excellent fibers.

Cotswold Facts

▶ **FLEECE WEIGHT**
8¾–20 pounds (4–9 kg), generally 12–15 pounds (5.4–6.8 kg); yield about 60 percent

▶ **STAPLE LENGTH**
7–15 inches (18–38 cm), generally 8–12 inches (20.5–30.5 cm)

▶ **FIBER DIAMETERS**
33–42 microns (spinning counts 36s–46s)

▶ **LOCK CHARACTERISTICS**
Heavy, lustrous, and hanging in ringlets.

▶ **NATURAL COLORS**
Traditionally white; also deep black and all shades of gray.

Using Cotswold Fiber

Dyeing. This fleece takes color beautifully, and the shine enhances the results. With the natural colors, overdye for subtly distinctive alternatives.

Fiber preparation and spinning tips. Spin the fleece from locks, pick and spin, flick, or comb. Keep in mind the length of the fibers, and keep your hands well separated when drafting.

Knitting, crocheting, and weaving. This fleece is good for heavyweight items that need to endure a lot of wear, like rugs or bags or outer garments. Spun fine, it can make a lovely lace, because the fiber's sturdiness makes the yarn-overs stand open distinctly. It is terrific for warps and for use in weft-faced textiles.

Best known for. Wavy, distinct locks; luster; length; and a range of colors from white through silvery gray to black.

Clothespin sheep by Jacqueline Ericson

Organic cotswold yarn from Garthenor

Clothespin sheep wrapped in Cotswold wool. Compare this sheep to the one made with Columbia, on page 244.

See for Yourself

Even though they are a quintessential British breed, Cotswold sheep are found in a number of petting zoos and historic centers in the United States.

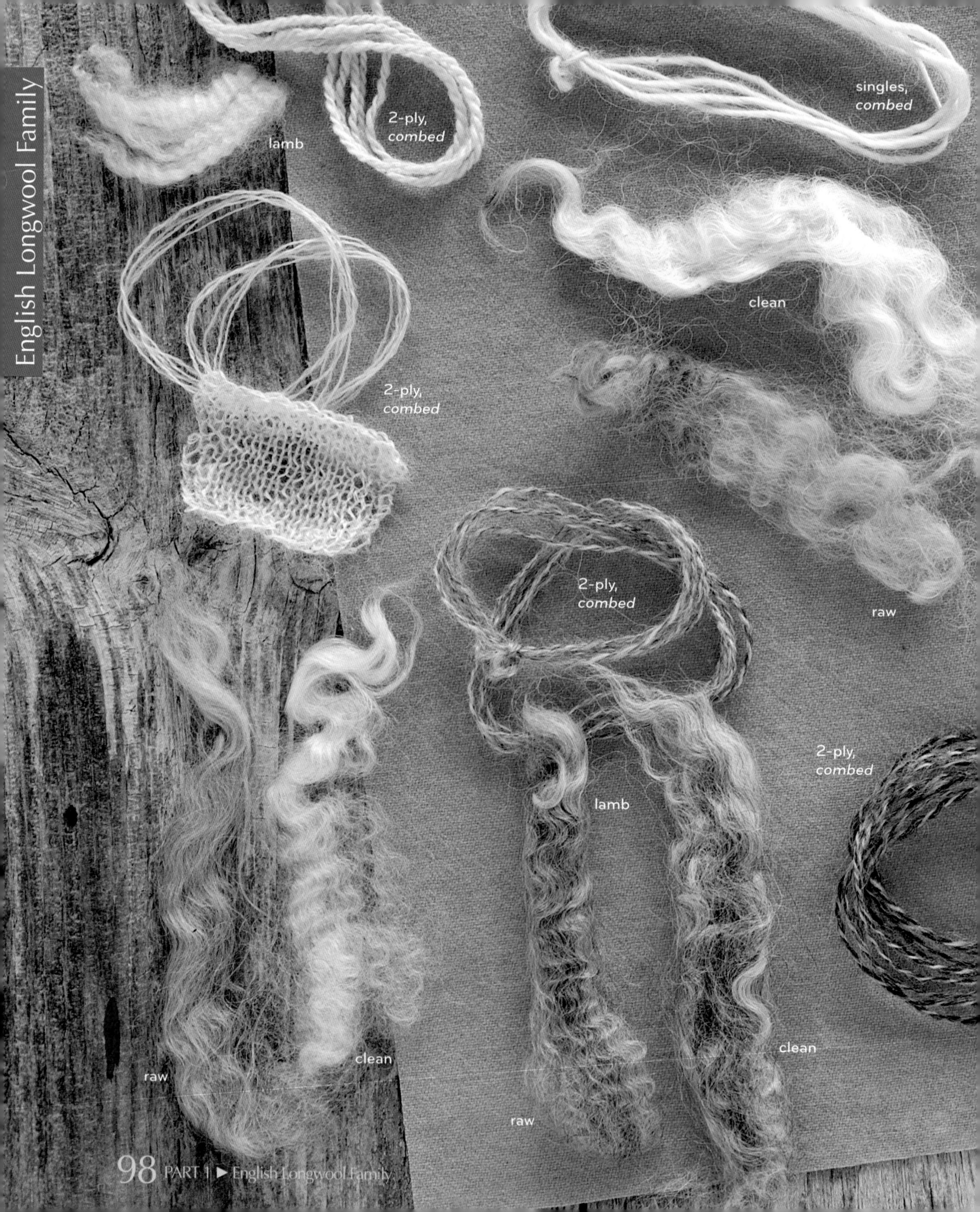
lamb
2-ply,
combed
singles,
combed
clean
2-ply,
combed
2-ply,
combed
raw
2-ply,
combed
lamb
raw
clean
clean
raw

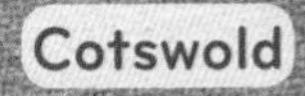

Cotswold

An array of colors, with options for spinning fine or thick, smooth or textured. When color variety exists in a fleece, either enjoy the shifts as they come or blend thoroughly to create a uniform shade.

The Dartmoors: Greyface and Whiteface

Conservation Breeds

DARTMOOR, now a national park in the county of Devon (southwest England), is known for its granite-topped hills, the highest at 2,040 feet (622 m) above sea level. It is also known for two breeds of sheep: One is simply known as the Dartmoor (though it can be referred to as the Greyface or Improved Dartmoor), and an older breed, the Whiteface Dartmoor, is the more endangered of the two, although both qualify as conservation breeds.

The early native sheep of southwestern England are referred to as West Country Moorland sheep. They were probably developed from the breeding of Roman sheep with the ancient primitive sheep of Britain. The range of the West Country Moorland sheep covered four or five counties. Among the breed's descendants in the world of modern sheep, the Whiteface Dartmoor most closely resembles those early sheep. These animals lived in just a small segment of Dartmoor, around the village of Widecombe-in-the-Moor. During the eighteenth and nineteenth centuries, when "breed improvement" was at its height, this rugged moorland area was off the beaten path, so Whiteface Dartmoors saw minimal change. There are written descriptions of the sheep in the earliest records of Widecombe that read very much

Whiteface Dartmoor

like the breed's description today. The horns of Whiteface Dartmoors are particularly reminiscent of those of their ancestors; the rams have impressively curling horns not often seen in more modern breeds, and the ewes also have small horns.

Dartmoor sheep are cute, small- to medium-sized animals that in their full fleece look a bit like shaggy sheepdogs. They are also a relatively old breed, developed in the 1800s by combining Longwools (Leicesters, Lincolns, and the now-extinct Notts) and perhaps some Cheviot and Southdown bloodlines. The Greyface moniker refers to the grayish-black mottled coloring around the animals' snouts.

Both Whiteface Dartmoors and Greyface Dartmoors grow outstandingly long, strong, curly fleeces, somewhat lighter per animal in the Whiteface. Few wools are so gleefully willing to be spun into highly textured yarns and still have adequate durability and strength. Yet that's not all the Dartmoors can let you do. Because of the fibers' length, low-twist yarns are an option. Because of their strength, the fibers can poke out of a yarn and not disintegrate at the first sign of wear. Soft? Not a chance. Sleek — if you want it. Robust — absolutely!

If you're looking for shine, length, and strength in your wool, the Dartmoors offer those qualities in abundance. As a bonus, you'll get curls. Dartmoor lambs are often shorn for their first, softer (yet still curly) fleece. Adult Dartmoor wool has traditionally been used to make sturdy fabrics, including hard-wearing blankets, rugs and carpets, and tweeds and similar cloth. The locks can be used unspun for pile or a fleece rug, or for added texture when woven into a base fabric. And if the length daunts you, you can cut the staples in half and tame them a bit.

Vive la Différence!

Although the two Dartmoors (Greyface and Whiteface) sound identical in almost all of their statistics, our samples brought out a curious difference that may just apply to the fleeces we were spinning, which did come from several different sheep. While *halo* is a word usually applied to yarns and fabrics made from mohair, Angora rabbit, and dog hair, check out the surfaces of those Greyface Dartmoor samples (see page 103). Most of the Whiteface samples, on the other hand, spun up sleekly. The coarseness of the individual samples didn't appear to explain completely the presence or absence of a halo.

There are a lot of subtle variations in wools that don't always show up in the numbers we tend to rely on (micron counts in particular). A lot more spinning of samples from both Dartmoor breeds, side by side, might reveal where this particular quality is coming from — or it might raise more questions! Put on your creativity goggles, and see what you can do with these wools' glorious presence.

Greyface and Whiteface Dartmoor Facts

▶ **FLEECE WEIGHT**
Greyface Dartmoor 13–20 pounds (5.9–9 kg), possibly up to 33 pounds (15 kg) for rams
Whiteface Dartmoor 10–16 pounds (4.5–7.3 kg), possibly up to 21 pounds (9.5 kg) for rams

▶ **STAPLE LENGTH**
6–12 inches (15–30.5 cm)

▶ **FIBER DIAMETERS**
36–40 microns, or a bit more (spinning counts 36s–40s)

▶ **LOCK CHARACTERISTICS**
Long, with big, deep waves — one every inch (2.5 cm) or so — and pointed tips; heavy and compact.

▶ **NATURAL COLORS**
White.

Using Greyface and Whiteface Dartmoor Fiber

Dyeing. The fleece's luster gives colors shine and depth; if the tips of the locks are slightly yellowed, they will give that undertone color to portions of the dye. Interesting!

Fiber preparation and spinning tips. Depending on length, strength, and cohesion of the locks, pick by hand; flick; or use combs with coarse teeth. Tease the locks apart before mounting them on the combs, don't overload the combs, and take very long strokes! Don't forget to keep your hands very far apart — farther than the length of a single lock. A light touch will benefit your hands, because this is a strong wool that can actually cut your fingers if you tug too hard against it; it's also slippery, so it will both clump up and drift apart if you cling to it.

Knitting, crocheting, and weaving. This is a sturdy wool that is best used for textiles that want some heft and durability, along with luster and, if spun in certain ways, texture.

Best known for. Producing large quantities of very strong, curly wool, without overgrazing.

Greyface Dartmoor

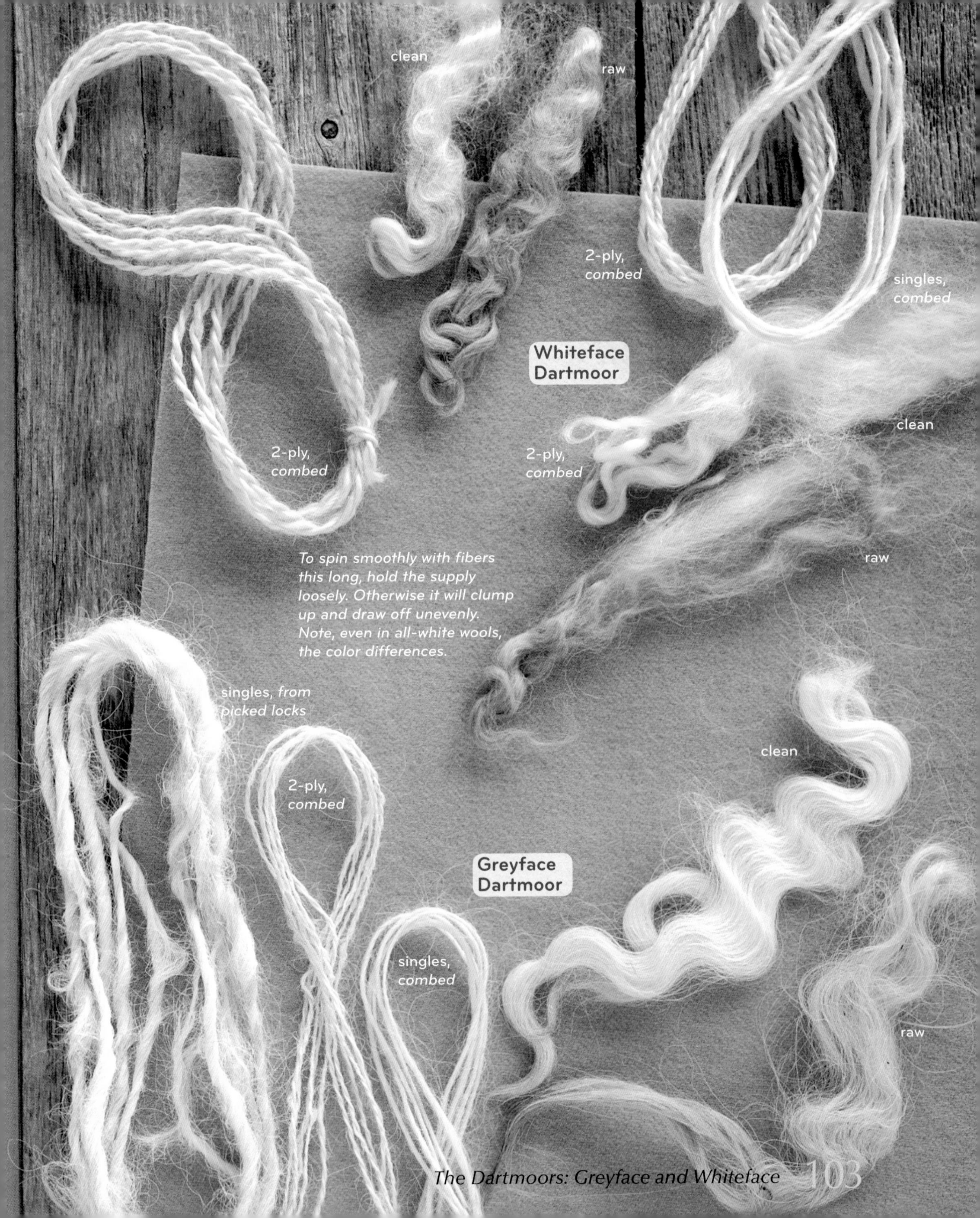

clean

raw

2-ply, *combed*

singles, *combed*

Whiteface Dartmoor

clean

2-ply, *combed*

2-ply, *combed*

raw

To spin smoothly with fibers this long, hold the supply loosely. Otherwise it will clump up and draw off unevenly. Note, even in all-white wools, the color differences.

singles, *from picked locks*

clean

2-ply, *combed*

Greyface Dartmoor

singles, *combed*

raw

Devon and Cornwall Longwool

Conservation Breed

DEVON AND CORNWALL are the southwesternmost counties of England, forming a peninsula of land that extends out into the Atlantic Ocean below the Irish Sea. In the eighteenth century, two recognized breeds, the Bampton Nott and Southam Nott, were known in the area. Breeders used Leicesters (see page 83), and later Lincolns (see page 106), to modify these two breeds, yielding the Devon Longwool in the northern parts of these counties and the South Devon in the southern range. These two breeds were similar; the South Devon was a bit larger, while the Devon Longwool had a slightly heavier fleece. Over centuries, the differences between the two breeds diminished, so in 1977 breeders combined their registries to form the Devon and Cornwall Longwool Association.

The Devon and Cornwall Longwools have not had introductions of new bloodlines since the early years, so they are considered somewhat primitive. They are raised as dual-purpose animals, producing both meat and fiber. Their wool gives new meaning to the word *substantial.* It's like a Cotswold's or Lincoln's big brother: the sort of fiber you'd want to have around for protection on the playground. Long, sleek, and strong, it requires a light hand of the spinner and rewards the knitter, crocheter, weaver, or other artisan with heft, drape, and presence. Felters will find it a suitable choice. Although the adult wool is classed as a carpet type, the lambs are usually shorn at six months; that wool is quite fine and can be used for next-to-the-skin projects.

Devon and Cornwall Longwool Facts

▶ **FLEECE WEIGHT**
13–22 pounds (6–10 kg) for ewes, 26 pounds (12 kg) for rams, with some fleeces as large as 44 pounds (20 kg)

▶ **STAPLE LENGTH**
8–12 inches (20.5–30.5 cm) normally, sometimes up to 18 inches (45.5 cm)

▶ **FIBER DIAMETERS**
36–40 microns or more (spinning counts 32s–40s)

▶ **LOCK CHARACTERISTICS**
Classic longwool lock form: long and wavy with curly, pointed tips.

▶ **NATURAL COLORS**
White.

Using Devon and Cornwall Longwool Fiber

Dyeing. The fleece will take colors clearly and brightly because of the luster.

Fiber preparation and spinning tips. Spin from the lock, pick and spin, or comb with sturdy combs that have widely spaced teeth; be careful not to overload the combs, and then practice taking very long strokes. The fibers are slippery, but when drafting, keep your hands farther apart than a staple length so the fibers will slip past each other; you don't want to be holding both ends of the same set of fibers. The fiber may also be flyaway, or have a tendency to be difficult to control, in low humidity.

Knitting, crocheting, and weaving. The fleece is suitable for use simply as locks, or for weaving (plenty strong for warps of all kinds), or to knit or crochet something sturdy. A great shiny rug yarn, and also possibly for tapestries, although not too forgiving if the sett (or the spacing of the warp threads) is off a bit.

Best known for. Massive, strong, lustrous, curly fleeces.

Both samples were prepared on peasant combs and required some added humidity to control the flyaway tendencies. That two-ply wants to be a fine rug warp!

Lincoln Longwool

Conservation Breed

THE NATIVE SHEEP of Lincolnshire evolved from the time when the Roman Empire extended its reach to the east coast of England. These native Lincoln sheep were tall, with long, coarse fleeces, but they were also thin animals that didn't provide good meat. Robert Bakewell (see page 86) used some of the native Lincolnshire sheep to develop the Leicester Longwool. Later, shepherds of Lincolnshire brought in some of Bakewell's improved Leicester rams to breed with the native ewes in order to improve carcass quality. Through continued selection, the Lincoln Longwool was developed as a recognizable dual-purpose breed by the late 1700s.

Although Lincolns are the largest breed of sheep in the British Isles, and most likely in the world, with rams exceeding 350 pounds (159 kg), these animals are rather docile individuals. Lincolns produce large, well-muscled lambs, although they grow slowly.

Not only are Lincolns physically big, they have "big wool." They grow lots of heavy, long, outstandingly lustrous fiber that handles much like mohair, although the comparison doesn't do justice to either fiber (each has its own personality). Where other breeds' genetic programs aim to produce finer and finer wool (which fits industrial processing best), Lincolns in North America are, by breed standard, supposed to grow wool "no finer than low quarter (46s)," which is about 33.5 microns and up. Sturdiness is one of Lincoln wool's defining qualities.

Handspinners have been responsible for the continued existence and breeding of colored Lincolns, which are registered separately from the white sheep. The fleeces on colored Lincolns are shaded on different parts of the body, with the darkest wool on the shoulders and legs.

For sturdiness in a wool, for length, and for large fleeces, it's hard (dare we say, impossible?) to beat Lincoln Longwool. A single Lincoln fleece has weighed in at 46 pounds (21 kg). That's not a typo. Lincoln fiber lengths have been recorded

Lincoln Longwool yarn from Wild Geese Fibres

in the vicinity of 30 inches (76 cm) or so. Tough, shining, and abundant — that's Lincoln. It will felt, but you'll probably need to work at it more than with some of the other English Longwools.

Lincoln Longwool Facts

▶ **FLEECE WEIGHT**
Generally 11–16 pounds (5–7.3 kg); yield 55–80 percent

▶ **STAPLE LENGTH**
Generally 7–15 inches (18–38 cm), average 7–10 inches (18–25.5 cm)

▶ **FIBER DIAMETERS**
U.S. 33.5 (minimum)–41 microns (spinning counts 36s–46s)
U.K. Not specific for white; colored, 41–45 microns (spinning counts 36s–40s)
New Zealand White, 36–41 (or more) microns (spinning counts 36s–40s); colored, 38–44 microns (spinning counts 36s and coarser)

▶ **LOCK CHARACTERISTICS**
Distinct, firm, heavy locks, with pointed and often spiraled tips and defined crimp; the locks can be used directly in projects such as making dolls' wigs or for weaving into "fleece rugs" (by adding raw locks to the base fabric during a rug's construction to form the textile's fluffy pile surface, which ends up looking like, but isn't, a sheepskin).

▶ **NATURAL COLORS**
White, silvery grays (light through dark), black, and possibly some moorits (a shade of brown; see page 157).

Using Lincoln Longwool Fiber

Dyeing. This fleece takes colors clearly and well because of luster.

Fiber preparation and spinning tips. Spin the fiber directly from locks, from picked locks, flick, or comb. Viking combs are great for this wool. Lincoln works well for making smooth yarns, or for textured yarns that still have structural integrity. This is because with the long staples, the curls can stick out at the same time that the fiber is well secured by twist. With really long fibers, the hands need to be kept far apart so drafting can occur.

Knitting, crocheting, and weaving. Look to Lincoln Longwool for hard-wearing, sturdy, gorgeous fabrics. Lincoln lamb's wool can be used for finer textiles, but the adult fiber should be allowed to do the things it does well: Upholster a chair with it; weave a fleece or flat-woven rug; make yourself a bag that you can use every day without wearing it out; weave tapestries that shine.

Best known for. Fiber volume, length, luster, and strength.

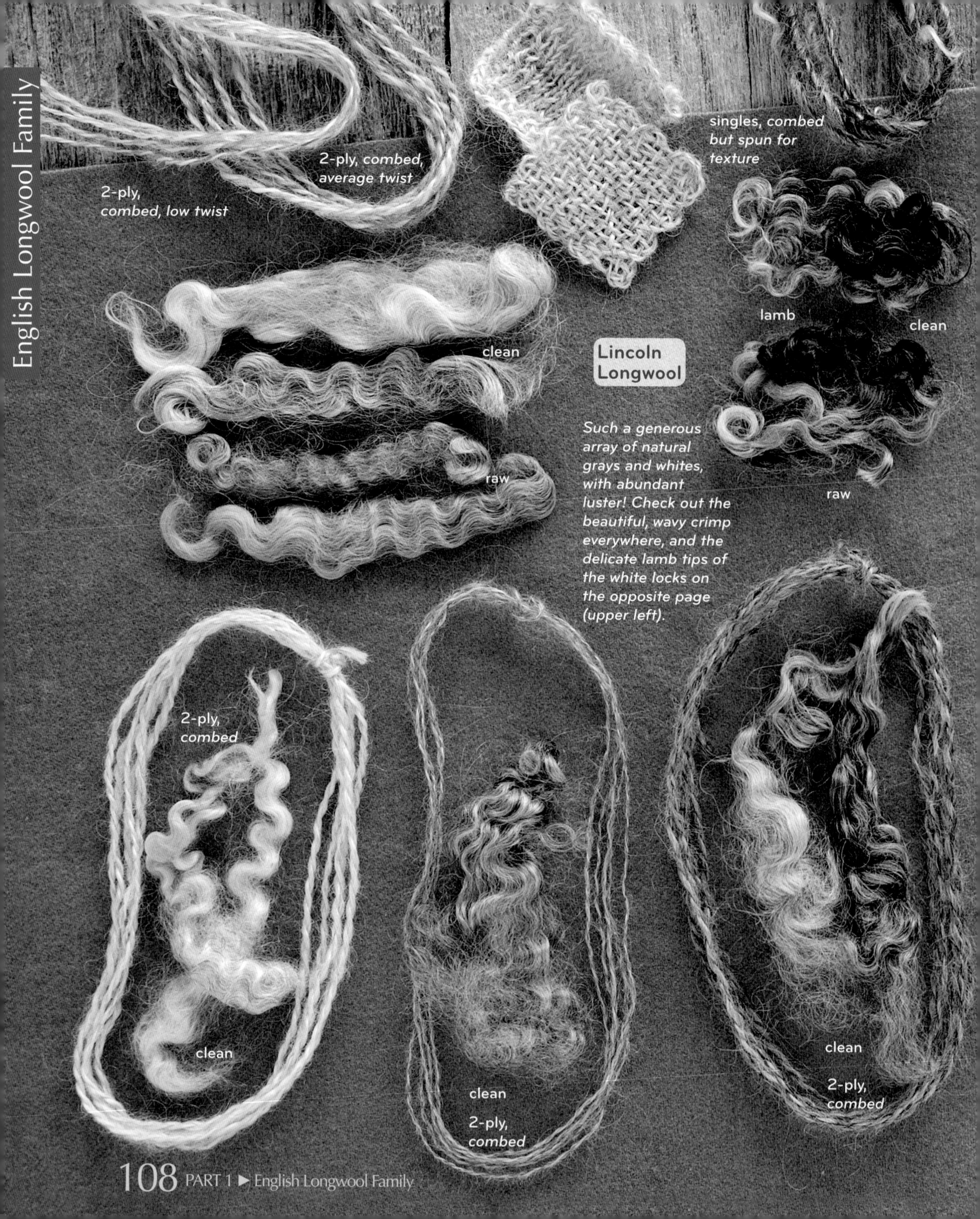

2-ply, combed, low twist

2-ply, combed, average twist

singles, combed but spun for texture

lamb

clean

clean

raw

raw

Lincoln Longwool

Such a generous array of natural grays and whites, with abundant luster! Check out the beautiful, wavy crimp everywhere, and the delicate lamb tips of the white locks on the opposite page (upper left).

2-ply, combed

clean

clean

2-ply, combed

clean

2-ply, combed

2-ply,
combed

clean

2-ply,
combed

lamb

raw

clean

raw

Both whites on this page and all of the grays across the bottom of both pages come from one flock: Christiane Payson's North Valley Farm in Oregon.

clean

clean

2-ply,
combed

2-ply,
combed

Romney

Conservation Breed

MUCH OF THE ROMNEY MARSH, a hundred-square-mile (259 sq km) coastal plain in southeast England, is below sea level. It is kept habitable by a centuries-old system of drainage ditches and seawalls, started by the Romans during their control of Britain (43–410 CE) and expanded during the Medieval period (1216–1484). The area started out as a huge salt marsh, with a few higher and drier areas interspersed among large swaths of permanently wet ones. As the drainage system and seawalls were constructed, opening more area to human habitation, the peaty soils of the marsh began to grow abundant and rich forage that was excellent for feeding livestock. However, much of the area remained boggy (especially during high tides) and sparsely populated.

Sheep were brought into the Marsh by its earliest inhabitants, but in the wet conditions, the animals usually yielded poor-quality fleece and suffered from myriad health problems. However, as the centuries passed, the sheep adapted to these conditions. Ultimately they were able to produce usable fiber in spite of the environment. In the mid-nineteenth century, shepherds in the Marsh improved the area's native sheep, which had been adapting to the locale for centuries, through a combination of intensive selection and the introduction of Dishley Leicester bloodlines (see Robert Bakewell's Sheep, page 86). This led to the modern Romney breed, which grows high-quality fleece even in soggy locations.

Romneys were first imported to the United States in 1904 by William Riddell. This Scottish immigrant farmer also raised Lincoln Longwool and Cotswold sheep on his Oregon farm. The Romneys' tolerance of humid climates works well in the Pacific Northwest and in most other humid areas of the United States and Canada, where they are found in farm flocks. Romneys have been exported around the globe and are now found in Eastern Europe, Africa, the Falkland Islands, New Zealand, and Australia. They are large sheep, known for calm and friendly dispositions, and are used for both wool and meat production. North American breeders can register colored Romneys and enter them in shows, though in other parts of the world only white animals are eligible for registration or showing.

Romney is the fleece most likely to be voted president of the Wool High School senior class. It can't do everything, but it's an all-around good citizen and extremely versatile, with personality and charisma. It's a classic, for many good reasons. Because Romneys are now found in many geographic regions, breeders can closely adapt their flocks to their environments. The wool reflects some of these environmental factors, but the fiber characteristics are similar enough for textile workers' purposes. The fleeces from animals grown in different locations will feel familiar once you've experienced a range of Romney fiber. Fiber diameter descriptions for Romneys appear to conflict a lot between the Bradford grades and the contemporary micron counts. The breed's history means that it has been traditionally evaluated by subjective means (with Bradford grades). Consider the inconsistent descriptions as an artifact of the Romney's range, in both geography and fiber particulars.

Romneys produce wool in a range from moderately coarse to fairly fine. The fiber belongs to the Longwool family, but it's enough finer than most of the other Longwools to feel like a cousin, rather than a sibling. (See Wensleydale, page 119, and Teeswater, page 116, for a pair of cousins of a different type.) The finest Romney fleeces might come from very fine adults or, more likely, lamb or hogget (first-clip, but not necessarily lamb) shearings. Romney can be soft enough to be worn next to the skin, although most Romney wool works better in garments that are one layer out from the skin.

If you have a relatively fine Romney and want to maximize its smooth qualities, spin it worsted style, so the ends of the fibers don't stick out of the yarn. The coarser Romneys make great outerwear — think of jackets and smoothly spun hats, mittens, and socks. They can also make rugs, mats, chair seat pads, and other items that require durability. The luster in a worsted-spun Romney yarn can be delicious.

The fleece of the Romney is large and dense, with well-defined crimp, and it hangs in separate locks. The wool should ideally be uniform in spinning count throughout the fleece, and the crimp should be consistent from the butt to the tip of each staple. Kemp is a demerit in judging a Romney fleece. Although the annual wool growth is specified as a minimum of 5 inches (12.5 cm), some animals are shorn twice a year, so actual spinnable length in an individual fleece can range from 3½ to 8 inches (9–20.5 cm), though 4 to 6 inches (10–15 cm) is a good, workable average. The raw fleeces contain relatively small amounts of grease.

If you want a fiber that will let you explore diverse preparation methods for handspinning, consider Romney. It can be spun from picked locks, flicked locks, carded rolags, drum-carded batts, or almost any other preparation you want

Romney yarns from eBay (left) and Solitude Wool (right)

to play with. Your resulting yarns will be different depending on the path you choose, but that's the point.

If nonspinning knitters, crocheters, and weavers could get their hands on reliable supplies of breed-specific commercial Romney yarns, they'd fall in love. Romney is currently available for sale from some handspinners and as custom-processed yarns offered by people who maintain farm flocks. Then again, Romney is a reason all on its own to learn to spin your own yarn, and the fiber is so amenable it will help you learn the craft.

Romney is generally a good felting wool, but because of the range of qualities possible within the breed, the speed of felting and the results depend on the individual fleece. The finer Romney fleeces are less likely to felt well.

Romney Facts

▶ **FLEECE WEIGHT**
8–12 (or more) pounds (3.6–5.4 or more kg); yield 70 percent

▶ **STAPLE LENGTH**
4–8 inches (10–20.5 cm)

▶ **FIBER DIAMETERS**
North American 29–36 microns (spinning counts listed at 44s–50s, but micron counts suggest 44s–54s)
New Zealand 33–37 microns (spinning counts 44s–52s, but micron counts suggest 40s–46s)
British 30–35 microns (spinning counts listed at 46s–54s, but micron counts suggest 44s–50s)

▶ **LOCK CHARACTERISTICS**
Hangs in distinct locks; crimp bold, uniform, and with consistent quality throughout from butt to tip.

▶ **NATURAL COLORS**
White, black, gray, silver, and brown.

Using Romney Fiber

Dyeing. This fleece takes dye well and clearly. Yellow staining or brittle sun-bleached tips may affect dyed colors. Consider overdyes on the natural colors.

Fiber preparation and spinning tips. Card, flick, comb, or spin from clean locks — or because of the low grease content, spin a freshly shorn fleece in the grease. On dark fleeces especially, watch for sunburned tips that can break off and make lumps in your yarn (choose another fleece, snip the brittle tips off the locks, or decide to make the most of the irregularities). Some white fleeces have yellow portions or "canary stain" that may or may not wash out. Romney is versatile, and the wool is as good-natured as the sheep who grow it. Play, and see what you come up with!

Knitting, crocheting, and weaving. Romney yarns can be used for a wide range of applications, from sweaters, mittens, caps, and shawls to chair cushions or rugs. Choose the finer fleeces for clothing and the coarser ones for upholstery or floor coverings. Romney is usually not a next-to-the-skin fiber, but a fine (perhaps lamb's), carefully prepared fleece might pass that test for some people. Before making a garment that will rub against bare skin, make a swatch and test it against the back of the neck or the wrist.

Best known for. High-yield wool clip, easy to spin, and very lustrous.

lamb
clean
raw
2-ply,
combed
2-ply,
combed
2-ply,
carded
clean
raw
Romney
Extremely versatile, with a nice range of colors and generally easy-to-open locks.
2-ply,
combed
2-ply,
combed
clean
clean
raw
raw

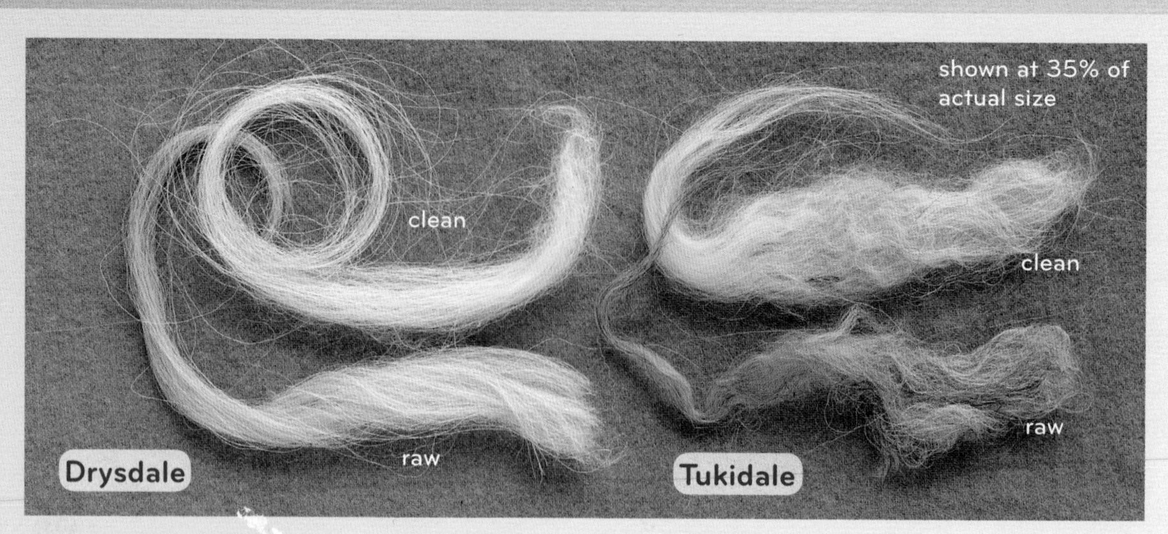

The Hairy Gene

Romney sheep have a singular gene, called the HH1 gene (for halo hair), which on rare occasions has shown up in the breed. When a sheep is born with the HH1 gene, it produces a coarser, hairier wool. In 1929, New Zealand professor F. W. Dry began researching the gene that was responsible for the isolated carriers of the hairy trait. By 1940, he'd isolated a gene, which he originally referred to as the N gene, and [illegible] stood its role in inheritance. For a long time [illegible] was considered a bit of genetically-interesting-yet-not-necessarily-practical research.

In the 1960s, the New Zealanders realized they had something of value — high-quality carpet wool — in the animals carrying this gene. They began breeding for it and named the breed Drysdale, in honor of the professor. Drysdales may produce wool in colors other than white, and their wool has been used to produce static-free carpeting for environments that house vulnerable electronic equipment.

Other Romney carriers of the hairy genes (there are now two known genes, HH1 and HH2) have been identified in Australia, leading to the development of three carpet-wool breeds: the Carpetmaster, Elliottdale, and Tukidale breeds. Here are some facts about these special Australian breeds.

▶ FLEECE WEIGHT

Drysdale 11–15½ pounds (5–7 kg)
Carpetmaster, Elliottdale, and Tukidale 6½–9 pounds (3–4.1 kg)

▶ STAPLE LENGTH

Carpetmaster 6 inches (15 cm)
Drysdale 3–6 inches (7.5–15 cm)
Elliottdale 4¾–8 inches (12–2[illegible]5 cm)
Tukidale 6 inches (15 cm)

▶ FIBER DIAMETERS

Carpetmaster 38–41 microns (spinning counts 36s and coarser)
Drysdale 40–42 microns (spinning counts coarser than 36s)
Elliottdale 38–40 microns (spinning counts [illegible] and coarser)
Tukidale 35–45 microns (spinning counts coarser than 36s–44s)

The Owlers of Romney Marsh

Romney Marsh was known through much of its history as an area inhabited by smugglers and brigands. Historians point to a tax on wool, implemented in 1275, as a catalyst for the illegal trade of wool. Tales abound of great battles, from then through the 1700s, between the king's men and hundreds of Marsh residents — also called "owlers," for their use of owl calls to communicate among themselves during shady nighttime activities.

If you are a child of the 1960s, you might remember a Disney program called *The Scarecrow of Romney Marsh*. Based on a 1915 novel by Russell Thorndike, it personified the Marsh's smugglers as Robin Hood–like characters who were only interested in justice for the area's citizenry — people beleaguered by a tyrannous King George, much like the American revolutionaries.

Romney

Teeswater

Conservation Breed

YOU'VE PROBABLY figured out that many British sheep breeds are named for a place, and the Teeswater is no exception, but with a twist: It is named for the River Tees. The river meanders about 85 miles (137 km) from Cross Fell (the highest summit in the Pennines) to the North Sea.

The original Teeswaters were probably similar to the old Lincolnshire and Leicester sheep that Bakewell used in the development of his New Leicesters. (See Robert Bakewell's Sheep, page 86.) Teeswaters received infusions of New Leicester blood fairly early after Bakewell developed his breed but then remained isolated and thus reasonably true to themselves. Their numbers dipped precipitously in the early part of the twentieth century, but after World War II, their remnants were recognized in a flock book. The breed's numbers picked up, thanks to its use as the primary sire breed of Mashams (see page 328).

Teeswaters never made an entry to North America, or if they did they left no trace. But today a group of breeders on the western side of the Atlantic is working on developing the breed through a system called *upbreeding* or *upgrading*. They are using imported semen to crossbreed to ewes of other (specified) longwool breeds. The offspring ewes from this cross are again inseminated with purebred Teeswater semen, and this process is repeated again, and again, and again. That last generation, the fifth, is considered to be more than 96 percent Teeswater and is classified as a purebred American Teeswater.

Long, lustrous, shiny wool: That's what it's all about with the Teeswater. The locks hang individually and don't clump together. There are no kemp or dark fibers, and the wool is uniform throughout the fleece. Because of the length, yarns with low twist work out well. It is a reluctant felting wool, at best.

On paper, Teeswater and Wensleydale wools can be described with similar numbers. The Wensleydale (see page 119) may grow a heavier fleece, while the Teeswater is considered to have the edge in meat production. Teeswaters have been bred for white wool, and our samples of Teeswater were a warmer white than similar Wensleydales. Based on what we've seen, Teeswater wool may have a slightly greater sense of body than Wensleydale, like a weighted silk, but they're similar enough that this may just be our personal fantasy.

Teeswater Facts

▶ **FLEECE WEIGHT**
7½–18 pounds (3.4–8.2 kg), often 12–15 pounds (5.4–6.8 kg)

▶ **STAPLE LENGTH**
12–15 inches (30.5–38 cm) first clip; thereafter (twice a year) about 6 inches (15 cm)

▶ **FIBER DIAMETERS**
30–36 microns (spinning counts 44s–50s)

▶ **LOCK CHARACTERISTICS**
Very long, wavy locks with brilliant luster and a smooth surface.

▶ **NATURAL COLORS**
Warm, clear white.

Teeswater yarns from Teeswater Wools

clean

Teeswater

2-ply

raw

Length means low-twist yarns can work well but also means previous combing experience helps.

Using Teeswater Fiber

Dyeing. This fleece makes brilliant colors because of the luster.

Fiber preparation and spinning tips

- The locks can be spun (or woven) as locks to give texture or used unspun to make a fleece rug (the same is true of any of the English Longwools with long enough staples).
- Open by flicking, although extra-long fibers require more work to flick than moderately long ones. Combing works very well, but because of the length and strength of the fibers, a beginning comber will need to remember to do *looooooong* strokes; draw off the top with a combination of firmness, widely spaced hands, and patience.
- Be sure to keep your hands far enough apart to draft the very long fibers effectively.
- Like Cotswold, Lincoln, and Wensleydale, among others, Teeswater is a great source of texture in yarns, or for use as dolls' hair.

Knitting, crocheting, and weaving. Teeswater is a great weaving yarn and is exceptionally durable without being stiff or scratchy. In knitting or crochet, it drapes well and won't incorporate much air, so it will be sleek and shiny and have terrific stitch definition. It can be laid directly (unspun) into woven, knitted, crocheted, or other ground fabrics.

Best known for. Luster and length, along with being fine for a longwool.

The Downside of Upgrading

The practice of upgrading is controversial in the case of the American Teeswater, American Wensleydales (page 119), and American Gotlands (page 163). Why? Our friends Phil Sponenberg and Don Bixby, of the American Livestock Breeds Conservancy, answer best in their book, *Managing Breeds for a Secure Future: Strategies for Breeders and Breed Associations* (American Livestock Breeds Conservancy, 2007).

"The importation of Wensleydale and Teeswater sheep semen has posed a direct short-term threat to other luster longwool breeds in this country, such as Cotswold, Lincoln, and Leicester Longwool. Ewes of these three breeds are designated . . . as the only recognized recipients of this semen to establish a base for upgrading to high-grade Wensleydales and Teeswaters. The irony . . . is that the breeds from which these ewes come are rare in their own right, and in some cases more than the imported resource. . . .

"The Wensleydales and Teeswaters in the USA, however, will be the result of an upgrading and will likely never be recognized as full Wensleydales or Teeswaters in their country of origin due to breed regulations in the United Kingdom. Upgraded sheep will therefore never be able to contribute to international breed numbers or conservation efforts for Wensleydales and Teeswaters. This situation has made one rare breed directly threaten others. . . .

"A further complication . . . is that these breeds do indeed tend to compete directly with one another for space on farms. A space taken by a Wensleydale or Teeswater upgraded sheep is very likely to have been filled by a purebred Cotswold, Lincoln, or Leicester Longwool rather than by a crossbred or non-longwool sheep. This becomes a zero-sum game in many instances, with the import displacing and preventing meaningful conservation of another breed in danger and already established in this country."

Controversy aside, the upgraded animals are providing North American fiber lovers with some great wool, and if all our spinners, knitters, weavers, and other fiber users start actively seeking out breed-specific yarns from all the conservation breeds, they will help protect their genetic resources around the globe. That is a good thing!

Wensleydale

Conservation Breed

WENSLEYDALE may be the only breed that can be traced directly to a single ancestor. In 1839, a ram lamb was born in North Yorkshire to a Mug ewe (an old-type Teeswater ewe that didn't show much of the New Leicester influence). The paternal ram was a New Leicester. The off-spring had the blue head and ears that show up as a recessive trait in Leicesters from time to time and was named Bluecap by its owner.

Bluecap grew up to be a potent ram and was leased by shepherds through a fairly wide area for a number of years. He was primarily used for breeding Teeswater ewes. His blue-headed trait passed to his progeny, and by the 1870s, these unique sheep (although closely related to the Teeswater) were recognized as a separate breed and called Wensleydales. Like the Teeswaters, there was never a population of Wensleydales in North America. However, breeders today are upgrading to produce them, using imported semen on ewes of a variety of related (and frequently also rare) breeds.

As befits their close relationship to the Teeswaters, Wensleydales grow long, lustrous, shiny wool. In spite of the many decades the two breeds have been considered to be distinct, the difference between them is hard to tell by looking at the numbers used to describe their fleeces. Statistically, they have nearly identical wools, and it's likely that their fibers handle similarly in industrial situations. Yet to a handspinner, there does seem to be a subtle yet definite difference between them.

Like Teeswater locks, Wensleydale wool hangs in distinct, curly ringlets and doesn't felt well. It also has no kemp, and the wool is uniform throughout the fleece. And because of the length, yarns with low twist work out well.

Among our samples, the white Wensleydale seems to be a cooler, brighter color than the Teeswater — and there are well-established strains of colored Wensleydales that produce gorgeous shades of gray and black. It's probably illusion, or an effect of the specific structure of the wool fibers, but the Wensleydales seem to be a bit more supple than the Teeswaters, while the Teeswaters have a slightly bolder presence.

Wensleydale yarns from Garthenor, Spinning Flock Farm, and somewhere in Scotland

Wensleydale Facts

▶ **FLEECE WEIGHT**
7–20 pounds (3.2–9 kg), often 10–15 pounds (4.5–6.8 kg)

▶ **STAPLE LENGTH**
7–12 inches (18–30.5 cm)

▶ **FIBER DIAMETERS**
30–36 microns (spinning counts 44s–50s)

▶ **LOCK CHARACTERISTICS**
Very long, wavy, distinctive, curling locks with brilliant luster and a smooth, supple surface.

▶ **NATURAL COLORS**
White, gray, black.

Using Wensleydale Fiber

Dyeing. The exceptional luster produces clear, bright colors.

Fiber preparation and spinning tips. See the notes for Teeswater (page 118).

Knitting, crocheting, and weaving. Wensleydale works exceedingly well as a weaving yarn, with unusual fineness for a longwool. Both Wensleydale and Teeswater nudge slightly more toward tapestry, upholstery, and fine worsted woolens than rugs, because of their relatively fine fiber diameters. Wensleydale brings drape and clear stitch definition to knitting, crochet, and other construction methods, unless it's spun into a dynamically textured yarn that becomes a statement in itself.

Best known for. Luster, long staple, and fine fiber diameter for a longwool.

clean

raw

2-ply, *combed*

Luster and length abound, and, like Teeswater, the fiber feels finer than the micron counts suggest.

clean

2-ply, *combed*

Wensleydale

singles, *lightly combed and spun for texture*

clean

raw

Gulf Coast Native ewe
ARapawa Island sheep

Feral GROUP

Cats probably come to mind first when people hear the term *feral*, but livestock breeds can go feral as well. Because sheep are vulnerable to predators, most feral sheep populations exist on islands, where predation is minimal. There are a couple of exceptions; of the breeds we cover here, the Gulf Coast Native is one.

When any livestock group goes feral on an island, it may cause dramatic changes to the ecosystem, including shifts in vegetation and erosion patterns or wildlife populations. As a result, government agencies and environmental organizations have often decided to eliminate or reduce the feral populations, frequently through wholesale slaughter. As this has gone on, groups dedicated to preserving agricultural biodiversity have worked to conserve at least some animals from these unique, potentially valuable, and irreplaceable populations. Fiber and products from many of these breeds are being marketed specifically to help with conservation efforts. Conservation needs to happen not just for the novelty of these sheep, but because they, like many other breeds of endangered animals, are the living repositories of genetic traits that we may need in the future.

Counter to the prevailing idea that sheep destroy natural environments, some breeds are being used to restore wildlife habitats. One example in the British Isles involves the dependence of the endangered Red-Billed Chough (a bird that has become extinct in most of the British Isles) on the presence of semiferal Manx Loaghtan sheep, which help maintain an appropriate environment for the choughs' feeding habits (see page 172).

While sheep of the feral group are often small compared to other, more refined breeds (and produce smaller quantities of both wool and meat), they are generally scrappier and very hardy, because they have had to survive on their own in harsh circumstances. They can make use of scarce or unusual food sources, such as seaweed, in the case of North Ronaldsays, and not much at all, in the case of the Soay, who made their home on wind- and sea-battered rocks in the north Atlantic Ocean. Most have good disease and parasite resistance.

Feral sheep grow black, brown, or gray wool more frequently than their tame relatives do. Horns are normal for feral rams and are not uncommon in ewes. Depending on their genetic background, sheep of feral and semiferal breeds may have a more wild-type fleece, with a soft wool undercoat of wool and a protective outercoat of hair, possibly containing kemp. Like their wild ancestors, these sheep tend to shed their wool naturally in warm weather. Often, the under- and outercoats loosen at separate times, providing a natural separation of the fiber types. Even feral sheep that have single-coated fleeces (including, interestingly, those with Merino ancestry) may shed their wool.

We have included two notable feral breeds, the Boreray (page 154) and Soay (page 194), with their relatives in the Northern European Short-Tailed family, rather than here, because of their history.

Gulf Coast Native

Critical Conservation Breed

This breed has a bit of an identity crisis. Depending on whom you talk to, you could hear it referred to as the Gulf Coast Native Sheep, Woods Sheep, Gulf Coast Sheep, Common Sheep, Pineywoods Sheep, Florida Native, Louisiana Native, or Scrubs Sheep. Sheep brought to North America by the earliest Spanish and French explorers evolved in response to natural selection and environmental pressures in wet, hot, and buggy regions that weren't well suited for the species. The animals that survived over time required little attention from people.

By the start of the twentieth century, hundreds of thousands of half-wild sheep were living in the southeastern United States. After World War II their numbers plummeted, when "improved breeds" were brought into the area, now able to survive the heat and parasites with the help of modern worming medications. Today, Gulf Coast Natives are so vulnerable to extinction that simply maintaining viable flocks of live animals amounts to a major achievement.

They are worth saving! Excellent foragers, Gulf Coast Natives will consume many noxious plants, such as kudzu and honeysuckle. Their lambs are small but hardy, and Native lambs survive more often than do the offspring of other breeds in the South. Their best trait, however, is critically important for Southern shepherds: they have outstanding resistance to many internal parasites and diseases (such as foot rot) that still plague sheep in the region.

Wool quality is variable. Care of the wool as a crop is coming along as a priority, and the breed can be a source of nice wool that sometimes falls within the next-to-skin comfort levels. Our sample came from a ewe who has lovely potential, but whose fleece for that year was affected by a series of nasty storms, which subjected her to physical, emotional, and nutritional stress. Stress causes weak spots in the year's growth of wool. In addition, a long period of steady humid weather affected the fiber quality. The wool was nonetheless a pleasure to work with. If you want to try your hand at felting, this is a fiber to look for.

Gulf Coast Native yarn from Running Moon Farm

Gulf Coast Native Facts

▶ **FLEECE WEIGHT**
4–6 pounds (1.8–2.7 kg)

▶ **STAPLE LENGTH**
2½–4 inches (6.5–10 cm)

▶ **FIBER DIAMETERS**
26–32 microns (spinning counts 48s–58s)

▶ **LOCK CHARACTERISTICS**
Single-coated fleece; open, wavy, and/or crimpy fibers; low in grease. .

▶ **NATURAL COLORS**
Mostly white; some tan, dark brown, black, and multicolors in patches.

Using Gulf Coast Native Fiber

Dyeing. This fleece is a fine candidate for dyeing.

Fiber preparation and spinning tips. The fleece is easy to wash, and the vegetable matter should fall out nicely because of the low quantity of grease. Short fleeces will card nicely, and the medium to longer ones will be happy with combing. The wool has nice, open, reasonably long staples. Use any prep and spinning techniques you feel like exploring.

Knitting, crocheting, and weaving. The fleece is crisp enough to work well to show stitch or weave textures.

Best known for. Ability to survive in a decidedly sheep-unfriendly environment.

Gulf Coast Native

2-ply

This fleece grew during a year of many hurricanes, and the effects of the weather can be seen and felt in the locks.

clean

raw

Hog Island

Critical Conservation Breed

HOG ISLAND is a 70-square-mile (181 sq km) barrier island located off the coast of Virginia, near the mouth of the James River. The island was settled almost four hundred years ago, and sheep were raised extensively there in communal fashion, since there was no need for fences. Owners marked their ewes with unique ear notches, much like cattle ranchers later used brands. In the spring, islanders would join together and gather the flock. Ewes and rams would be sheared, and lambs received the same ear-notch patterns as their dams.

The sheep went feral in 1933, when the human inhabitants abandoned the island after a devastating hurricane. In 1974, the Nature Conservancy acquired the island and removed the sheep to improve survival of the native grasses. Some of the sheep found new homes at historic sites in Virginia, such as Gunston Hall Plantation (which no longer has its Hog Island flock) and Mount Vernon (which continues to provide a number of Hog Island sheep with a home). A handful of people, including a few dedicated shepherds on private farms, are playing critical roles in preserving the Hog Island breed.

There are so few Hog Island sheep that it's difficult to make generalizations about the wool, except that it is medium in quality and comes in a beige-toned white, some grays, and a nice black. We had computerized analyses done on our samples to establish a few data points. The white averaged 30.5 microns, although

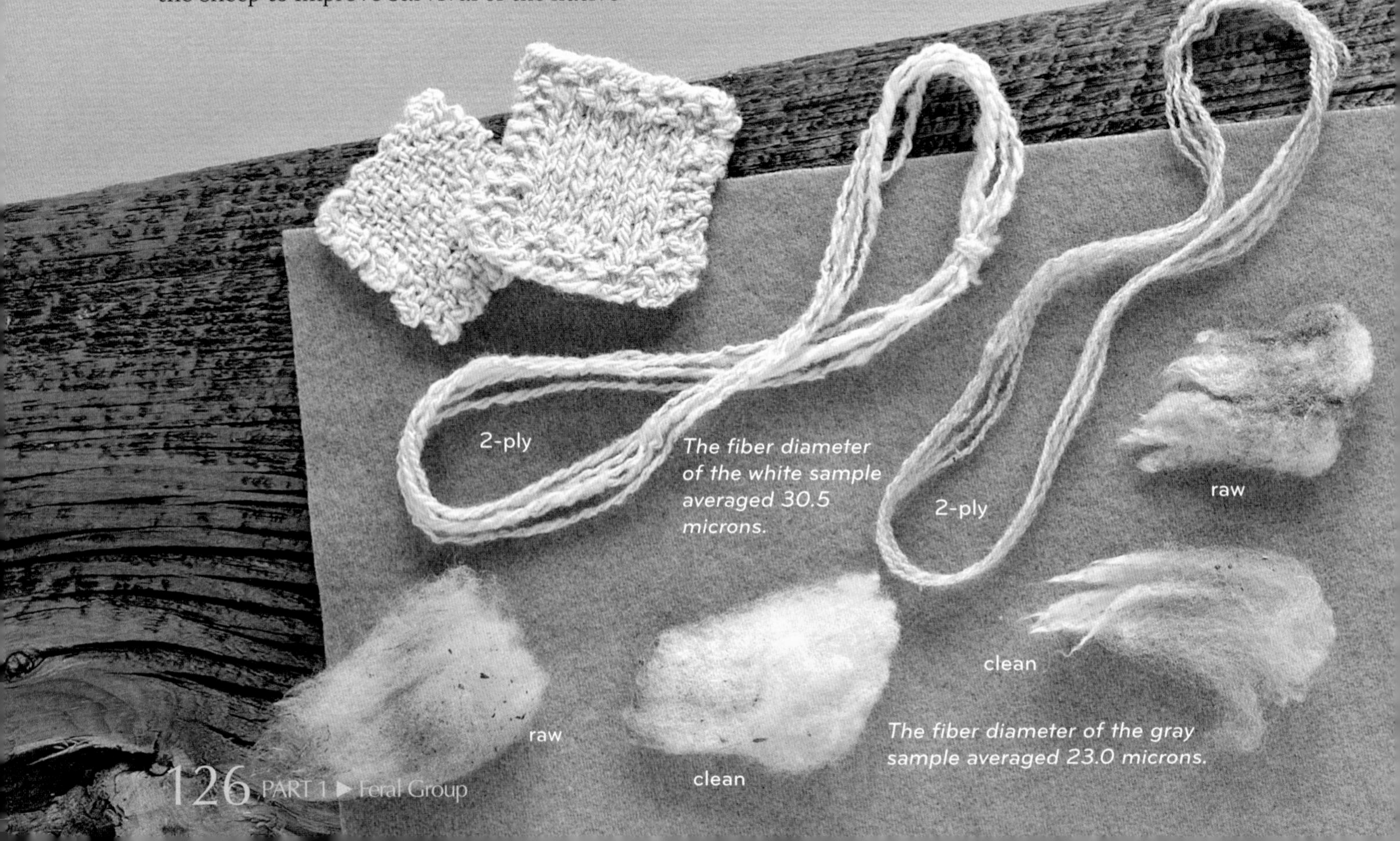

2-ply

The fiber diameter of the white sample averaged 30.5 microns.

2-ply

raw

clean

raw

The fiber diameter of the gray sample averaged 23.0 microns.

clean

more than half of its fibers were larger than 30 microns. This means it was not an obvious candidate for next-to-the-skin wear, although it would make good hats, mittens, and outer-wear sweaters — or sturdy blankets. On the other hand, the black wool we tested averaged 22.6 microns, with more than 90 percent of its fibers at or under 30 microns, so a fair number of people would likely find it appropriate for use next to the skin (5 percent of fibers at or above 30 microns is the suggested cut-off for manufacturers' next-to-the-skin use).

As a feral breed, Hog Island sheep have developed the ability to gradually shed their wool, although they are commonly shorn now that they are in off-island flocks maintained under closer shepherding supervision. The fleeces are unusually high in lanolin, which protects the wool from harsh weather (a plus for residents of a storm-battered coastal island) but lowers the clean yield of fiber.

Hog Island

The fiber diameters of both black samples averaged 22.6 microns.

raw

clean

2-ply

2-ply

raw

clean

Hog Island Facts

▶ **FLEECE WEIGHT**
2–8 pounds (0.9–3.6 kg), usually in the range of 3–5 pounds (1.4–2.3 kg)

▶ **STAPLE LENGTH**
Insufficient data; samples we examined ranged from 1½ to 2½ inches (3.8–6.5 cm)

▶ **FIBER DIAMETERS**
Inconsistent and insufficient data; fine to medium (estimated average range 22–32 microns)

▶ **LOCK CHARACTERISTICS**
Single-coated fleece, high in lanolin, with locks somewhat rectangular and dense; disorganized crimp.

▶ **NATURAL COLORS**
White, sometimes black (1 or 2 of every 10 sheep).

Hogg Island

On old maps the island's name was sometimes spelled with two Gs. In Britain, *hogg* (or *hogget*) means a sheep under one year of age.

Using Hog Island Fiber

Dyeing. A matte surface and natural pigment means that colors may be muted.

Fiber preparation and spinning tips. The generally short length and disorganized crimp mean that machine preparation may lead to neps and other irregularities that will make it difficult to spin an even yarn. Hand carding or combing, depending on length, is likely to be the best approach. The biggest challenge in spinning will be fiber length; for yarn integrity, keep the singles reasonably fine so that the fibers are well caught in the twist.

Knitting, crocheting, and weaving. The fleece makes a yarn that traps a lot of air and has a slightly crisp feel. It can be used for sturdy garments (sweaters, socks, blankets) that may have next-to-the-skin qualities, dependent on the specific fleece.

Best known for. Not well known at all! Enjoyable and satisfying to work with, if you are a spinner with some experience in fiber preparation and can get your hands on good-quality fleece — or if you are a nonspinner lucky enough to find some already spun into yarn.

New Zealand Ferals

Conservation Breeds

ALTHOUGH CAPTAIN COOK brought sheep with him to New Zealand in the eighteenth century, those animals did not survive. Sheep found in the islands of New Zealand today were brought in primarily during the nineteenth century, and some arrived during the twentieth century. Husbandry practices were relatively lax in New Zealand when flocks were established, and a number of animals began to live under feral conditions on islands large and small. Some of these flocks remain, although they have been at risk because of conflicts with more controlled livestock management and wildlife priorities. The feral breeds on Arapawa Island and Pitt Island have some supporters working on their behalf, and you may run across their fiber someplace. We were able to get samples of both. Some of the fiber we obtained at a major wool festival, all prepared and ready to spin.

Arapawa Island is a small speck of land in the South Pacific, located off the northeastern coast of New Zealand's South Island. No one is sure exactly how, or when, the sheep (or goats and pigs, for that matter) got to the island. During early European exploration and colonization of the region, livestock was carried shipboard; probably some of these animals were given to native Maori tribes as gifts or simply dropped on various islands to create feral populations that voyagers on future ships could hunt. There was a whaling station on Arapawa Island in the 1830s, so there surely would have been some populations of European livestock in residence by then; however, the first written documentation of sheep on the island was for a flock of Merinos introduced in 1867.

Various feral critters thrived on the island for over a century, but in the 1970s, the government of New Zealand planned to eliminate the livestock. All feral animals were to be shot. When making their plans, though, they didn't count on Betty Rowe. Betty and her husband Walt were American expatriates who settled on Arapawa Island in 1972. The couple and their kids fell in love with the island's wildlife, including the wild sheep, goats, and pigs. Betty spearheaded a movement to save the feral breeds of Arapawa, creating a 300-acre (121 hectare) sanctuary and raising awareness. Walt has now passed away, but Betty has stayed the course and is still a feisty advocate who has managed to shine an international light on the island's feral breeds.

About 500 miles (804 km) east of the main islands of New Zealand, Pitt Island is part of the Chatham archipelago. Due to its proximity to the international date line, it is the first inhabited place in the world to experience each new day. (Or century! At the end of 1999, many people traveled there to watch the new millennium dawn.) For over a thousand years, the island was home to a small population of Moriori people (of similar heritage to New Zealand's Maori), who called it Rangiaotea. The island has fewer than 40 human residents today, so the island's feral sheep far outnumber the people. (This is also true on the main islands of New Zealand, where there are 13 times as many sheep as there are people! Although of course not all these sheep are feral.)

Many of the New Zealand feral sheep have considerable amounts of Merino blood, resulting in fine fleeces. Like most feral animals, they often shed their wool. Locks are fine, crimpy, and relatively short stapled. An unusual aspect of these sheep is the color range of their wool. The Arapawa Island and Pitt Island flocks are almost all black. Colored fleeces also occur in the other feral populations, including the Diggers Hill, Hokonui, Mohaka, Omahaki, and Stewart Island flocks. Still, among the fine-wooled feral groups, the Clarence Reserve sheep have white wool, and most of the Chatham Island sheep (which otherwise resemble Pitt Island animals) are white. Woodstocks are basically white-fleeced, but some animals have a badger-faced pattern of tan or creamy stripes and splotches reminiscent of a badger's coloring.

Two of the identified rare feral flocks have stronger (or coarser) wools. The Campbell Island sheep are only one-quarter to one-half Merino (the remainder is Longwool) and are predominantly white. The Raglan flock more closely resembles early Romneys (see page 110), and the appearance of these animals suggests that there may be some Cheviot (see page 54) influence as well.

Feral Sheep of New Zealand

NAME	THEORETICAL SOURCE	NOTES	DATE INTRODUCED
Arapawa Island	Merino origins	Most are colored, commonly black. There are also grays and white to cream-colored animals. Some have spots, or have white face markings. The wool is fine (as delicate as 11 microns) and high bulk, and felts well. The sheep have a tendency to shed.	Livestock, including sheep, may have been dropped off by whaling expeditions as early as the 1500s; a documented introduction of Merinos occurred in 1867.
Campbell Island	Merino + Longwool (like Lincolns or Leicesters)	No sheep are currently on the island. Some are being bred on the New Zealand mainland. Wool studied has measured 18 to 32 microns, with most between 26 and 29 microns. Fleeces weigh 2¾–4 pounds (1.3–1.8 kg).	1895 (Merinos), 1901 (Merinos), 1902 (probably Longwools). The island sheep were abandoned in 1931 and had all been exterminated or removed by the late 1980s.
Chatham Island	Merino characteristics	May have originated from Saxon Merinos taken to nearby South East Island in 1841. Similar to Pitt Island, but Chathams are white.	Possibly 1841.
Clarence Reserve	Merino origins	All are white; fine but light fleece.	Origins unknown; may have escaped from European settlements in the late nineteenth century.
Diggers Hill (Deans Forest, Fiordlander, Takitimu)	Basically Merino	Tendency to shed; fleece fine but chalky. A newly established breed standard suggests 16–20 microns for fiber diameter.	Late 1880s, from escaped and abandoned livestock.
Herbert (Kakanui, Waianakarua)	Unknown	Few have been domesticated, and little information has been gathered.	Escaped from domestication at unknown times.
Hokonui	Merino type	Foundation animals were likely Saxon Merino. A newly established breed standard calls for fine, kemp-free wool. The sheep come in a number of colors.	Escaped and abandoned sheep from about 1850.
Mohaka	Merino type	Feral Merinos with black or white wool.	Well established by the 1880s.
Omahaki (Ngaruroro)	Merino origins	Percentage of black sheep has been increasing; similar to Mohaka sheep.	Likely became feral during the nineteenth century.
Pitt Island	Merino type	See notes for the origins of Chatham Island sheep. Of the Pitt Island group, about 85% are solid black, 10% white, and 5% brown. They shed their fleeces.	Possibly 1841.
Raglan	Resemble early Romney imports to New Zealand, with some Cheviot-like characteristics	Not much information is available.	Unknown.
Stewart Island	Merino type	Similar to Arapawa Island; mostly black, with white patches on face and on tip of tail.	Escapees from established farms, beginning in 1874.
Woodstock	Merino origins	Most have white wool; some are badger-faced, and this is the only New Zealand feral group with this characteristic.	Escapees from established farms, beginning in the 1890s.

Santa Cruz

Critical Conservation Breed

THESE SHEEP are named for Santa Cruz Island, the largest in the Channel Islands chain off the California coast. How and when the sheep got to the island is a mystery, though researchers suspect that Merinos or Rambouillets may have been among the parent stock. In the 1890s, Santa Cruz — a fairly large island measuring almost 100 square miles (259 sq km) — was a ranch owned by Justinian Caire. Sheep were an important commodity during Caire's tenure, with the flock reaching a peak of 50,000 head. After Caire's death, his children fought over the disposition of the estate. By the 1930s, when several of the children sold their shares of the island, the sheep population had fallen dramatically, and the remaining animals had gone feral.

In 1978, the major portion of Santa Cruz Island was acquired by the Nature Conservancy (TNC), and the remainder was turned over to the National Park Service in 1997. TNC began a program to eradicate the sheep by slaughtering them on site, but the American Livestock Breeds Conservancy intervened and managed to bring some of the sheep off the island, with the help of five California shepherds. Today several of those shepherds have retired, so there are only a couple of flocks of Santa Cruz sheep. The SVF Foundation in Newport, Rhode Island, is helping to protect the fewer than 150 animals that are left. SVF's mission is to help maintain some of the most critically threatened livestock breeds through a program in which they gather embryos and semen from these breeds to act as a gene-bank collection.

Because they are one of the critically endangered island sheep breeds, there isn't enough data on Santa Cruz fleeces to fill in the information profile. But the SVF staff, Jennifer Heverly, of Spirit Trail Fiberworks, and Sonja Straub, of Legacy Farm, provided us with samples; we've also seen and felt (although not spun) wool from this breed in the past and remember it very fondly. The existing Santa Cruz animals are for the most part being cared for with a view to their physical survival, not to the production of the best fiber for hand crafters. Nonetheless, we'd be willing to go out on a limb and say that Santa Cruz is one of the softest and pleasantest wools we've ever encountered, with astonishing potential for elasticity in the yarn. Whatever else the breed's survival stands to offer, the fleece should be factored in with a big plus.

Like many other feral breeds, Santa Cruz sheep shed their fleeces if they don't get shorn. This is a rarity for sheep with Merino and Rambouillet in the family tree unless those sheep have been feral for quite a while.

We pulled a lock of our white sample and sent it off to the lab. The results came back with an average micron count of 23.2 and a comfort factor of 97.3 percent — which means almost all of the fibers measured 30 microns or less. This is firmly in the realm of next-to-the-skin soft. Some of the fibers in the array measured as fine as 14 microns. The black represents a very nice depth of color in a soft, unusual wool. Santa Cruz wool is another ready felter, if you have the chance to find some of this very rare fiber.

Santa Cruz Facts

▶ **FLEECE WEIGHT**
Insufficient data; small animals, so fleece weight will be relatively light

▶ **STAPLE LENGTH**
We're guessing 2–4 inches (5–10 cm)

▶ **FIBER DIAMETERS**
Insufficient data; fine; on pure speculation, estimated average range 18–26 microns (approximate spinning counts 58s–80s)

▶ **LOCK CHARACTERISTICS**
Insufficient data; samples here are very finely crimped but disorganized in the staple.

▶ **NATURAL COLORS**
Mostly white; a few are medium brown, and some are a dark brown that appears nearly black.

Using Santa Cruz Fiber

Dyeing. The fleece has a matte surface, so dyes will be clear but not shining.

Fiber preparation and spinning tips. Wash as for a fine wool to remove the lanolin and suint, then it's happy to be carded at shorter staple lengths or combed when it's longer. Our samples were quite short, at about 1½ inches (3.8 cm); we don't think they're representative of the breed's potential. Experiment with techniques to elicit the extraordinary elasticity.

Knitting, crocheting, and weaving. This fleece is beautifully soft. Keep that in mind, and put this wool in the comfort corner, instead of asking it to hold up to use in a rug.

Best known for. Being extremely rare. Delightful softness when husbandry and environmental conditions provide the animals with wool-oriented nutrition and lack of stress. Astonishing potential for elasticity.

Saxon MeRino
Debouillet

Merino FAMILY

Merinos are truly the royal family of the wool world. They may also bear the only breed name that is commonly recognized, because it so often appears in catalog descriptions and on yarn labels. The family also happens to have a royal history!

Spain, separated from Morocco by a mere 8-mile (13 km) gap of water known as the Strait of Gibraltar, has long been a point of ready exchange between Europe and Africa. The history of the Merinos is one that crosses both sides of this blue ribbon between the Atlantic Ocean and the Mediterranean Sea. In the twelfth and thirteenth centuries, Spanish royalty began importing rams from the Beni-Merines (members of a Berber tribe centered in the area of present-day Morocco) to cross with their best sheep. This was a match made in heaven. The Spanish ewes, when bred to the African rams, yielded fine-wool sheep like none other. They were named Merinos after the African tribe.

By the Middle Ages, the Spaniards had refined these sheep to a point where Spain held a corner on the European wool market. They outlawed the export of Merino sheep, with the penalty of death to anyone found breaking the law. Then in the 1700s, some Spanish monarchs began giving breeding stock to their relatives in other European courts, ultimately leading to the Merinos' expansion across Europe and around the globe. Starting in 1793, Merinos were exported to North America from Spain in several small lots. In 1809, William Jarvis, a Vermont native and then American consul to Portugal, brought over a ship full of Portuguese Merinos (more than 3,500 animals), providing the real foundation for Merinos in America.

The ubiquitous nature of Merino wool in the marketplace, and the fact that it is processed industrially in such massive quantities, might give the impression that Merino is always Merino, that it's a single, consistent type of soft, fine fiber. Not so. All Merino falls within what is generally considered the fine range of wools, and it can all be used for next-to-the-skin wear, although the softness levels vary noticeably; the span from the finest Merino wools to the coarsest is wider than the range of most other breeds, running from about 11.5 microns to perhaps 26 microns at the upper end. There are ultrafine Merinos, strong Merinos (coarse for this family, and the most durable Merino fiber), and every step in between. Unless you seek out finer fibers — either in the fleece or by seeking labeled yarns from, say, Tasmanian Merino (see page 139) or Sharlea (see page 146) — the commercially prepared yarns and fabrics you see will mostly include Merino in the 20- to 25-micron part of the breed's range.

Merinos grow large quantities of dense, fine wool with regular crimp patterns. The high density of Merino wool is due to the large number of wool follicles in the skin, compared to other breeds. In round numbers, a fine Merino has 72 wool follicles per square millimeter, a medium Merino has 65, and a strong Merino has 57. Compare that to a Corriedale, with

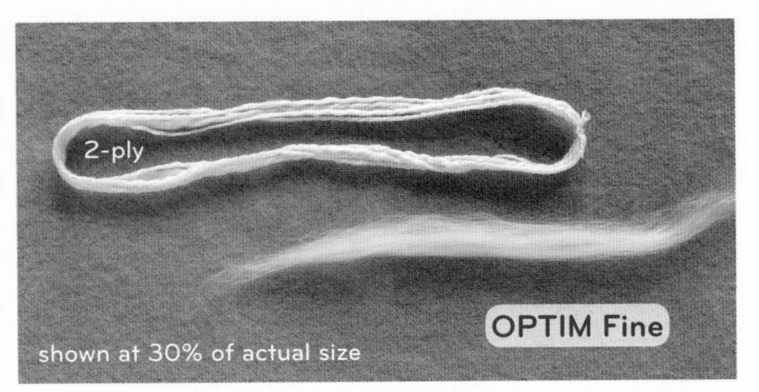

shown at 30% of actual size

OPTIM

CSIRO, a worldwide scientific research institution, has developed two ways of treating medium-fine Merino wools to make them behave differently.

OPTIM Fine. OPTIM Fine, found as a handspinning preparation and in ready-to-use yarns, involves stretching 19-micron Merino fibers 40 to 50 percent and then permanently, chemically setting them in that condition. Stretching reduces the fiber diameter by 3 to 3.5 microns, yielding diameters of 15.5 to 16 microns. It also eliminates some crimp. The resulting fiber is strong and silky, still absorbs water, and drapes well — but lacks Merino's characteristic crimpy resilience. Stretching increases the fiber's luster as well.

Spinners can obtain OPTIM Fine as sliver. One end may draft more easily than the other. Static may be an issue during spinning, and you'll probably need to use hand cream to keep the fiber from snagging.

OPTIM Max. OPTIM Max is used only in industry to make cloth. The fibers are temporarily stretched 30 to 40 percent, blended with regular wools, spun along with those wool fibers, and then restored to their original length through washing in hot water. This produces an unusually high-bulk, lightweight yarn that can be made mechanically — and machines are not normally good at spinning yarns of this type.

just 29 follicles per square millimeter, or a Suffolk with 20. Lincolns and Cheviots only have around 15. In addition, Merinos have far more secondary follicles than other breeds. (Remember, these are the kind of follicles that grow the finer fibers in a fleece.)

Closely packed fibers like these mean that dirt can't easily get down into the staples, which are also protected by a heavy coat of grease. As a result, the tips are frequently weathered, but the bulk of the wool stays clean. Getting the grease fully removed becomes one of the tricks to working with Merino, and the same is true of all similarly developed fine wools. If the grease partially dissolves in inadequate washing and then is redeposited on the fiber, it can be a bear to ever get off and makes the wool hard to spin; it also contributes to a grubby look in the yarn and fabric. Secrets to success include really hot water that is not allowed to cool off, a reliable cleansing agent (an effective scouring aid), and no agitation at all, because Merino loves to felt. While it's best to store all wools clean (moths enjoy greasy wool even more than clean wool), fine wools like Merino are much easier to wash if the job is done as soon as possible after shearing.

Merino is a near-miraculous wool. Yet its easy availability means it is often used to make textiles for which another breed's wool would be much more appropriate. Select one of the finest Merinos to make an exquisite, delicate, extraordinarily soft garment (and wash it oh so carefully). Use a comparatively sturdier Merino for plenty-soft fabrics that will wear better (and still be gentle with your washing techniques). For substantial durability, look to another breed.

Types and Strains of Merinos

Merino flocks are historically managed to maximize their productivity. What that means depends on the human goals behind the particular flock of sheep. There are easily dozens of types, strains, and identified breeds of Merinos found around the world. You are unlikely to come across fiber for every one, and most are not even marketed by their unique names.

In short: Our information on Merinos is incomplete; the breeding targets move constantly, seeking newly defined production-oriented targets; and we're still committed to sharing with you whatever we've managed to learn about this dominant breed's varieties of wool.

It can be either very easy to find data on types of Merino wool or nearly impossible. For example, Sharlea and Tasmanian Merino breeders aim for extremely precise and predictable fiber diameters in their flocks. Yet when we talked with an expert in Australia about Saxon Merino, we were advised that there's so much crossbreeding of the strains that the words *Saxon Merino* now refer more to a style of wool than any arbitrary descriptive qualities, like fiber diameter or staple lengths. (Our adviser was, nonetheless, able to give us an idea of where most wool that's called Saxon Merino is likely to fall in terms of fiber diameters.)

While we can't cover every contemporary strain of Merinos here, we have done our best to highlight the types you may see or hear about. Keep in mind that in this context the words *fine*, *medium*, and *strong* are relative to the range of Merinos, not the full scope of all wools.

Booroola Merino (Critical Conservation Breed). Booroolas were developed in Australia for high fertility rates and year-round breeding. The breed was once imported to North America, but in their pure form these sheep may now be extinct on this continent. That said, the Booroolas' B gene, a single gene that is responsible for their high birth rates, is found in some North American flocks of Merinos. Fleece weights range from 9 to 15 pounds (4.1-6.8 kg); yield is 55 to 70 percent. Staples are 3 to 4 inches (7.5-10 cm) long with fiber diameters of 18 to 23 microns (spinning counts 62s-80s).

Debouillet (Conservation Breed). See pages 142-143.

Delaine Merino. See pages 144-145.

Est à Laine Merino. A polled Merino of French origin that grows a fleece described as "very fine." The name translates to "wool of the east," reflecting the breed's development in northeastern France, on the border with Germany. There have been a few importations to North America in the last few decades, and our samples came from a North American flock, but this breed is mostly found in Europe. Fleece weights range from 10 to 14½ pounds (4.5-6.6 kg), and staples are about 4 inches (10 cm) long.

Fonthill Merino. An Australian type that was developed from a cross of American Rambouillet rams and Saxon Merinos, intended for both wool and meat purposes. The goal has been to keep wool production the same as for other Merinos while increasing the amount of meat produced per animal. We found no specific information on fleece weights or staple lengths, other than that they are like those of other Merinos raised under similar conditions. Fiber diameters range from 20 to 22 microns (spinning counts 64s-70s).

Peppin Merino. This strain is of paramount importance to the Australian wool industry. In fact, almost 70 percent of Australian sheep are Peppins, and people will refer to their Merinos as either Peppin or non-Peppin.

Although we located no individual information on staple lengths, the statistics we did find for this breed do shed light on its dominant

Merinos Compared to All Other Wools

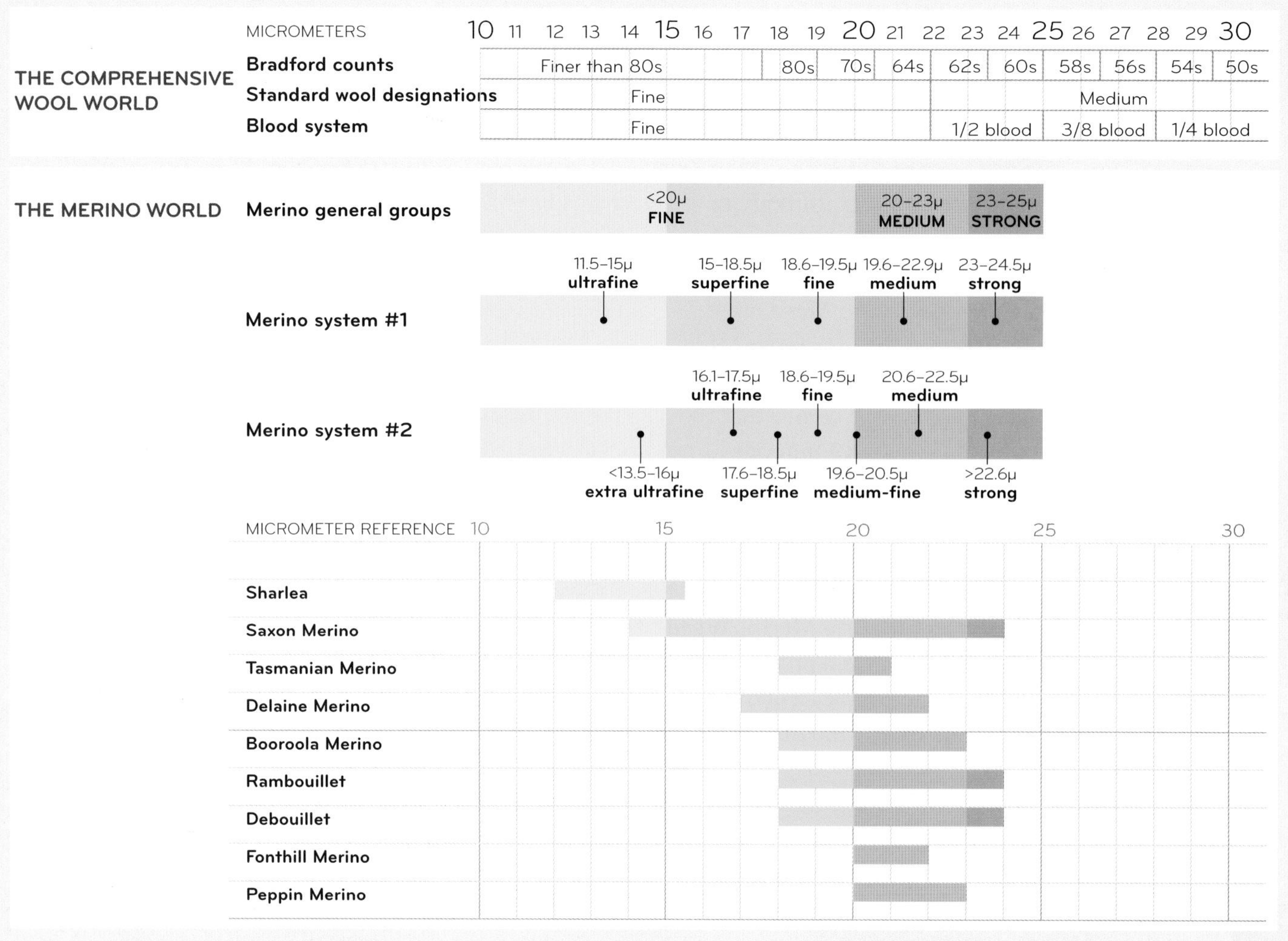

status. A single sheep can grow up to 40 pounds of wool (18 kg), with diameters of 20 to 23 microns (spinning counts 62s-70s). Those are potentially massive quantities of soft yet not overly fragile fiber.

Poll Merino. The polled trait (without horns) has shown up from time to time in Merino sheep due to a recessive gene. The polled trait has now been selected for in Australia, starting in the last decades of the twentieth century. Because the animals used in developing the Poll Merino came from within a variety of different flocks, there are fine-, medium-, and strong-wool Poll Merinos. The wool characteristics will follow those of the flock from which the polled strain has been developed.

Rambouillet. There are so many Rambouillets, derived from French refinements that began in the 1700s, that we've listed this breed separately, starting on page 148.

Saxon Merino. See pages 146-147.

Sharlea. See Saxon Merino, pages 146-147.

South Australian Merino. Also known as SA Merinos, these are a type of strong-wool

Merino. They thrive in dry areas of southern Australia, where rainfall can be as low as 5 inches (12.5 cm) per year. They are a dual-purpose Merino. We were not able to locate detailed information on individual wool characteristics, except that fiber diameters are up to 23 microns (spinning counts up to 62s).

Tasmanian Merino. Raised on the Australian island state of Tasmania, this strain is known for the production of fine Merino fiber, with most of its wool in the range of 18 to 21 microns (spinning counts 64s–80s). It is also extremely high yielding, at around 70 percent average, and noted for exceptional whiteness. You can buy yarns identified as Tasmanian Merino.

Type A Merino (Conservation Breed). Now comparatively rare and raised primarily in historic settings, this type was developed in Vermont with the intention of increasing the surface area of the skin, and therefore the wool yield, through breeding for wrinkles, like those seen on a Shar-Pei dog. Although the goal was reached, the skin folds make these sheep very difficult to shear. We found no individual wool information on this breed, which has also been known as the Vermont Merino.

Type B Merino (probably extinct). These were developed in Ohio to maintain Merino-quality fleece while moving the body toward higher meat productivity. We found no individual wool information on this breed.

Type C Merino. See Delaine Merino, pages 144–145.

More about Merino

If you are interested in Merino, be sure to find and read Margaret Stove's *Merino: Handspinning, Dyeing, and Working with Merino and Superfine Wools.*

Shrek the Sheep

Sheep are normally shorn each year, but a certain wether (castrated male) living in the mountains of Otago (in New Zealand) evaded the shearing squad for six years. When he was finally caught in 2004, his fleece weighed 60 pounds (27 kg)! He looked like some kind of strange woolly monster, with an uncanny resemblance to Shrek, the movie ogre, and thus he was named.

Shrek's owners, John Perriam and his wife, Heather, run more than eight thousand sheep on Bendigo Hill Station, a 27,000-acre (10,925 hectare) ranch — that's about 42 square miles (109 square km) — in a mountainous area of the South Island. It was this vast, craggy landscape that allowed Shrek to elude the shearers for so long. Once Shrek was captured, John knew he would have to clip Shrek with old-fashioned hand shears (which resemble a giant set of scissors) rather than using modern shaver-type clippers, because the fancy equipment wouldn't go down far enough into the wool without stalling out on every attempted pass. He arranged to have a champion shearer give Shrek his haircut before a New Zealand TV news crew and to auction fiber from Shrek for a New Zealand charity, Cure Kids, via the Internet.

Because of the Internet, Shrek's story spread quickly, with media from all over the world running stories about the fugitive sheep, who turned out to be quite gentle. The fleece raised $50,000 for Cure Kids, and the whole event not only helped make Shrek a celebrity around the world but also raised awareness about sheep production in New Zealand. A little over two years later, Shrek was sheared again — on his tenth birthday — to raise more funds for Cure Kids. For publicity's sake, this second shearing took place on an iceberg that was thought to have calved from the Ross Ice Shelf in Antarctica six years earlier!

General Merino Fact Sheet

For most Merino wool that we come across in yarns or spinning fiber, the strain or type isn't recorded. Exceptions include the occasional labeling of known-to-be-fine varieties like Saxon, Sharlea, and Tasmanian Merinos. Ultrafine Merino and lamb's wool will also be on the finer end of the overall range.

► **FLEECE WEIGHT**
Fleece weights range from 6½ to 40 pounds (3–18 kg), averaging 9–14 pounds (4.1–6.4 kg); yields from 35 percent to (very rarely) around 80 percent; as a rule of thumb, think 50 percent

► **STAPLE LENGTH**
2–5 inches (5–12.5 cm)

► **FIBER DIAMETERS**
11.5–25 microns overall (spinning counts 60s to finer than 80s), with most falling between 20 and 22 microns (spinning counts 64s–70s)

► **LOCK CHARACTERISTICS**
Dense, blunt locks, often weathered just at the tips, with a heavy, protective layer of grease. Crimp fine, tight, and well defined in both fibers and locks.

► **NATURAL COLORS**
Predominantly white because of the vast quantities of wool that are destined for industrial processing, where colored fibers become a major liability. Some strains are known for clear white wool, while some of the finest have straw-colored wool. Uncommonly, there are blacks, grays, and moorits (a shade of brown; see page 157).

Merino yarns from Koigu and from eBay

Using Merino Fiber

All of the Merinos have similar usage guidelines. The individual fleece or yarn you are evaluating will be somewhere within this breed's range, and its individual qualities will guide your decisions about how to use it best.

Dyeing. The whites take clear colors well.

Fiber preparation and spinning tips

- Getting the grease out is the biggest concern in preparing Merino for spinning. It requires hot water, a good scouring agent, and keeping the washing solution from cooling off so the grease and suint don't get redeposited on the fiber. Although washing needs to be thorough, it's also important to avoid agitating (and felting) the fiber.
- Some spinners wash individual locks. For a fiber that is likely to be spun very fine, this isn't unreasonable.
- You can spin from the lock, comb, or card. Because of its fineness, Merino wool tends to form neps. Use your fine-toothed combs, cotton carders, or special fine-fiber carding cloth on a drum carder.
- Merino wants to be spun fine. If you go for a thick yarn right off the bat, the wool will tend to draft in clumps, the ends of the fibers may not be secured well within the yarn, and your yarn will have a short lifespan and pronounced tendency to pill. To produce a thick yarn, spin several thin, nicely twisted strands and then ply them together.

Knitting, crocheting, and weaving. A wonderful fiber for lightweight, soft, and often delicate fabrics: Merino is a classic fiber for shawls, camisoles, and (with enough twist to wear well) baby garments.

Best known for. Large quantities of fine, soft, highly crimped wool with elasticity and resilience.

2-ply

2-ply

Merino

fiber length

top

Although most Merino wool is white (because of the breed's commercial focus), lovely colors do exist.

clean

raw

raw

2-ply

fiber length

roving

Of our prepared fibers, the white is very smooth and even. The black has many neps, and it will make a textured yarn no matter how it is spun.

Debouillet

Conservation Breed

HAILING FROM NEW MEXICO, this breed was developed in the 1920s by rancher Amos Dee Jones from crosses of Delaine Merinos and Rambouillets. Not many breeders outside the arid southwestern United States raise Debouillets, and they are hard to find, though the Jones family still maintains a flock today (most of their wool clip is sold directly to fiber mills, such as the Pendleton Woolen Mills — which has also been a primary destination for the white Romeldale clip). For a member of the Merino family, Debouillet has very long staples.

Debouillet Facts

▶ **FLEECE WEIGHT**
9–18 pounds (4.1–8.2 kg); yield 35–55 percent

▶ **STAPLE LENGTH**
3–5 inches (7.5–12.5 cm)

▶ **FIBER DIAMETERS**
18–24 microns (spinning counts 60s–80s)

▶ **LOCK CHARACTERISTICS**
Dense, blunt locks, often weathered just at the tips, with a heavy, protective layer of grease. Crimp fine, tight, and well defined in both fibers and locks.

▶ **NATURAL COLOR**
White.

Using Debouillet Fiber

All of the Merinos have similar usage guidelines. The individual fleece or yarn you are evaluating will be somewhere within this breed's range, and its individual qualities will guide your decisions about how to use it best.

Dyeing. Clear colors work nicely with whites.

Fiber preparation and spinning tips

- To remove the grease, use hot water and a good scouring agent. Wash thoroughly without agitating (felting) the fibers. It is important to keep the washing solution hot, so that grease and suint don't blend back into the fiber.
- For a fiber that is likely to be spun very fine, it isn't unreasonable to wash individual locks.
- Clean wool can be spun from the lock, comb, or card. To best prevent neps, use your fine-toothed combs, cotton carders, or special fine-fiber carding cloth on a drum carder. Work deliberately and with patience.
- It is easiest to spin Debouillet into fine yarn. If you want to produce thick yarn, ply together a number of thin, nicely twisted strands. If you try to spin a thick yarn, the wool will tend to draft in clumps, the ends of the fibers may come loose within the yarn, and the work is likely to have a short lifespan and pill easily.
- The crimp can also be challenging to handle, although it's also a nice thing to have because it contributes to elasticity.

Knitting, crocheting, and weaving. A great fiber for lightweight, soft, and often delicate fabrics, with enough twist to add durability.

Best known for. Large quantities of fine, soft, highly crimped wool with elasticity and resilience.

Debouillet

2-ply

Once the sticky tips were opened up, this fleece felt very fine and springy and was relatively easy to spin.

clean

raw

Delaine Merino

ALSO KNOWN AS the Type C Merino, this variety has smoother skin than some other Merino strains, which makes shearing easier. Delaines are practical, sturdy sheep that do well on the western ranges of the United States, although they can adapt to all types of terrain, from deserts to mountains. Like other Merinos, Delaines have very long productive lives, running to a decade or more.

Delaine Merino Facts

▶ FLEECE WEIGHT
9–14 pounds (4.1–6.4 kg); yield 45–55 percent

▶ STAPLE LENGTH
2½–4 inches (6.5–10 cm)

▶ FIBER DIAMETERS
17–22 microns (spinning counts 64s to finer than 80s)

▶ LOCK CHARACTERISTICS
Dense, blunt locks, often weathered just at the tips, with a heavy, protective layer of grease. Crimp fine, tight, and well defined in both fibers and locks.

▶ NATURAL COLOR
White.

Using Delaine Merino Fiber

All of the Merinos have similar usage guidelines. The individual fleece or yarn you are evaluating will be somewhere within this breed's range, and its individual qualities will guide your decisions about how to use it best.

Dyeing. Clear colors will dye nicely with white fleeces.

Fiber preparation and spinning tips

- Remove the grease with hot water and a good scouring agent, maintaining a hot temperature so that grease and suint won't blend back into the fiber. Do a thorough job of washing, but be careful not to agitate (felt) the fiber.
- Since the fiber is likely to be spun very fine, it isn't unreasonable to wash individual locks.
- You can spin from the lock, comb, or card of clean wool. As with all Merino wools, you will need to take measures to best manage neps. Use your fine-toothed combs, cotton carders, or special fine-fiber carding cloth on a drum carder, and work patiently and deliberately.

◎ The character of Delaine Merino fleece lends itself to being spun fine. To produce a thick yarn, ply together a number of thin, nicely twisted strands. If you try to spin a thick yarn, you will likely be disappointed: the wool may draft in clumps, with loose fiber ends poking through the yarn, and the work will pill easily and have limited durability.

◎ The crimp has the benefit of contributing elasticity to your yarn, but it can be challenging to handle.

Knitting, crocheting, and weaving. Merino is a classic fiber for shawls, camisoles, and baby garments.

Best known for. Having wool that is fine, soft, and highly crimped, with elasticity and resilience.

Notice how much the apparent staple length shortened in washing. This is characteristic of fine, crimpy wools. The length is still there; it's just contracted, like a spring.

Saxon Merino

ACCORDING TO RENOWNED New Zealand spinner and fiber artist Margaret Stove, Saxon Merino, an Australian type, grows in a wide range of fiber diameters, from less than 14 microns ("extra ultrafine," no matter who's calling the shots) to 24 microns (strong-wool Merino turf). Yet most Saxons fall in the finer end of the spectrum (15-19 microns), and the most delicate Saxon Merino is the fiber that has allowed Stove to create masterpieces like the shawl, in her original lace design, presented to Prince William at the time of his birth in 1982. That sizable shawl could be passed through a wedding ring, in the tradition of shawls made in Shetland of wool from the Scottish breed associated with those northern islands.

In the early 1990s, Catskill Merino Farm in Goshen, New York, imported some Saxon Merino rams and has been upbreeding an American Saxon line. The farm sells wool products at an online store and at the Union Square Greenmarket in New York City.

Saxon Merino Facts

▶ **FLEECE WEIGHT**
6½–13¼ pounds (3–6 kg); no specific information on yields

▶ **STAPLE LENGTH**
Probably 2½–4 inches (6.5–10 cm)

▶ **FIBER DIAMETERS**
14–24 microns (spinning counts 60s to finer than 80s), most are 15–19 microns (spinning counts 80s to finer than 80s)

▶ **LOCK CHARACTERISTICS**
Dense, blunt locks, often weathered just at the tips, with a heavy, protective layer of grease. Crimp fine, tight, and well defined in both fibers and locks.

▶ **NATURAL COLORS**
Predominantly white.

Sharlea

Sharlea is a registered name for a type of Saxon Merino that has been carefully bred and raised with specific husbandry practices that result in very fine, exceptionally clean fiber without any weather damage to the tips. Sharlea fiber diameters run from 12 to 15.5 microns (indisputably ultrafine), and they are also the highest-yielding of the Merinos, with more than 75 percent clean weight derived from the raw fleece. Almost all Sharlea goes to textile manufacturers for the production of luxury woolens.

Saxon Merino yarns from Catskill Merino Farm

Using Saxon Merino Fiber

All of the Merinos have similar usage guidelines. The individual fleece or yarn you are evaluating will be somewhere within this breed's range, and its individual qualities will guide your decisions about how to use it best.

Dyeing. White fleeces will produce nice results with clear colors.

Fiber preparation and spinning tips

- As with all of the Merinos, extra care must be taken to remove grease. Use hot water and a good scouring agent, being careful not to let the wash solution cool. Wash thoroughly without agitating (felting) the fibers.
- Because the fleece is likely to be spun very fine, you may want to wash individual locks.
- When the wool is clean, you can spin from the lock, comb, or card. Like all Merinos, the fine wool tends to form neps. To best prevent neps from forming, use your fine-toothed combs, cotton carders, or special fine-fiber carding cloth on a drum carder. Be deliberate and patient while you work.
- It is best and easiest to spin the wool fine. If you want a thick yarn, ply together a number of thin, nicely twisted strands.
- The crimp will contribute to elasticity, but it can be challenging to handle.

Knitting, crocheting, and weaving. For making lightweight, soft, and delicate fabrics, Merino is a superb choice. It is the classic fiber for shawls, camisoles, and, because it has enough twist to wear well, baby garments.

Best known for. Having an abundance of fine, soft, highly crimped wool with elasticity and resilience.

Saxon Merino

2-ply

clean

raw

In very dense fleeces, there's often dirt at the tips of the locks, while their bases are clean. The outer end of the wool was the first to grow and has had the longest exposure time.

Rambouillet

In 1786, when the Spanish began allowing Merinos to be exported, Louis XVI (soon-to-be-beheaded husband of Marie Antoinette) imported 359 of the Spanish sheep to his estate at Rambouillet in northern France, where breeding modifications resulted in the development of a separate breed. In 1840, just over a half-century later, Rambouillets were first introduced to the United States. The breed quickly spread throughout North America, particularly leaving its mark on western range operations. There it was valued for its hardiness and ability to thrive on somewhat sparse native grasslands; Rambouillets perform especially well in that regard, thanks to their very strong flocking instinct. This basically means that while these sheep will spread out somewhat during the day while they graze, at night they bunch up tightly to sleep. This offers them a higher degree of protection from predators than

breeds consisting of individuals that spread out to sleep. Over the years, a fair amount of adaptation has occurred within the North American flock of Rambouillets.

Within the extended Merino family, the American Rambouillets are sort of like a first cousin; the family resemblance is strong, but the features are slightly different. Rambouillet sheep produce nice, soft fibers that trade a little of Merino's elegance for extra doses of warmth and resilience. Compared to most Merinos, Rambouillet has a bit less luster, more disorganized crimp (which means more loft and elasticity), and longer fibers. Where Merino wool can produce sleek yarns, a collection of Rambouillet fibers naturally incorporates more air (which can translate into warmth) and bounce.

If you've been looking at the special designations of fiber fineness used for Merinos, you'll discover that Rambouillet doesn't go into the ultrafine range (or very far into superfine). Instead, it remains concentrated in the medium and strong segments of that particular scale — still in the fine to medium-fine portions of the full spectrum of wools. Rambouillet is a great wool to blend with luxury fibers because it contributes bounce and flexibility without compromising softness. Like its Merino cousins, it is a readily felting wool.

Rambouillet Facts

▶ FLEECE WEIGHT
8–18 pounds (3.6–8.2 kg), averaging 10 pounds (4.5 kg); yield 35–60 percent

▶ STAPLE LENGTH
2–4 inches (5–10 cm), occasionally as long as 5 inches (12.5 cm)

▶ FIBER DIAMETERS
18–24 microns (spinning counts 60s–80s)

▶ LOCK CHARACTERISTICS
Square, flat-tipped, dense staples with well-defined crimp, probably somewhat less organized than in Merinos. The tips often collect dirt and need to be carefully cleaned.

▶ NATURAL COLORS
White is by far the most common color, although some small flocks produce other colors, especially blacks and grays.

Using Rambouillet Fiber

Dyeing. The fiber has a more matte surface than Merino. The wool will take colors clearly and well, but with a softer overall effect than in more lustrous wools.

Fiber preparation and spinning tips

- Be sure to clean the wool well, while being careful not to encourage its tendency to felt. The high grease content takes both hot water and a good cleansing solution to remove, and you want to be sure to keep the temperatures high enough that the dissolved lanolin and suint don't get redeposited on the wool.
- It's also best not to store Rambouillet in the grease, because it is likely to become very hard to clean.
- It can be spun from flicked locks, combed, or carded. In carding, the wool will tend to form neps if not carefully managed. Use fine-toothed carders or carding cloth, or comb or flick it.
- This wool will be better spun fine than coarse, because of the slender fibers. To benefit from the fiber's elasticity, straighten but don't stretch the fibers while spinning.

Knitting, crocheting, and weaving. The crimp patterns of Rambouillet give it more loft and elasticity than Merino. It's very soft, with good insulating qualities. It is superb for next-to-the-skin fabrics at the luxury-fiber softness level and for baby garments and other special items.

Best known for. Being a member of the Merino family with huge fleeces; wool has great crimp and bounce and is somewhat loftier than other wools from Merino ancestry.

Rambouillet undershirt from Rambler's Way

A New Business Model

In 1968, Tom and Kate Chappell moved from Philadelphia to Kennebunk, Maine. They started a line of all-natural personal care products and dubbed their company Tom's of Maine. Fast-forward to 2006: Tom and Kate sold their company to Colgate-Palmolive and focused on their farm. They looked for a new creative niche to provide natural products and came up with Rambler's Way. This company produces next-to-the-skin garments from their own Rambouillet sheep and from the wool grown by other U.S. Rambouillet producers. Their products are fabricated in mills in New England and the Carolinas. Both their farming and their manufacturing processes are done with an eye toward sustainability.

Springy, with good fiber length, and lovely colors that can be spun for variety or blended for uniformity.

HebRidean
Soay

Northern European Short-Tailed FAMILY

Probably the most famous member of the Northern European Short-Tailed family found in North America is the Icelandic breed, which has enjoyed growing popularity since its introduction to Canada and the United States in the mid-1980s. But there is a plethora of interesting critters in this crowd. As the name implies, these breeds originated in the northern latitudes: in Scandinavia, on islands scattered throughout the North Atlantic, and across into Northern Russia. In some of the remote and rugged island locations where these breeds are found, some feral populations still exist, such as the Soay and the Boreray from the archipelago of St. Kilda. A number of these breeds still provide cultural cornerstones in the lives of their people.

The family name also suggests one of the distinguishing traits of this clan: Naturally short tails are the norm. When it comes to fiber, this is an interesting family. Many of its members are dual-coated, offering both coarse hairy fiber and fine downy fiber from the same animal, and grow fleeces in a range of natural colors.

Out of the wild. The mouflon is the wild sheep from which domesticated sheep were developed around 9,000 years ago. The Northern European Short-Tailed breeds are generally primitive breeds, meaning they retain characteristics similar to those of their wild progenitors.

Boreray

Critical Conservation Breed

St. Kilda is a group of four small islands (Boreray, Dùn, Hirta, and Soay) and a number of sea stacks (rocky outcroppings that jut from the sea), located about 112 miles (180 km) west of mainland Scotland. Today the islands are managed by the National Trust for Scotland and have been recognized by the United Nations as a World Heritage Site, thanks to their significant cultural and scientific values.

The history of sheep and these islands is fascinating, albeit a bit perplexing and knotty to untangle — but we are game to give it a try! St. Kilda was probably occupied permanently about two thousand years ago, although Bronze Age tools show that humans built structures on the islands four thousand or more years ago. Stuck out so far in the Atlantic, the islands were inhospitable, and life was extraordinarily hard. The population, which counted on sheep and fishing for its livelihood, peaked at fewer than 200 residents on Hirta in the 1600s and was down to just 36 residents by 1930. Then, after a particularly hard winter in which everyone nearly died of starvation and a young mother perished from appendicitis, the St. Kildans petitioned the Scottish government to relocate them. They were soon evacuated.

Most of the St. Kildans' sheep, at the time of evacuation, were descendants of the now extinct Scottish Tan Face, with an infusion of Hebridean Blackface bloodlines; these sheep had been introduced to St. Kilda in 1872. (There was also a flock of Soay sheep on Soay Island. Read more about these animals on page 194.) The residents' sheep and their few cattle were to be moved on the mail boat to the mainland of Scotland. The seas were rough that day, so the mail boat *Dunara Castle* moored in a bay. Then 527 sheep were ferried to it in small boats (two previous shiploads of sheep had already been removed), while 13 cattle swam to it, towed behind boats on lead lines. The *Dunara Castle* departed the island at noon, and the people of the island left the next day on the *Harebell*.

The St. Kildans had always kept a second flock of their domestic sheep on Boreray

The Edge of the World

In 1936, the acclaimed British movie director Michael Powell filmed a story inspired by the evacuation of Hirta. The black-and-white movie, called *The Edge of the World*, is a work of fiction and was actually filmed on Foula, an island in the Shetland archipelago. It's available for rental and is quite worth seeking out. Our favorite part was watching people capture the sheep and *roo* (or pluck) their fiber. The movie also highlights the importance of wool in an island's economy. In one scene, the governing lord comes to the island and takes the surplus wool, as well as the island-spun-and-woven tweed fabric.

Island as backup, should something befall the main flock. Traditionally, villagers would go to Boreray once a year, capture the sheep, and harvest the fiber. This was a dangerous trip, and one year the group of St. Kildans who made the voyage was stranded on Boreray over the winter. For at least three years prior to the evacuation, there had been too few able-bodied men left on the island to handle that flock, so the sheep on Boreray returned to a feral state. Those animals were simply too wild to remove during the evacuation, so they became the fully feral population that's still found on Boreray Island today. In 1970, six animals from that flock were moved off the island as part of a conservation effort to protect their bloodlines, and those gave rise to a few small flocks of these sheep in other areas of Britain.

You might think you've got this all straight — but we need to warn you that it gets even more confusing. Animals of the breed known as Hebridean sheep are also sometimes called St. Kilda sheep. The name seems to have its origins in the late 1800s, when wealthy landowners in Scotland and England kept the Hebridean breed on their country estates — even though these sheep are not directly related to the domestic or feral sheep of the St. Kilda archipelago.

Now, to add one final complication to the naming of these sheep, there are actually two Boreray Islands off the coast of mainland Scotland. One is the Boreray Island of the St. Kilda archipelago; the other is in the Outer Hebrides archipelago. Today one crofter, or small-scale farmer, named Jerry Cox lives on the Boreray Island in the Sound of Harris, tending a flock of feral Hebridean sheep, marketing their fiber, and operating a remote and rugged tourist house that requires a 4-mile (6.5 km) sea journey in an open boat to reach. Most of this fiber goes to a young English clothing design company called Makepiece, run by partners Beate Kubitz (who raises Shetland sheep) and Nicola Sherlock, who specialize in sustainable clothing from natural yarns that are produced on sustainable farms in undyed and natural-dyed colors. For the record: In this write-up, we are talking about the Boreray sheep of the St. Kilda archipelago, not about Jerry Cox's Hebridean sheep (see page 166 for our discussion of the Hebridean breed).

Early in this project we sent a first batch of 30 wool samples off to a lab for analysis. The samples all represented breeds for which we had little or no information. We included locks from several rare breeds. When we received the results, the first item on the list, the Boreray, showed an average micron count of 17.9 for the bit we'd sent. Our first thought was that there'd been a mistake and that, despite extreme care, we'd mislabeled a bag, or the lab technicians had made an error. That micron count is cashmere territory (see page 348), or the realm of the finest Merinos (see page 135). Could this result possibly be true? When we got to spinning the wool from which the sample was taken, it became apparent that the analysis is correct. This sample of Boreray wool *is* that fine.

Will all Boreray surprise you this way? Probably not, and later samples came up with different, yet not incompatible, results. But as our lab snapshots affirm, the fiber is quite variable, and you'll need to take each fleece as a unique representative of its range of possibilities.

Boreray Facts

▶ **FLEECE WEIGHT**
2–3½ lbs (0.9–1.6 kg)

▶ **STAPLE LENGTH**
2–6 inches (5–15 cm)

▶ **FIBER DIAMETERS**
Not enough data; what we found in other sources suggests 23–32 microns (spinning counts 48s–62s), which is a very wide range, yet one sample we tested (on the left below) was much finer, averaging 17.9 microns (spinning count finer than 80s)

▶ **LOCK CHARACTERISTICS**
The locks are indistinct and quite open, have slightly pointed tips, and may have the "sticky" base ends characteristic of a fleece that sheds: When the wool prepares to shed, the loosening ends often get stuck together.

▶ **NATURAL COLORS**
Creamy white, tan, gray, sometimes dark brown.

Using Boreray Fiber

Dyeing. Colors will be influenced by the underlying tone.

Fiber preparation and spinning tips. Because of the variability, preparation and spinning approaches depend on the fleece profile. Open the butt ends, then choose your next step based on the length and coarseness of the individual fleece: Spin from the lock, or card, flick, or comb.

Knitting, crocheting, and weaving. The fleece is mostly suitable for tweeds, rugs, hard-wearing pillows, and other sturdy applications, although our especially fine sample would be great for garments that combine sturdiness and softness.

Best known for. It's not well known at all!

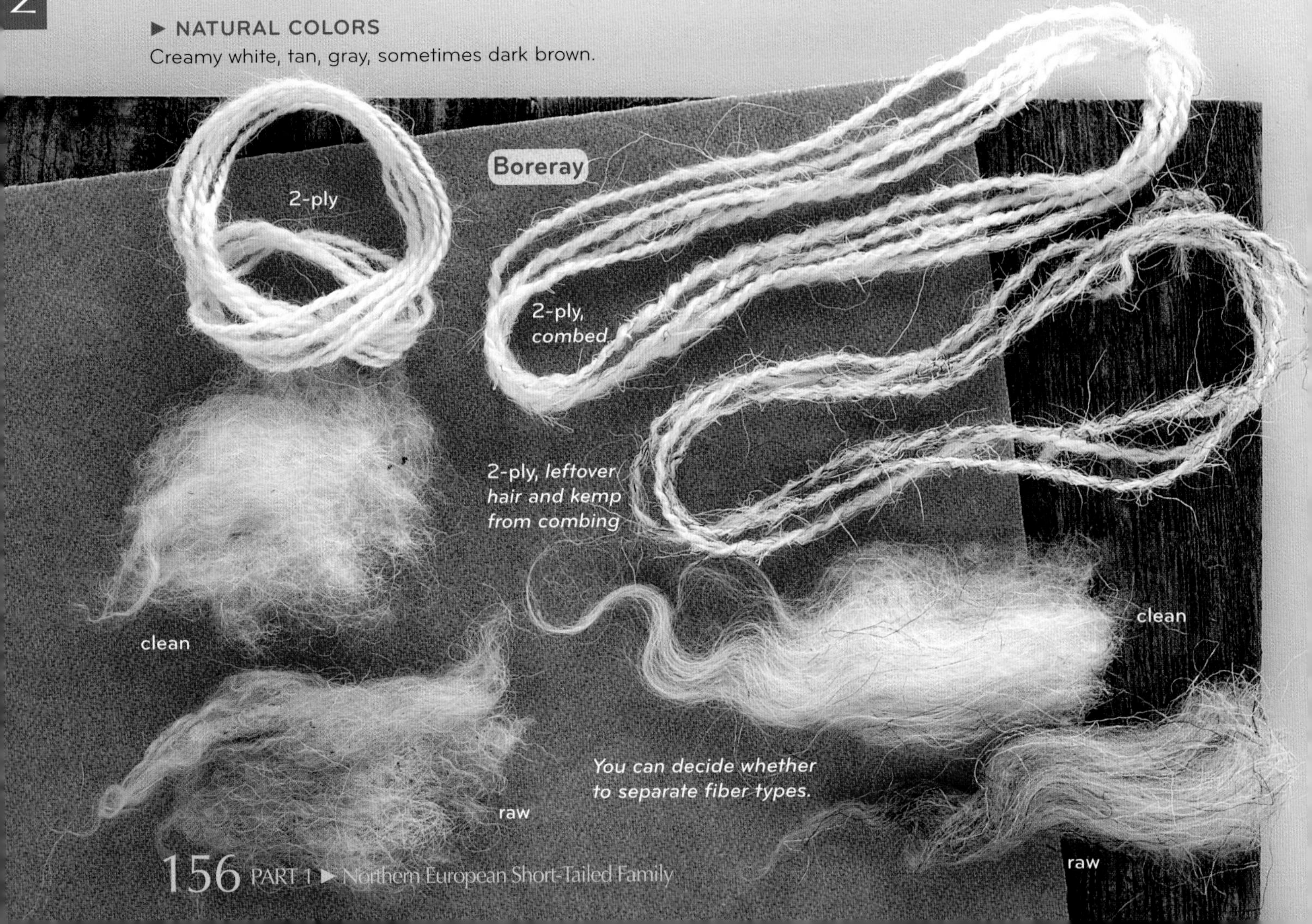

Castlemilk Moorit

Critical Conservation Breed

Sir Robert William Buchanan Jardine (1866-1927) was a Scottish aristocrat whose family acquired its wealth through trading, particularly of opium to and from China. He made his home at the family's Castlemilk estate in Scotland, where his pastime was breeding animals: hounds, cattle, horses, and sheep. One of Jardine's oddities as a breeder was that he was fascinated with breeding for shades of brown. He bred his sheep primarily to be beautiful lawn ornaments and named them for his estate plus the term *moorit*, for their color. Moorit, which means reddish brown, is derived from the old Norwegian-influenced language of the Shetland and Orkney islands. Essentially, it translates "as red as the moors."

Jardine used moorit Shetlands, Manx Loaghtans, and a mouflon ram to come up with an easy-care brown sheep with short, tight wool suitable for handspinning by the family. The Jardine family continued raising the breed until 1970, when they dispersed the flock. Cotswold Farm Park, established in the 1960s to protect rare breeds of livestock, acquired six ewes and a ram. These few animals are the foundation for all the Castlemilk Moorits today.

Although raised in large part for appearance, nonetheless the breed is known for high-quality wool. Spinners we spoke with have gone into raptures over this wool. There aren't many Castlemilk Moorits; the first small samples we obtained were not consistent enough for us to spin a small demonstration skein. Previously printed sources on the wool quality offered us no help; the micron counts and Bradford designations are not even in the same neighborhood. One set of locks we sent for lab analysis tested out at an average of 26.5 microns, which is much finer than the prevailing suggested range of 33 to 35 microns; it is also more in keeping with what we'd expect from the breed's ancestry. We were given a spinnable sample near the end of this project. When tested at the lab, it had a fiber diameter of 28.5 microns, again finer than we were led to believe this wool would be.

Because of the small number of sheep, this breed rarely shows up in skeins ready for nonspinners to experience. When it does, it is worth trying out!

Castlemilk Moorit Facts

▶ **FLEECE WEIGHT**
2–3 pounds (0.9–1.4 kg)

▶ **STAPLE LENGTH**
1½–3 inches (3.8–7.5 cm)

▶ **FIBER DIAMETERS**
According to sources we located, listed as 30–35 microns or Bradford 48s–50s (which roughly corresponds to 29–32 microns). Our money's on the 29–32-micron range, because both samples we sent to the lab tested even finer than that.

▶ **LOCK CHARACTERISTICS**
Like Manx Loaghtan (see page 172), a Castlemilk Moorit fleece tends to hold together as a single unit, from which the locks can be separated. They are short, blocky, and often sun-bleached at their outer edges. They occasionally have slightly pointed tips.

▶ **NATURAL COLORS**
Light to medium reddish brown, often with sun-lightened tips.

Using Castlemilk Moorit Fleece

Dyeing. The fleece will accept dye, of course, but choose colors that will look good when overdyed on brown.

Fiber preparation and spinning tips. The fleece is short enough that most will want to be carded. It could be spun from loosened locks. Handle like a short, moderately fine wool; most comparable, perhaps, to Manx Loaghtan.

Knitting, crocheting, and weaving. Best for mid-range garments and fabrics, like outerwear sweaters, hats, and blankets.

Best known for. A very rare breed with a distinctive light-brown fleece and coloring related to wild mouflon, which, to oversimplify, means that the body is darker on its upper parts and lighter on the lower bits, including the neck and belly.

Castlemilk Moorit

Fiber is variable between fleeces and often short. Some fleeces are delightful for handspinning. Commercially, the wool is often blended to make it easier to process.

2-ply

clean

clean

clean

Finnsheep

ALSO KNOWN AS the Finnish Landrace, or just Finn for short, Finnsheep are an ancient Scandinavian breed, though there is little knowledge about their early ancestry. However, they are known to be one of the most prolific sheep in the world. As with the Romanovs (see page 183), litters can have as many as nine lambs born at a time, though three to four lambs is more common.

Finns' fleeces are light, averaging around 5 pounds (2.3 kg) when cleaned, but their wool, often described as silky, is more sleek than fluffy. It has a nice amount of crimp that gives yarns spun from it a pleasant resilience, although it can't be called springy, like wools with tighter crimp patterns can. Most Finn fiber that gets into general circulation is clear white, with a very slight warm cast to it — clean, but not eye-popping bright (some of us think this is a fine place to be on the color spectrum). Like many of the other Northern European Short-Tailed breeds, this one can produce a full array of colors, including blacks, grays, browns, fawn colors, and some spotted (or piebald) individuals. However, in Finns the colors are far from common, especially in North America. Almost all Finns are single coated, although their heritage means a double-coated fleece may show up now and then.

Knitters can find Finn and Finn-blend yarns fairly easily, and spinners can obtain prepared top and roving. For a transcendent experience, though, comb your own top from freshly shorn and washed Finn wool and spin your own. There's not a lot of grease, so it's easy to clean and prepare. Because of this wool's propensity for felting, though, watch the water temperatures and don't agitate it.

Finnsheep Facts

▶ **FLEECE WEIGHT**
4–8 pounds (1.8–3.6 kg); yield 50–70 percent, often on the high end

▶ **STAPLE LENGTH**
3–6 inches (7.5–15 cm); sometimes shorn twice a year, yielding 3–4 inches (7.5–10 cm) each time

▶ **FIBER DIAMETERS**
24–31 microns (spinning counts 50s–60s); Canadian Finns may run a little finer (average 22–24 microns; spinning counts 60s–62s), and New Zealand colored Finns a little coarser (25–33 microns; spinning counts 46s–58s)

▶ **LOCK CHARACTERISTICS**
An open fleece, from which you can easily separate out locks that have slightly pointed (occasionally sun-damaged) tips.

▶ **NATURAL COLORS**
White, black and gray (less common), shades of brown (least common).

Using Finnsheep Fiber

Dyeing. In white Finns, the very subtle warm undertone will add an appealing unity to colors (you won't notice it, it'll just be sweet), while the fibers' gentle sheen emphasizes the shades' clarity. With the colors, overdyeing can be fun.

Fiber preparation and spinning tips. Shorter locks can be carded, but Finn has perfect staple lengths for combing; keeping the fibers parallel makes the most of the luster. You can fluff out the locks and spin straight from them, or with one of the longer fleeces, you might want to spin from the fold. Finn is a lovely, easy wool to spin. Commercial top may become compressed enough to be hard to draft, so loosen the fibers before you head for the wheel or spindle.

Knitting, crocheting, and weaving. Finn is extremely versatile. It's sturdy enough to wear well, and some is in a fineness range where it won't be irritating when worn next to the skin; it's an easy choice for sweaters, blankets, and other garments to snuggle into, although there are better selections for camisoles. The fiber's body and luster make it a great choice for textures, like knit/purl patterns or woven laces, as well as for crisply defined color patterns.

Best known for. Reliability, flexibility, nice combination of softness and durability, silky feel, and look. Prolific mamas!

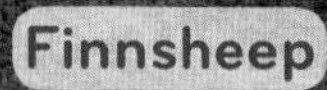

Finnsheep

clean

2-ply

raw

This raw yearling fleece was silky, shiny, and sweet.

Commercial preparations of fibers, especially top, may look the same but feel very different. The fiber in the top to the right was finer than the fiber in the top to the left.

2-ply

2-ply

top *(fiber length)*

top *(fiber length)*

clean

Gotland

Gotland in this context refers to the modern Swedish breed developed in the early twentieth century from the more primitive Goth, Gute, or Gutefår (see page 317), which sometimes are also referred to as Gotland sheep because of the geographic coincidence of the related breeds' origins on the largest island in the Baltic Sea.

Gotland is an unusual wool, resembling a fine mohair or an English luster longwool more than the other Northern European Short-Tailed breeds. It's also more often comfortable in next-to-the-skin garments than would ordinarily be expected from the characteristic fiber diameters. It can be spun into either smooth or heavily textured yarns, and the unspun locks themselves are beautiful and can be used to decorate other textiles or form a sheepskin-like surface on a woven or knitted fabric. It does felt. The predominant colors are a range of grays, from pale silvery through a charcoal that's close to black, although blacks, whites, and a few browns can also be found.

Different strains of Gotlands have different fleece characteristics, but they have in common length, luster, and well-developed wavy crimp. We found that commercially processed top was matte, rather than shiny, and felt coarser than the fiber we processed

from raw wool. It really felt like a different fiber, even though both our processed and our glossy from-the-fleece samples came from Gotlands raised in the British Isles.

The Stansborough strain of Gotlands in New Zealand is being bred for increased fineness, and theirs is the wool used to make the Elven cloaks worn by the main characters in all three *Lord of the Rings* movies — an ideal choice for garments that need to be lightweight and durable, drape well, and possess a subtle gleam that suggests their magical qualities. Gotlands in North America are being developed through upgrading and include more brown, especially moorit, colors.

Gotland Facts

▶ FLEECE WEIGHT
5½–11 pounds (2.5–5 kg)

▶ STAPLE LENGTH
Variable, 3–7 inches (7.5–18 cm); often shorn more than once a year to produce a 3–4 inch (7.5–10 cm) staple, which is optimally workable for a variety of processing methods

▶ FIBER DIAMETERS
Lambs from as fine as 18 microns (spinning count 80s) to low to mid-20s (spinning counts 60s–70s); adults in North America and New Zealand, 27–34 microns (spinning counts 46s–54s); in Britain, average 35 microns (spinning count 44s)

▶ LOCK CHARACTERISTICS
Relatively long, high-luster, dense, astonishingly wavy — even curly — wool, and staples that feel soft and silky. The locks are like shiny springs.

▶ NATURAL COLORS
Predominantly grays, from light silvery gray through charcoal to near black, often with appealing subtle variations in shading. Occasionally white, and a few browns. The grays and blacks often have brown tips on the staples.

Gotland yarn from Blacker Designs

Using Gotland Fiber

Dyeing. The natural grays are exquisite on their own, and the underlying color will influence any superimposed dyed hues; the results can be subtle and, because of the fiber's luster, glisteningly beautiful.

Fiber preparation and spinning tips. For both preparation and spinning, Gotland's relatively fine and lustrous fibers are slick and flyaway, especially in a dry climate. Open the locks slightly to allow the twist to get traction and spin directly; or flick the locks and spin; or comb; or, depending on fiber length, card. Gotland is easier to spin into fine, rather than bulky, yarns. When you are preparing the fleece, it will appear very fluffy. Yet when twist is added, the fibers nestle down together and make a yarn that feels heavier and has better draping qualities than you've been led to expect it will produce when you saw it unspun. It's slippery, too, yet very rewarding. Gotland can be spun woolen (from rolags or batts) for texture, or worsted (from top) for smooth results.

Knitting, crocheting, and weaving. The fleece is suitable for use in any technique. Worsted-spun yarns will emphasize the shine and drape, while woolen-spun varieties can incorporate an unusual amount of air and can be used to add texture, if spun for that. Gotland is happy to oblige with felting.

Best known for. A range of lovely grays, luster, supple fibers, often surprisingly soft.

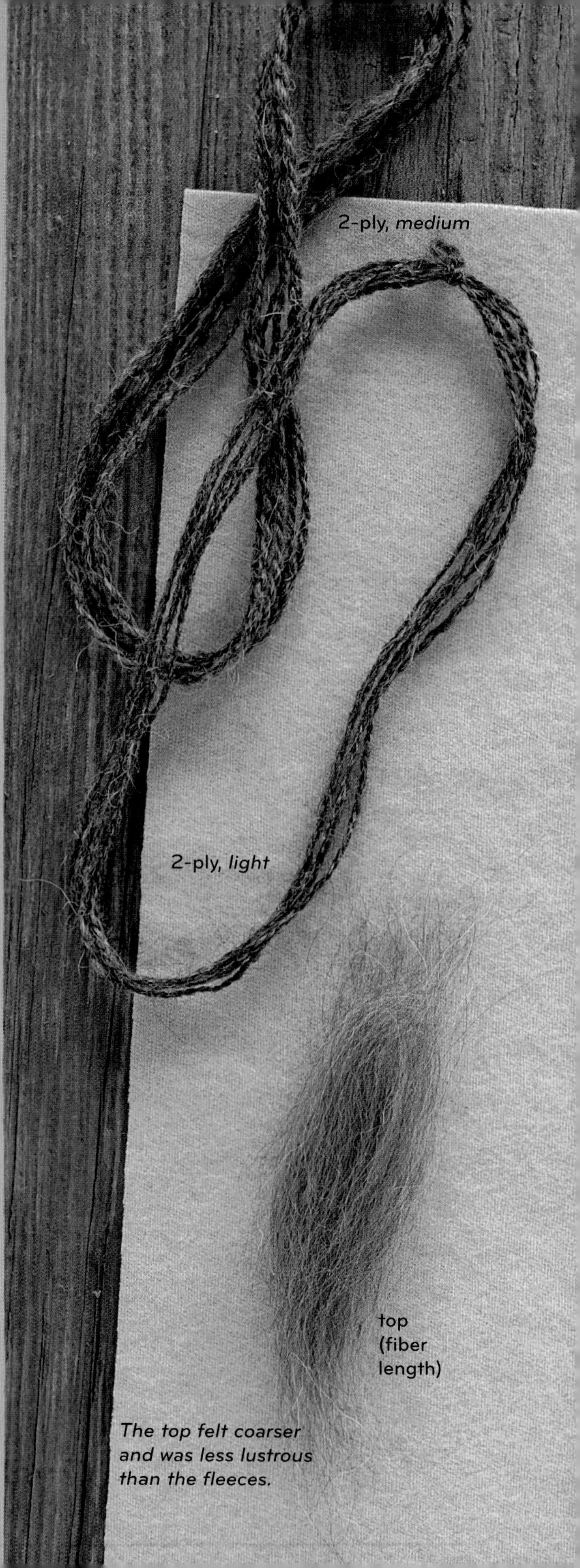

The top felt coarser and was less lustrous than the fleeces.

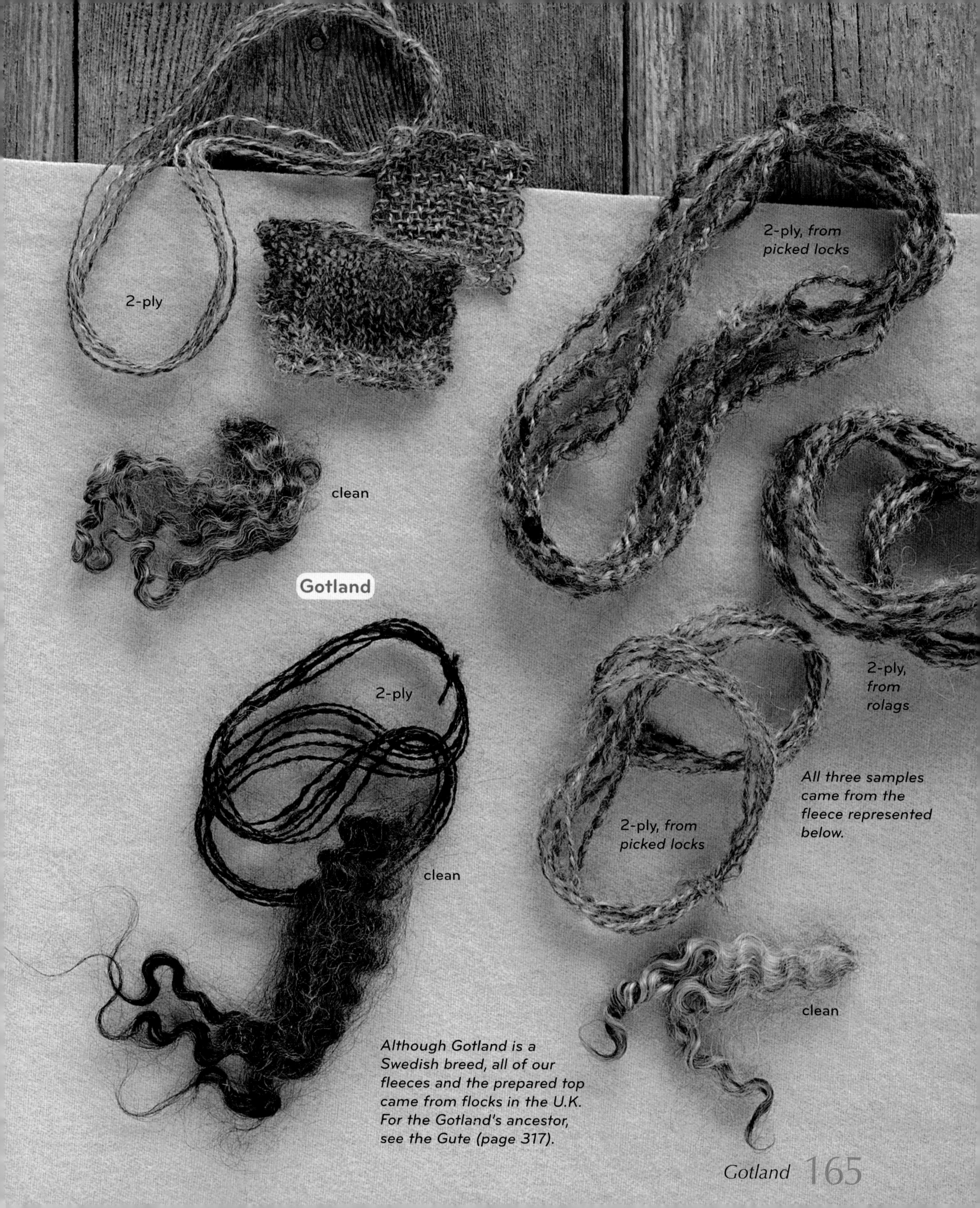

All three samples came from the fleece represented below.

Although Gotland is a Swedish breed, all of our fleeces and the prepared top came from flocks in the U.K. For the Gotland's ancestor, see the Gute (page 317).

Hebridean

Conservation Breed

THE HEBRIDES, a large archipelago off the west coast of Scotland, consist of over five hundred islands, though most are uninhabited rock spits. There are two major groups of islands: the Inner Hebrides (closest to the mainland of Scotland) and the Outer Hebrides (which lie farther to the west in the Atlantic). The warm Gulf Stream plus the varied mix of landforms and waters yield an amazingly rich diversity of marine life. Nearly a third of the world's species of whales, dolphins, and porpoises inhabit the seas surrounding the Hebrides!

Today's Hebridean sheep come from a handful of flocks that were relocated to estates in England during the late 1800s. Though they are probably related to the ancient primitive sheep that the Vikings brought to the Hebrides a millennium or so ago, these Hebrideans aren't necessarily true representatives of the sheep that once lived on the islands. Their ancestors on the islands died out as the crofters moved to raising "improved" breeds, such as Cheviot (see page 54) or Scottish Blackface (see page 47).

The current Hebrideans are frequently multihorned and always have black wool, which lightens to gray or reddish brown on the tips (with both age and exposure to sun). The original Hebridean sheep occasionally had multiple horns but far more frequently had only two, and they showed a range of colors, including blue-gray, brown, black, and russet. These original animals also sported white faces and legs. It is possible that the Hebrideans we know today are the result of specific selection by the aristocrats who chose these animals as lawn ornaments, or it's possible that the Hebridean sheep brought to the estates may have been crossed with Black Welsh Mountain sheep (page 207). Either way, today's flock of Hebrideans is consistently black, and the multihorned trait runs strongly through the breed.

The Hebridean is a breed with exceptionally variable, dense wool that can stand up to weather and wear. Although double coatedness is not noted in other information sources, our sample fiber could be separated into two coats. The demarcation between the two was not absolutely clear, as it sometimes is. There was a gradual shift, rather than a gap, between the two types of fiber, but separation was relatively easy.

The deep color and sturdy texture are givens. The softness may be suited to either everyday or outerwear garments, or to blankets or sturdy mats.

Sea World

To learn more about the sea life of the Hebrides, visit the Hebridean Whale and Dolphin Trust website at www.whaledolphintrust.co.uk.

Hebridean Facts

▶ FLEECE WEIGHT
3½–5½ pounds (1.6–2.5 kg); ram fleeces may be larger

▶ STAPLE LENGTH
2–8 inches (5–20.5 cm), generally 2–6 inches (5–15 cm)

▶ FIBER DIAMETERS
29–38 microns (spinning counts 40s–50s) average; our sample averaged 40.9 microns, obviously much coarser than we were led to expect, although not altogether surprising given the breed's rarity and the general lack of information on its wool qualities

2-ply, *undercoat*

2-ply, *outercoat*

clean

Hebridean

Spun the two coats separately, for two completely different yarns.

▶ **LOCK CHARACTERISTICS**
Lustrous. Triangular. May contain hair and/or kemp, especially on the hindquarters.

▶ **NATURAL COLORS**
Black; possibly sun-bleached or turning to gray with age; sometimes very dark brown.

Using Hebridean Fiber

Dyeing. This is generally pointless for Hebridean fleece.

Fiber preparation and spinning tips. Preparation and spinning approaches depend on fiber length and whether the fleece is double coated or not and on whether you want to separate qualities of fiber or spin them together. Depending on individual fleece characteristics, the spinning may be a little challenging.

Knitting, crocheting, and weaving. Use Hebridean where you want the luscious dark color and a lot of durability. It is great for outerwear.

Best known for. Dark, lustrous color.

Hebridean serape from Hebridean Wool House

Hebridean yarns from Garthenor

Icelandic

VIKINGS SETTLED Iceland between 870 and 930 CE. They brought sheep with them and added a few more a short while later. Since then, it's been illegal to bring in more sheep. Developed in almost total isolation, the Icelandic breed is one of the world's purest livestock populations. These sheep provide 25 percent of the island nation's agricultural revenue. Although sheep are grown primarily for meat in Iceland, the breed is best known elsewhere for its fleece. Certainly the Icelanders have established successful and distinctive markets for their wool.

In 1985, the first Icelandic sheep came to North America through the efforts of Stefania Sveinbjarnardottir-Dignum and Ray Dignum, of Yeoman Farm in Ontario. They imported a second batch in 1990, and all the Icelandic sheep in North America derive from their animals.

Leadersheep

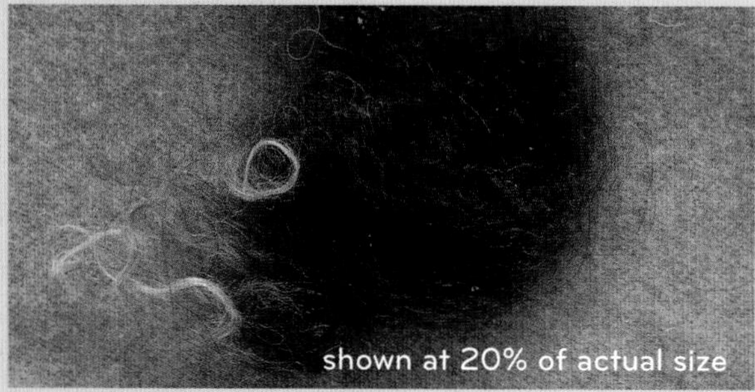

Icelandic Leadersheep fiber

Within Icelandic flocks, especially intelligent individual sheep play important social and protective roles in the flock, alerting the others to hazards like predators and storms. Called Icelandic Leadersheep, they are specially identified and bred. In 2000, the Leadersheep Society of Iceland was founded to conserve them.

Icelandic sheep grow dramatically double-coated fleeces. The strong outercoat is called *tog*. It sometimes has a texture like mohair and sometimes a more subtle, silky quality. The fine undercoat is called *þel*, usually typeset in English as *thel* (the initial letter is a character known as a *thorn*, pronounced with a *th* sound). The words *tog* and *thel* are occasionally applied to the fiber types from other two-in-one fiber sources, whether sheep, camelids, or others, but they originated with Icelandic wool.

Icelandic fleeces shorn at different times of the year feel different in handling and spinning. The coats can be spun separately or together. If you buy commercially prepared Icelandic wool yarn or spinning fiber, it usually contains both coats. When you begin with a raw fleece, getting the fiber types to draft evenly can be a challenge. To maximize the strength of the outercoat and the softness of the inner coat, separate the types and spin the long fibers worsted and the short ones woolen (remembering, however, that there are no rules in spinning). Tog may or may not be okay for next-to-the-skin wear, while thel should be delightfully soft.

Icelandic Facts

▶ **FLEECE WEIGHT**
4–7 pounds (1.8–3.2 kg); yields range from 50 to 90 percent; fleeces are light in lanolin and suint; high yields come from using all the fiber types

▶ **STAPLE LENGTH**
Outercoat 4–18 inches (10–45.7 cm)
Undercoat 2–4 inches (5–10 cm)

▶ **FIBER DIAMETERS**

Outercoat 27–31 microns (spinning counts 50s–56s)
Undercoat 19–22 microns (spinning counts 64s–70s)

▶ **LOCK CHARACTERISTICS**

In double-coated fleeces, staples tend to be triangular. The bulky undercoat spreads the staple's base, and the long outercoat puts a point on its tip. The longer tog may be loosely curly, while the shorter thel has some true crimp.

▶ **NATURAL COLORS**

A full range, including white, tan, brown, gray, black, and mixtures. If you want an exercise in studying color genetics, Icelandic sheep will give it to you.

Using Icelandic Fiber

Dyeing. Whites will produce clear colors; overdyeing natural colors results in interestingly subtle shades.

Fiber preparation and spinning tips

- The two coats can be spun separately or in combination.
- The fibers are easiest to separate when clean. Separate either by holding a lock in your hand, or by catching the base of a staple on a hand card or flicker, and pulling free the long tog fibers, leaving the thel behind in your hand or on the carder.
- Because Icelandic sheep naturally shed their wool (although they are most often shorn), the bases of the staples may clump together and require special care to open out. Try spinning from the lock (both coats), drum-carding, or using Viking-style combs.
- If you want to spin both coats together, work from the lock or combine the fibers by carding, and be careful to draft them evenly. .

Knitting, crocheting, and weaving. The undercoat is an obvious choice for knitting and crocheting, while the outercoat lends itself easily to weaving, needlepoint, and other stitchery. Experiment, however, because these fibers are versatile.

Best known for. Double-coated fleece, with a strong outercoat and a fine undercoat; a broad array of natural colors and patterning on the sheep.

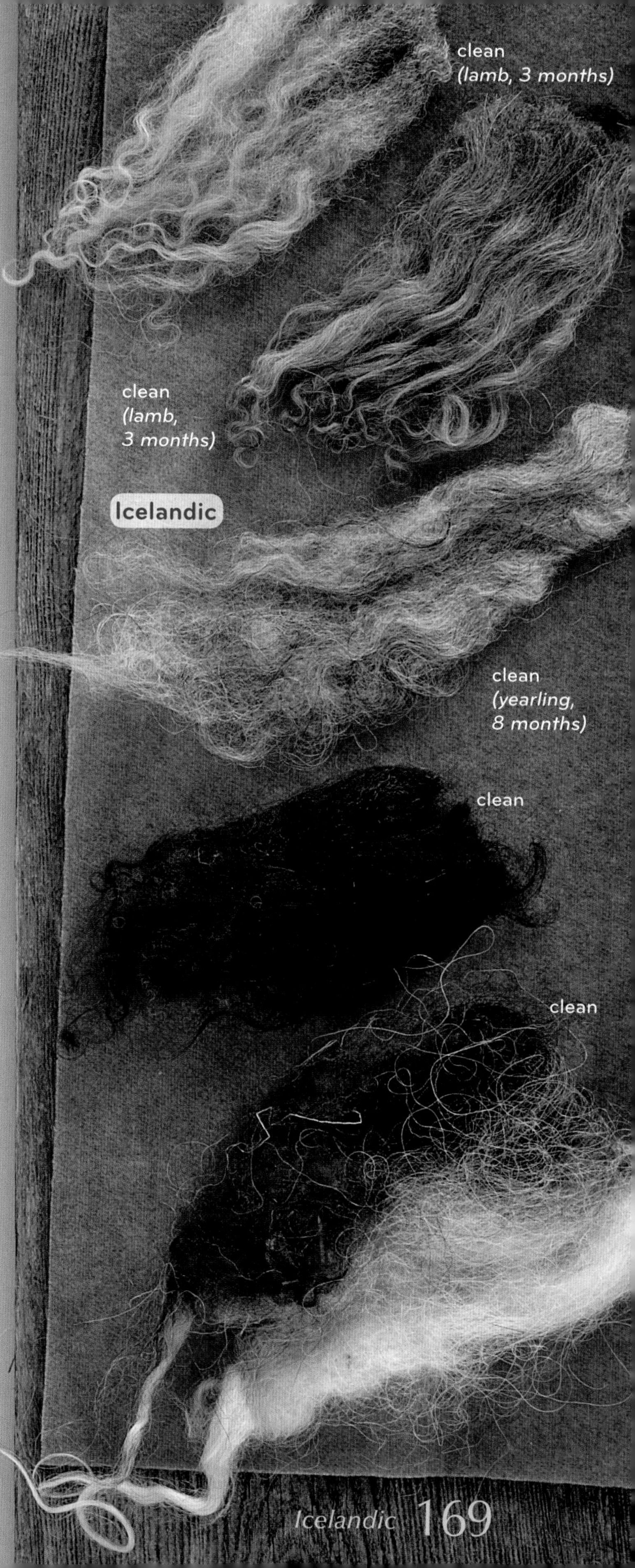

clean *(lamb, 3 months)*

clean *(lamb, 3 months)*

Icelandic

clean *(yearling, 8 months)*

clean

clean

Northern European Short-Tailed Family

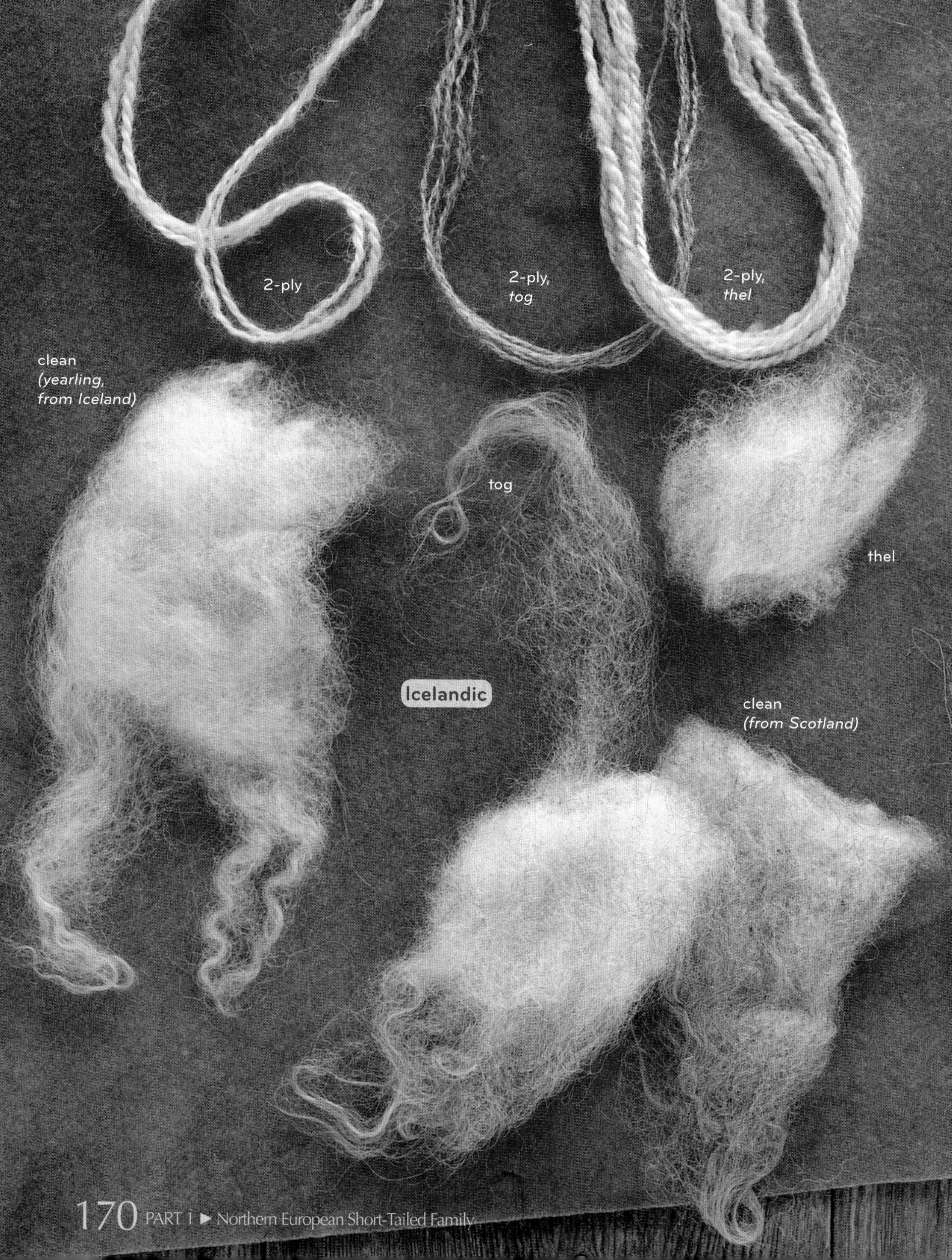
2-ply
2-ply, *tog*
2-ply, *thel*
clean (*yearling, from Iceland*)
tog
thel
Icelandic
clean (*from Scotland*)

The Lopi Story

Due to the nature of Icelandic fleeces, knitters in Iceland discovered they could knit with roving, without spinning first. The untwisted rovings are called *lopi*, which means simply "unspun wool." Although lopi is a bit fragile while being worked, a bit of moisture will stick it back together if it pulls apart, and it produces a durable knit piece.

The original unspun version is referred to in the textile world as *plötulopi*, meaning "a disk (or plate) of lopi," because the fiber is wound into flat, circular bundles for storage. Plötulopi is hard to find outside of Iceland (although it can be obtained through the Internet). Commercial lopi yarns are usually lightly spun, and three weights are available under two labels, Álafoss and Reynolds:

- an Aran weight that works up at 18 stitches to 4 inches/10 cm
- the baseline bulky weight that works up at about 14 stitches to 4 inches/10 cm
- a super-bulky version that works up at 10 stitches to 4 inches/10 cm

Icelandic yarn from Local Harvest/Foggy Sheep Ranch

Manx Loaghtan

Conservation Breed

THE ISLE OF MAN (or Ellan Vannin in the Manx Gaelic language) lies in the geographic center of the British Isles but has never been part of the United Kingdom. Just 33 miles (53 km) long and 13 miles (21 km) wide, the island has been inhabited for 10,000 years and is self-governing, with the world's oldest continuous parliament. The island also has its own very distinctive breed of sheep, the Manx Loaghtan.

Up until the eighteenth century, the native sheep, thrifty and hardy souls simply called Manx mountain sheep, lived throughout the island. They provided the human inhabitants with meat, and with fleeces in a range of shades including white, gray, and black, as well as a rare and valued soft shade of brown. Today brown is by far the dominant color within the breed, an artifact of its near extinction and the breeding preferences of the individuals who brought these sheep back from the brink. The lambs are born black and turn brown within weeks.

Manx Loaghtans have value for conservation grazing. One flock lives on an island known as the Calf of Man, off the southwest tip of the big island. The Calf is a bird sanctuary; in the first decade after the sheep were added to the small island's ecosystem, the number of nesting pairs has doubled for an alarmingly endangered member of the crow family, the Red-Billed Chough. The sheep are credited with grazing down hawthorns and coarse grasses, which encourages the native plants and wildflowers while creating a favorable environment for the insects and worms that the choughs feed on.

Different standards for the breed have been set by the Manx Loaghtan Sheep Society and the Rare Breeds Survival Trust, and some breeders prefer to maintain their independence and not participate in any registration program. The breed does carry horns in both sexes,

How Did You Spell That Name?

There is no standardized spelling for the second word in the name of this breed (pronounced *LOCH-tun*). It appears in at least seven forms (listed alphabetically): *laughton, laughtown, loaghtan, loghtan, loghtyn, loughton,* and *loughtyn*. Whatever its spelling, the word is believed to have come from one of two combinations of Gaelic words: *lugh dhoan* (mouse brown) or *lhosht dhoan* (burnt brown). *Manx* is also sometimes spelled *Manks*.

which is not particularly common in sheep, although you may not get that impression from the Northern European Short-Tailed group or a survey emphasizing the rare breeds. Some people argue that true Manx Loaghtans must have four horns (or two pairs) each, while others think that two are sufficient. Occasional animals have as many as six horns.

Today the breed is still raised largely for its meat, which is considered a delicacy. It is darker than most lamb meat and quite flavorful, yet low in fat and cholesterol.

Manx Loaghtan Facts

▶ **FLEECE WEIGHT**
3–5½ pounds (1.4–2.5 kg)

▶ **STAPLE LENGTH**
2½–5 inches (6.5–12.5 cm)

▶ **FIBER DIAMETERS**
Average 27 (woolly type) to 33 (hairy type) microns (spinning counts 46s–56s). There is a broad range of acceptable fleece qualities, from coarser to finer, which means that some fleeces are relatively soft and others are more durable.

▶ **LOCK CHARACTERISTICS**
Locks are close textured and uniformly brown throughout, with some luster. There are two types of fleece within the breed (not on the same animal), one hairy and one woolly. The wool overall tends to be soft and contains enough grease to protect it, more so than in some of the other Northern European Short-Tailed fleeces. The crimp is bold, uniform, and of consistent quality throughout, from butt to tip.

▶ **NATURAL COLORS**
Soft brown; the tips tend to sun-bleach to a lighter shade.

Although it's all moorit, the brown tones vary.

Using Manx Loaghtan Fiber

Dyeing. The natural brown can be overdyed to produce a subtle range of colors.

Fiber preparation and spinning tips. Pick and card, or comb fibers that are long enough. Between the fingers, Manx Loaghtan comes across as a relatively soft fiber with an exceptional amount of bounce, a quality that can be maximized through woolen-style preparation and spinning. Prepared and spun worsted-style, it can be very durable.

Knitting, crocheting, and weaving. Manx Loaghtan works best when made into relatively lightweight fabrics. It's a great wool for making knitted or crocheted sweaters, as well as socks. Traditionally it's been woven into Manx tartans and fabrics for tailored suits and vests. Before making a garment that will rub against bare skin, make a swatch and test it against the wrist or the back of the neck.

Best known for. Unusual color and softness. Should be better known than it is, because of its resilience and all-around appealing character.

Manx Loaghtan yarn from Blacker Designs

To Save a Rare Breed

The Manx Loaghtans were almost extinct 50 years ago, and though breed numbers have increased and stabilized since the 1970s, these sheep are still considered rare and at risk of extinction. So in 2001, at the height of the last outbreak of foot-and-mouth disease in the United Kingdom, the island's parliament (the Tynwald) took action. They cancelled the Tourist Trophy motorcycle races, a series of events that were started in 1907 and had previously only been cancelled during World War II. The 37-mile (59.5 km) Mountain Circuit race typically brings 40,000 visitors to the Isle of Man and is enormously important to the economy. Yet that year, the island's government felt the risk to the sheep was just too great to hold the event!

In 2009, the Manx Loaghtan breed received a different form of protection — the European Union recognized it with the Protected Designation of Origin. The designation is designed to help reconnect farmers with consumers and raise awareness of a unique product, crop, or animal.

North Ronaldsay

Conservation Breed

TEN MILES (16 km) from the north coast of the Scottish mainland lies an archipelago of approximately 70 islands called Orkney. The largest island is about 202 square miles (523 sq km); the smallest is less than a tenth of a square mile (0.25 sq km). North Ronaldsay, the northernmost island in the chain, is one of the larger islands, even though it is just 2.7 square miles (7 sq km). At its high point, this spit of land rises a mere 54 feet (16.5 m) above the waters of the North Sea. For an odd comparison of landscapes, the island of Hirta, now home of St. Kilda's Soay sheep, is about the same size, at 2.6 square miles (6.7 sq km). Yet Hirta has three peaks that rise from the water in cliffs that are more than 1,300 feet (396 m) tall.

In the early 1800s, the residents of North Ronaldsay made a good bit of their livelihoods by burning and drying the abundant seaweed that covered their shores, and sending the iodine-rich kelp and ash south to England. But then the kelp industry waned, and the residents had to return to subsistence farming.

In 1832, to maximize the production from their limited uplands, they built a 13-mile-long (21 km), 6-foot-high (1.8 m) dry stone wall around the island and placed their rugged little sheep outside the wall. These sheep had always consumed a good bit of seaweed, but once banished to the shore, they had to survive almost exclusively on their seaweed diet. Today, the island boasts a human population of 60, folks who largely make their livings from tourism, from lighthouse and communication system operations, and from the island's other residents: the extremely independent North Ronaldsay sheep.

Historically and from a fiber perspective, it's hard to get a handle on North Ronaldsay sheep, because their origins are obscure and because their niche in the landscape is so unusual. Their closest relatives are perhaps the Soay (see page 194), which have been around even longer, and the Shetlands (see page 184), which have been developed through centuries of systematic breeding efforts that have emphasized wool quality. Attempts to "improve" the North Ronaldsays with crossbreeding have not succeeded because the resulting animals have not been able to survive on the island.

Although no one knows their exact origin, North Ronaldsays are undoubtedly one of the most ancient sheep breeds we have access to

A 21st-Century Industry for a Stone Age Sheep

It's not surprising that North Ronaldsay, with its key position at the outermost reach of the islands of Orkney, hosted one of the earliest lighthouses in Britain. Built in 1789, the original North Ronaldsay lighthouse was close to shore and contained a lighting system composed of oil lamps and copper reflectors, which, though state-of-the-art at the time, required careful burnishing with a linen cloth and fine chalk to remain bright. Perhaps because of these painstaking and possibly impractical maintenance requirements, the light was sometimes mistaken from the sea for a ship's mast instead of a warning light and thus it didn't always help navigators. Called the Old Beacon, this lighthouse was closed in 1809, three years after another lighthouse was built on nearby Sanday Island.

It turned out that a light on North Ronaldsay was a good idea after all, due to the dangerous waters in the area. By the middle of the nineteenth century, better mapping of the sea and improved lighthouse technologies aided both the placement and the construction of the island's second lighthouse, completed in 1854 and still in service.

However, this new lighthouse, known as Dennis Head, was fully automated in 1998. The switch from human-powered to computerized operation left several buildings around the lighthouse unused — buildings that had both water and power supplies available, features not to be taken for granted on North Ronaldsay. A number of islanders, including Jane and Peter Donnelly, envisioned a new opportunity for their small, remote community. In 2003, an empty lighthouse-keeper's home was redeveloped to house North Ronaldsay's latest industry — a mini-mill that processes fleece into roving and batts, yarn, and locally produced knitwear for a cooperative, value-added company: A Yarn from North Ronaldsay.

today. Like the Soay, North Ronaldsay sheep seem to date back deep into the early history of domestication. They have frequently been called an Iron Age (about 1200 BCE) breed, but recent evidence takes them back even farther. According to the Rare Breeds Survival Trust, DNA studies of the North Ronaldsay sheep "have shown a close relationship to sheep in the Stone Age village of Skara Brae on mainland Orkney, which dates from 3000 BC."

Color, texture, and fiber-handling adventures reward the patient spinner who encounters North Ronaldsay fleece. Some fleeces are Shetland-like. Others contain mixes of fiber types that will put you through your paces and teach you things you never thought you'd learn. The yarns you produce can be delightful, and your journey in making them will certainly be interesting. The undercoat is most likely fine enough to be used next to the skin. The guard hairs, which help protect the sheep from the weather, are significantly coarser. As with other double-coated sheep, the fiber types can be spun separately or together.

As primitive sheep, North Ronaldsays previously shed their wool, so the crofters (a British name for small-scale farmers) *rooed*, or pulled, the naturally loosened wool off the animals, instead of clipping it. This process removed the soft undercoat and left the guard hairs behind because the different coats loosened at different times. Now the sheep are gathered in stone pens along the shore (called *punds*) and sheared, with the whole community participating.

North Ronaldsay wool is sometimes called coarse and sometimes fine. Those who call it coarse are evaluating on the basis of the guard hair; those who call it fine are examining fleeces that consist mostly of wool, rather than hair, or are looking at the undercoat portion of a mixed fleece. White and brown fleeces tend to have the least guard hair, and that includes the dark browns that are almost black. The gray or black fleeces get their color at least in part from guard hairs, which can be predominant. Rams have hairier fleeces than ewes; they also develop coarse beards and manes.

Although the wool tends to be low in grease, you need to count on a loss of 50 percent or so from raw weight to usable fiber. Because the sheep spend most of their time on a seashore in wild weather, up to half of the weight of a fleece may consist of sand and sea salt. The wool felts easily.

North Ronaldsay Facts

▶ **FLEECE WEIGHT**
About 2–2½ pounds or a little more (0.9–1.1 kg)

▶ **STAPLE LENGTH**
Generally 2–4 inches (5–10 cm) for the finer coat, but the guard hairs on some of our samples measured up to 6 inches (15 cm)

▶ **FIBER DIAMETERS**
Variable; for the undercoat, think in the range of 23–28 microns (spinning counts 56s–62s)

North Ronaldsay gloves by Jane Donnelly

▶ **LOCK CHARACTERISTICS**
Triangular locks with wide bases, as is characteristic of double-coated fleeces. Crimp varies; the undercoat tends to have fine, even crimp.

▶ **NATURAL COLORS**
Most are whites and grays (mixes of white with coarse black guard hairs), but the full range includes blacks and browns.

Using North Ronaldsay Fleece

Dyeing. Consider overdyes on the lighter natural colors. The guard hairs (mostly black anyway) don't shift colors much with dyeing, but on light grays the applied color in contrast to an underlying natural shade, accented by the guard hairs, can produce stunning effects.

Fiber preparation and spinning tips. You'll almost certainly want to wash the fleece well before spinning. The fiber tends to mat at the base of the lock, so it can be difficult to just card; the matted clumps won't work out very quickly. Combing removes the guard hairs, which may be exactly what you want; carding tends to leave them in, adding visual texture and also potentially some scratchiness. Examine the specific fleece carefully, and consider whether or not to separate the guard hairs. They add a lot of texture and color, but they also radically reduce the softness of the yarn.

Knitting, crocheting, and weaving. Yarns with guard hairs are best for making textiles where durability is a plus. Yarns spun from fleeces that don't contain guard hairs will be like Shetland and suitable for fine textiles.

Best known for. Having a broad range of colors; the quality varies according to the amount of hair and kemp, which may range from almost none to significant. North Ronaldsays are also famous for their ability to live by eating seaweed (they prefer dulse, a red seaweed). This has altered their metabolisms, so successfully moving them out of their natural environment requires specialized nutritional knowledge and care. They are very independent sheep.

North Ronaldsay

Fascinatingly diverse fleeces — some single coated and some strongly double coated — yet all our samples averaged 22.5 to 23.8 microns, including the hair components.

If you sort the fibers by their qualities, you can make anything from baby clothes to rope.

Ouessant

NAMED FOR THE SMALL and rocky French island in the English Channel known as Ushant or Ouessant, where the breed was developed, Ouessant (weh-SAHN) sheep are extremely unusual. The breed is one of the smallest in the world — even smaller than the Soay (see page 194) — reaching no more than 20 inches (51 cm) at the withers (or shoulders) in a mature ram. Not only did the island's shepherds traditionally select for small sheep that could produce a bit of meat and fiber on limited feed, they also strongly selected for black sheep. Although the breed can throw an occasional white animal, most are very dark.

In spite of their tiny size, Ouessants have big fleeces. Every year they grow wool equal to about 4 to 4.5 percent of their body weight. That's like a 150-pound (68 kg) person hauling around a 6-pound (2.7 kg) jacket. They do retain the Northern European Short-Tailed breeds' tendency to shed in the spring, giving them the option of lightening up even if someone forgets to relieve them of their winter clothes.

Ouessant fleeces are double coated, possibly even triple coated. There may be coarser fibers, including kemp, around the neck area. The breed's shedding tendency can make the butts of the shorn locks a little hard to deal with when it's time to prepare the wool for spinning, as they can be gummy or matted together (the fiber does felt). On most of our samples, we trimmed off the very bases of the locks, sacrificing a bit of fiber length for easier workability.

Today some North American breeders are using imported semen to crossbreed Ouessants from Shetland sheep, so sometime in the future you may see an American Ouessant.

Ouessant Facts

▶ **FLEECE WEIGHT**
Ewes 2¼–3¼ pounds (1–1.6 kg); rams 2½–4 pounds (1.1–1.8 kg)

▶ **STAPLE LENGTH**
There's not much data on this, but the information we found said 3¼–4 inches (8.5–10 cm); one of our samples was 5½ inches long (14 cm), which is a lot of length in comparison to the size of the sheep; one sample was 2 inches (5 cm) and two were 4 inches (10 cm)

▶ **FIBER DIAMETERS**
The undercoat can be quite fine, with micron counts in the mid- to lower 20s (spinning counts about 58s–62s); the outercoat can be quite a bit coarser, in the mid-30s (spinning counts about 40s–46s). We sent three samples to the lab, which came back with averages of 25, 26, and 38 microns (spinning counts about 60s, 58s, and 40s).

▶ **LOCK CHARACTERISTICS**
Typical of double-coated fleeces, the staples are triangular, with wide bases consisting of the finer undercoat and very pointed tips formed by the coarse guard hairs.

▶ **NATURAL COLORS**
Primarily black and brown; some whites and grays, possibly through infusions of other breeds' genetics.

Ouessant

2-ply

2-ply

clean

These two fleeces (above and directly below) were basically single coated, with a few coarser hairs.

2-ply, undercoat, carded

clean

raw

2-ply, outercoat, combed

This fleece was strongly double coated.

clean

Using Ouessant Fiber

Dyeing. The dominant natural colors are dark enough that dyeing is likely to be a waste of effort. If you have some of the white or gray fleeces, they can be dyed like other wools. Expect the different qualities of fibers in the fleece to take the color differently.

Fiber preparation and spinning tips. The coats can be separated or mixed and spun together. Combing or flicking will push you into a separate-the-fibers situation (although the long hairs can also be pulled out if you are just holding the butts of the locks). Carding will work if the fibers are not too long, although keeping the varied textures mixed will be a challenge. If there are midrange fibers, it'll be your judgment call about when to stop separating, knowing that your yarn will only be as soft as its coarsest components. The guard hairs will be most cooperative if spun worsted style, keeping them lying parallel to each other. If you separate out the undercoat, you can spin it either worsted (if it's long enough to help you out with this technique) or, more likely, in a woolen style.

Knitting, crocheting, and weaving. Any yarn with the outercoat mixed in will have pronounced texture and some scratchiness that will need to be taken into account when planning how to use it. Yarns made completely of the outercoat can be very strong, suitable for use as warp yarns in weaving or for texture and strength in knitted or crocheted items. A crocheted basket would be an interesting project. If the undercoat is completely separated from the outercoat, it could be soft enough for socks, mittens, and other garments.

Best known for. Tiny sheep with very dark, and relatively abundant, double-coated fleece.

Look how tiny Ouessant sheep are! Most of the photos in this book don't include people because we wanted to focus on the animals. Here's an exception: the person in this picture helps us show something about the sheep that wouldn't otherwise be apparent.

Romanov

THE ROMANOV BREED originated in the Volga valley near Moscow more than two hundred years ago and was named for the last imperial family of Russia. Romanov sheep began to be raised in North America relatively recently. They are still rare west of the Atlantic, though the breed is gaining interest among shepherds due to its prolificacy. Romanovs are said to lamb in litters; the record is nine lambs to a single ewe at one time, though typically Romanov ewes have three or four lambs per breeding. The lambs are black but fade to gray with age, and they sport white face markings (often badger-faced). These sheep are strongly double coated, with a surprisingly fine undercoat, and they do shed their wool if they aren't shorn. The fiber should be fun to experiment with if you are lucky enough to come across some.

Romanov Facts

▶ **FLEECE WEIGHT**
6–13 pounds (2.7–5.9 kg); yield 65–80 percent

▶ **STAPLE LENGTH**
4–5 inches (10–12.5 cm)

▶ **FIBER DIAMETERS**
Strongly double coated; outercoat 40–150 microns (mean about 72 microns; spinning count coarser than 36s) and undercoat 16–22 microns (mean about 21 microns; spinning counts 64s–80s and finer)

▶ **LOCK CHARACTERISTICS**
It's unusual that a double-coated breed's two coats are of similar length (a sheep's outercoat is almost always longer), but this is the case with Romanovs.

▶ **NATURAL COLORS**
Gray that is a combination of black and white fibers; rams have long, hairy manes as well.

Using Romanov Fiber

Dyeing. This might produce some interesting results, although the natural color of the wool will dominate.

Fiber preparation and spinning tips. Because the two coats are the same length, separating will not be as easy as with some other double-coated breeds, although the right set of combs might help you accomplish the task. This wool calls for creativity in preparation, spinning, and use.

Knitting, crocheting, and weaving. We were unable to obtain a sample of Romanov fleece to spin and assess. Traditionally, Romanovs' wool has been valued not for spinning and the construction of textiles from yarn but for pelts, or sheepskins, much as has been true for Karakuls and Gotlands, although those breeds' wools have made their way into the yarn world as well (see pages 276 and 162).

Best known for. Lots of lambs, and a natural-colored, unusual type of fleece.

Shetland

Conservation Breed

The northernmost islands of Britain are the home of, and source of the name for, Shetland sheep. The hundred-plus-island archipelago is located at a latitude about equal to that of Fairbanks, Alaska, and, along with the Orkney islands to its southwest, forms the break between the North Sea to the east and the Atlantic Ocean to the west. Fifteen of Shetland's islands are inhabited — by people, that is. Many more are home to Shetland sheep and ponies.

Shetland sheep have evolved in this harsh island environment for more than a thousand years. The roots of the breed's origin have been the subject of much study, some debate, and, as of yet, no conclusive agreements. Sheep, including types most closely represented today by the Soay and North Ronaldsay breeds, have lived in the vicinity of Shetland for several thousand years. From two thousand years ago to the present, genes from Roman-influenced types of sheep, moving north from the southern parts of the British Isles, have modified many of the northern sheep populations, including the Shetlands, to greater or lesser degrees. About fifteen hundred years ago, seafaring Scandinavian people brought other sheep from farther east to add to the island mixes. In addition, economic pressures in favor of either wool or meat production have nudged the gene pool back and forth, most dramatically in the eighteenth through twentieth centuries. DNA testing should help us unravel the complicated stories that have brought us today's Shetlands, but for the time being we don't know enough to explain how these sheep got to be the way they are.

Despite or because of all this shifting about, Shetlands offer fiber lovers the equivalent of a sheepy smorgasbord of wool options, with a wide variety of colors, textures, and styles of fleeces. Because of the variety, there's a whole vocabulary that goes with Shetland sheep and their fiber. There are 11 defined colors, which

Shetland yarn from Garthenor

Woven Shetland scarf from Blacker Designs

function as identifiable stopping points along a lush natural rainbow, including whites, grays, beiges, moorits (reddish browns), medium to deep browns, and blacks. The names assigned to the colors have a partial connection to tradition and are partially a delightful romantic overlay. They include terms like *emsket*, *shaela*, and *mioget*. (See The 11 Classic Colors of the Shetland Rainbow, on pages 186-187.) In some breeds, grays and light or medium browns occur through a visual blending of dark (black or deep brown) and lighter fibers. In the Shetland, there are true, solidly pigmented grays and soft browns.

Although the full color range still exists in the breed, it is threatened. According to our friend Jan Dohner, author of *The Encyclopedia of Historic and Endangered Livestock and Poultry Breeds*, "Nine of the rarest colors [out of the eleven identified shades] totaled just 214 sheep in 1994." That's not very many sheep for each of those shades, although since that time more breeders have realized the potential for loss and are breeding Shetlands specifically for color.

The animals themselves also have color names based on the patterning of their coats, listed in the table on page 193. There are 30 names, which, although not entirely traditional or used consistently, offer one way to make sense of a breed with extremely complex color alternatives. The body patterns are not mutually exclusive, meaning you can potentially have a *bersugget yuglet* or a *snaelit bielset*. Shetland sheep colors could fill a book in their own right.

Sometimes Shetland fiber is described as the finest wool grown by any sheep breed in the British Isles. And it can be remarkably fine, as seen in Shetland lace ring shawls, spun from carefully selected locks of neck wool from the finest-fleeced sheep; these shawls are so delicate that a large one can be drawn easily through a wedding ring. At other times, the fiber is too coarse for use in next-to-the-skin garments or baby clothes. We've even heard people refer to Shetland yarn as scratchy, thinking that was true of all of the breed's wool. What gives?

From the early history of the breed, there have been two distinctive types of Shetland fleeces: *kindly*, which is fine, soft, and more single coated and also has well-defined crimp; and *beaver*, a coarse style that contains less crimp and more hair. Double coatedness may be part of the original genetic composition of Shetlands, or it may have been introduced, or reinforced, by later contributions from other breeds.

In 1927, breeders who were concerned that the heritage traits of the Shetlands were about to vanish formed the Shetland Flock Book Society and defined breed standards. Discussions of what the Shetland breed should be like are ongoing, in Shetland and everywhere else that these sheep are now grown. Agnes Leask, a lifelong Shetland crofter (small farmer), is quoted thus in *Shetland Breeds — "Little Animals . . . Very Full of Spirit"* (edited by Nancy Kohlberg and Philip Kopper): "I don't agree with the flock book of 1927. . . . It's an engineered breed. You won't get modern flock book sheep surviving in the hills without a bit of pamper. That's speaking from experience." There is no question that Shetland sheep have experienced breeding pressures toward "improvement" in production values, whether toward wool or meat. In what form they originated and where breeding should go in the future are hotly debated among those who care passionately about the breed.

What we can do, as fiber lovers, is understand the history, recognize that breeders look for different strengths and qualities in their flocks, and understand that some Shetland fleeces (and therefore some Shetland yarns) will be exceptionally fine and soft, while others will be rough and sturdy.

Shedding vs. shearing. The ability to shed is present to a greater or lesser degree in different strains of Shetlands, both single and double

The 11 Classic Colors of the Shetland Rainbow

The Delicious Heritage of Color

Although white is preferred by mass producers of Shetland yarn, the unusually wide range of natural colors makes this breed highly valued by fiber lovers.

Traditional/historic		Modern/production-oriented
Full array of 11+ colors and 30+ marking patterns and variations, including whites, creams, shades of brown, shades of gray, and blacks (*some colors are exceptionally rare*)	← Increasing color variety	White wool
More likely to require (and reward) hand processing		Can be more easily processed commercially and made available in a wide range of dyed colors

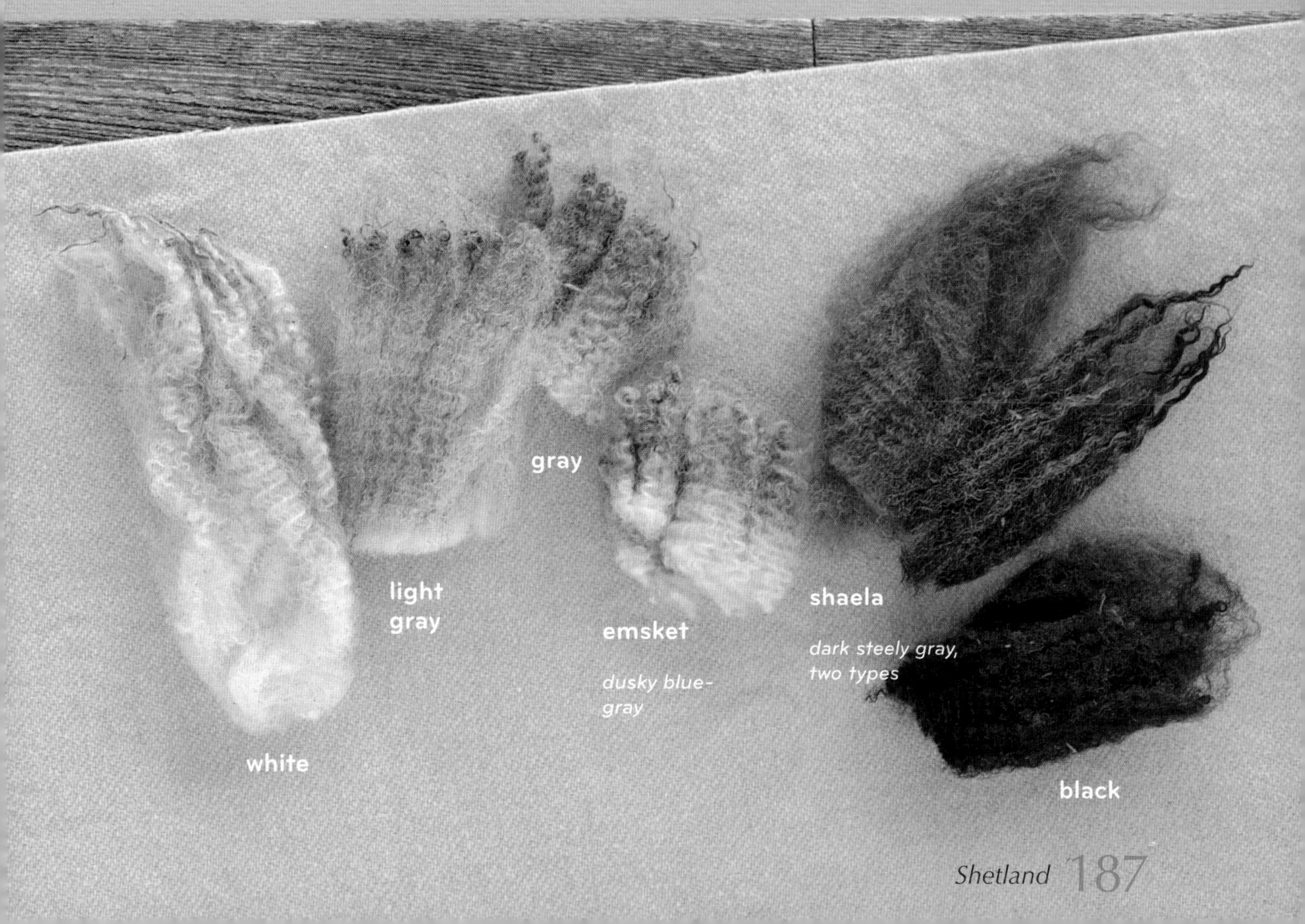

white

light gray

gray

emsket

dusky blue-gray

shaela

dark steely gray, two types

black

coated. In strongly double-coated animals, the undercoat and outercoat may shed at different intervals, making it relatively easy to separate the fibers as they come off the sheep by plucking, or *rooing*, as was traditionally done. Double-coated fleeces with high contrast between the fiber types may be a relatively recent development in the breed. Then again, they may represent the earliest types of Shetlands, which are believed to have been similar to — or even part of the same population as — today's North Ronaldsay sheep (see page 175), found on a neighboring chain of islands.

Most Shetlands today are sheared rather than plucked, because most Shetlands are white (so wool from a number of fleeces can be mixed); much of the wool is processed mechanically; and with shearing the fiber is available all at once, rather than in bits and pieces over a period of weeks. While shearing simplifies harvesting, it makes the separation of the fiber types in preparation for spinning (if the fleece is double coated and the spinner wants to divide the coats) more challenging.

Describing varying characteristics. The immense variability of the wool makes it hard to tell what to expect of fleece or yarn from Shetland sheep, although the several breed societies have similar standards for what the sheep should be like. These standards are evolving into more agreement on fleece qualities without significantly narrowing the possibilities.

In any case, Shetland fiber is a whole lot more predictable than that of the North Ronaldsays, which are thought to resemble what Shetland sheep may have been like a few thousand years ago. More predictable through the entire population? No. More predictable within flocks or breeding groups? Yes. One experienced breeder described the perfect Shetland fleece to us as being single coated with a slight "feathery" look and having an average micron count in the mid-20s, a staple length between 3 and 5 inches (7.5-12.5 cm) long, a very soft and silky feel, and nicely defined and even crimp. Other breeders have different opinions. A website with superb information on the world of Shetland sheep has been established by Linda

Characteristics of Shetland Fleece Types

This chart does not represent breed standards or anyone's idea of the perfect Shetland fleece but reflects the types of fleeces spinners and knitters may encounter labeled as purebred Shetland. In addition, any of these fleece types may have hair fibers (scadder) around the neck and along the back. This is characteristic of breeds from similar ancient backgrounds but is considered a fault and discriminated against.

True double coated	Short, fine undercoat and long, hairy outercoat	More likely to require hand processing
Transitional double coated	Even, and difficult to separate, mix of fine, short fibers and somewhat longer and coarser fibers	
Single coated	Short-stapled, fine, crimpy wool, fairly even throughout fleece	Can be more easily processed commercially, but also lovely to process by hand

Wendelboe and Kathy Baker at Shetland Sheep Information: www.shetlandsheepinfo.com

Effects of wool source. When discussing Shetland wool, it's useful to understand the distinction between wool grown by Shetland (breed) sheep and wool from Shetland (the islands). They are sometimes, but not always, the same. This produces some additional confusion that fiber folk may experience when trying to figure out what Shetland wool is like.

Wool from the purebred Shetland breed of sheep may be grown in Shetland or elsewhere. As with other breeds, environment and husbandry strongly influence the wool, and it's not easy to generalize about Shetlands if you are working only from a knowledge of wool grown in a particular region. For example, the Shetlands from which we spun samples all came from North American sources. They were all basically single coated, although they did not represent the finest wools available within the breed, and one had some coarser fibers from what was likely the britch area (the upper legs and buttocks).

Shetlanders also market wool that has been harvested on the islands, whether it came from purebred Shetlands or from other types of sheep kept in the island flocks. These wools are permitted to be labeled in the marketplace with terms like "100 percent Shetland wool" or "pure Shetland wool" whether or not they are from the Shetland breed. We applaud place-based identifiers, because they can be a crucial component of supporting local economies, but they can also leave one a bit baffled. Jamieson and Smith, the famous source of Shetland yarns for knitters and other fiber crafters, buys more than 80 percent of the wool produced on the islands, whether grown by purebred Shetland sheep or not. The company reserves the finer qualities for its own yarns and sells the other grades to other firms.

A wool for every taste. Ultimately, Shetland wool can provide whatever you need. You just need to know what you're looking for and, if you are working from fleece, how to select with your own goals in mind. If you are choosing yarn that's already been spun, feel the fiber and make samples before deciding what you will use it for. Some Shetland wool, especially the silky-feeling variety, will felt very quickly; those that are matte and more spongy than silky will resist felting. Putting a Shetland sweater in the washer and dryer is a really unfortunate idea unless you want it permanently shrunk.

Fleeces on this page are single coated. Note that the tips of the long-tipped locks are coarser than the rest of the length of fiber.

clean
raw
2-ply
clean
lamb
Variable textures and lengths
2-ply
raw
raw
clean
2-ply
raw
clean
Part of this fleece was single coated, and part was double coated.

Shetland Facts

▶ **FLEECE WEIGHT**
2–5 pounds (0.9–2.3 kg), usually in the range of 2–4 pounds (0.9–1.8 kg); yield 65–80 percent, usually on the higher side

▶ **STAPLE LENGTH**
Varies, depending on type of Shetland. Think 2–4½ inches (5–11.5 cm) in general; some fleeces, including those from many North American Shetlands, have a range of 4–6 inches (10–15 cm), while others are as long as 6–10 inches (15–25.5 cm). While some people question whether the longest fleeces (those with staple lengths of 6–7 inches [15–18 cm] or more) remain true to the unique qualities and history of Shetland wool, others prefer to breed their animals to produce this long fiber. The ultimate question for the fiber user is whether a given fleece meets your needs.

▶ **FIBER DIAMETERS**
Covering the full span of possibilities, most fibers are likely to fall in the range of 20–30 microns (spinning counts 54s–70s), with some as fine as 10–20 microns (neck wool from the finest fleeces: spinning counts 70s–finer than 80s) and some as coarse as 30–60 microns or more (outercoat on a double-coated sheep: spinning counts coarser than 36s–50s). The Shetland Sheep Society gives an average fiber diameter of 23 microns, reflecting the breed's bias toward the fine end (spinning count 62s).

▶ **LOCK CHARACTERISTICS**
Regardless of style, the fleece is dense with locks that are usually wider at the base and somewhat pointed, so the essential shape is triangular. The fine fibers are nicely crimped, and the longer, coarser fibers are wavy to nearly straight.

▶ **NATURAL COLORS**
The widest array of any breed (see The 11 Classic Colors of the Shetland Rainbow on pages 186–187). Our Shetland samples all washed up considerably lighter and whiter than they looked in the lock; the grease seems to affect color by darkening and yellowing it. That's true of all wools but seemed to be dramatically so for the Shetlands. If you are starting with grease fleece, you won't know what color you've got until you have washed it.

Using Shetland Fiber

Dyeing. The whites are frequently dyed. The natural colors are most often used in their original state, but they can be overdyed.

Fiber preparation and spinning tips. Depending on the characteristics of the fleece, spin from the lock (any type); card (short, fine fleece); comb or flick (longer fine, medium, or mixed fleece). Where several types of fiber are present, they may be easy to separate, or they may not be clearly different enough and may need to be spun together. Regardless of the type of Shetland fleece, the spinning should be comparatively smooth and easy. The finished yarn will lean toward the feel of its coarsest component fibers.

Knitting, crocheting, and weaving. Shetland lends itself to varied construction techniques, ranging from lace shawls (both the fine sort and a heavier, everyday variety) to color work in crochet and knitting to woven tweeds. The fine to medium fleeces make exceptional sweaters. The fiber tends to combine light weight with unusual durability for its range of fineness. For example, the finest fibers, when processed parallel (worsted) in a cobweb yarn, will make an ethereal shawl that is sturdier than it looks. Commercially spun Shetland yarns are often bulkier and feel a bit sturdier than their handspun counterparts.

Best known for. Exceptionally wide array of natural colors; variable fleece; when selected and used carefully, can produce everything from the sheerest and most delicate shawl to sturdy rope. Fleece closely connected to development of Shetland ring shawl and hap (everyday, practical) shawl traditions and to the original Fair Isle color-patterned knitting, worked with natural or dyed colors. (Fair Isle is part of Shetland, located about 24 miles, or 39 km, from the main group of islands.)

Naming Shetland Colors and Patterns

Names for the marking patterns on the sheep are traditional names, based on a list found on the website of the North American Shetland Sheepbreeders Association. The patterns are not mutually exclusive: a sheep can have more than one name.

bersugget	having irregular patches of different colors; variegated
bielset	having a complete circular band of a different color around the neck
bioget	having a white back and darker sides and belly, or the reverse — darker back with white sides and belly
blaeget	having a lighter shade on the outer portions of the wool fibers, especially in moorit and dark brown sheep
blaget	white with irregular dark patches, so it looks like the ground partially covered in snow
bleset	dark colored with a white stripe down the forehead, or the reverse, white with a dark-colored stripe down the forehead
blettet	having white spots on the nose and the top of the head
brandet	having stripes of a different color across the body
bronget	dark colored with a light-colored breast, or the reverse, light colored with a dark-colored breast
flecket	white with large black or brown patches (less well defined than in Jacob sheep, which also have a different overall body shape and wool quality)
fronet	black spotted, with a white head and black spots around the eyes
gulmoget	having light underparts with a dark-colored body; the opposite of katmoget
ilget	white with spots of a different color (usually gray or black)
iset	dark colored with many white fibers, so the overall appearance from a distance is bluish
katmoget	having a light-colored body with a dark belly and legs, along with moget facial markings; the opposite of gulmoget
katmollet	having a light-colored nose and jaws
kraiget	having a neck that is a different color from the rest of the body
kranset	dark colored with white around the eyes and the neck
krunet	dark colored with a white patch on the top of the head
marlit	various shades of different colors, mottled
mirkface	white with dark patches on the face
moget	characteristic dark and light patches around the mouth, eyes, and ears
mullit	white with a dark nose and jaws, or the reverse, dark with a white nose and jaws
sholmet	of any nonwhite color, with a white face
skeget	having stripes of different colors on the sides
smirslet	dark colored with white around the mouth, head, or neck
snaelit	light colored with a snow-white face
sokket	having legs of a different color from the body
sponget	dark colored with small white spots, or the reverse, white with small dark-colored spots
yuglet	having color around the eyes that is different from the color on the remainder of the body

Soay

Critical Conservation Breed

For the Soay, we return to the islands of the St. Kilda archipelago (see Boreray, page 154), particularly the islands of Soay and Hirta. Pronounced "soy" or "SO-ay," the word *Soay* is derived from the Norse words for "sheep" and "island." The Soay breed is feral and closely resembles the earliest domesticated sheep of Neolithic times. No one is sure when these sheep arrived on Soay island, but over the centuries, the St. Kildans — who lived on Hirta, another island in the cluster — would go to the island once a year and gather wool from these small, wild sheep, who retain the primitive characteristic of shedding their fleeces seasonally. The St. Kildans made their living in part through spinning (done by the women) and weaving (done by the men) and used wool from all of the sheep, feral and domesticated, within the small group of islands.

The residents of St. Kilda never owned any of their islands. The land, such as it was, was sharecropped for a nonresident lord who was the actual owner, and ownership changed over the years (mostly among the MacDonalds and MacLeods of Skye, an island in the Inner Hebrides). In the very early years of the twentieth century, the lord took small numbers of Soay sheep to England and gave some to his patrician friends as ornamental additions to their estates and parklands. The British elite bred these animals for a dark mouflon color, and they became known as Park Soay.

Then, in 1932, after the evacuation of St. Kilda (see page 154), the lord sent some of the St. Kildan men back to the islands to capture and move a flock of 107 of these primitive sheep from Soay to Hirta. This was to expand their population as well as control the vegetation on the relatively much larger island, which had previously been kept in check by the islanders' primary sheep, a localized type of Blackface cross.

Lord Dumfries, the fifth Marquess of Bute, owned St. Kilda since shortly after the evacuation. Upon his death in 1956, the islands were taken over by the National Trust for Scotland. The National Trust subsequently encouraged scientists to study the ancient Soay sheep. Scientists brought a flock to England for additional study, and that flock helped expand Soay populations to other parts of the British Isles and later, through exports, to North America.

Today Hirta is home to seasonal staff from the National Trust and a small year-round military installation. It can be visited during the summer through volunteer work programs operated by the National Trust and specially arranged briefer trips, but as in the past, the journey to the island is made with no guarantee that the weather will permit a landing!

At least four lineages of Soay sheep have been described over the years, two in Great Britain (the Park Soay and the St. Kilda, or Island, Soay) and two in North America. All of these Soay are rare. The Soay in North America derive from two importations to Canada. The first, in 1974, saw four lambs imported to Assiniboine Park Zoo in Winnipeg; animals from that group were crossed with some other breeds (possibly including Barbados Blackbelly, wild mouflon, and other hair sheep). Their offspring are now known as American or North American Soay. A second importation to Canada, in 1990, to Athestan, Quebec, of six sheep registered with Britain's Rare Breeds Survival Trust, gave rise to a population

of purebred Soay commonly known in North America as "British," distinguished by the fact that their ancestors trace exclusively to the 1990 importation. There are more than five hundred British Soay in North America (more than three hundred ewes), held by conservation breeders. In addition, over the years many American ewes have been bred to British rams.

The American and British Soay have many aspects in common, most notably color, which runs from light brown through near black. There is a range of fiber-quality possibilities, from very fine wool to kemp. The wool sheds naturally and can be plucked, or *rooed*, when it loosens up in late spring or early summer. Some shepherds raising American Soay have breeding programs designed to produce fleeces that will be more familiar feeling to contemporary handspinners than those grown by the wild island animals.

What does all this mean for textile people? Exploring Soay wool can be a worthwhile adventure in which you won't be sure what you're getting. Or when you'll get it — the wool can be challenging to locate, although we had an easier time finding Soay fiber for our samples than we did for a number of other breeds. Soay yarn isn't readily available, however, though you may obtain some directly from breeders. Some Soay fiber is spun at Still River Mill in Eastford, Connecticut, and woven into scarves, wraps, bookmarks, and table runners by Margaret Russell of Antrim Handweaving in Byfield (in Newbury), Massachusetts.

Soay Facts

▶ FLEECE WEIGHT
¾–2 pounds (0.3–0.9 kg)

▶ STAPLE LENGTH
1½–4 inches (3.8–10 cm), usually in the 2-inch (5 cm) range

▶ FIBER DIAMETERS
Most often listed as a spinning count of 44s–50s, or approximately 29–36 microns, Soay fleece actually ranges from about 9 to 48 microns (spinning counts finer than 80s to coarser than 36s), with some outliers (kemp) as great as 100 microns. The multiple samples that we had analyzed, from animals in three lineages, had average micron counts in the low to mid-20s (spinning counts 58s–70s), yet all also contained fibers in the high 30s and 40s (spinning counts 36s and coarser than 36s). Because of the way wool classing is done, the coarser fibers would significantly lower the assigned spinning count. Rams may have a coarser "mane" on the neck and shoulders and more coarse britch (upper legs and buttocks) fibers than ewes.

▶ LOCK CHARACTERISTICS
Locks tend to be indistinct, a little blocky, sometimes with very slightly pointed tips. Sometimes distinctly double coated, sometimes more consistent in fiber quality. There are two traditional types of fleeces, woolly and hairy, although variety is the name of the game overall. All of the samples we scanned were of the woolly type, although a couple had scatterings of very coarse fibers.

▶ NATURAL COLORS
Most frequent is a medium to dark brown, with some animals growing much lighter brown wool (tan) and others producing an exceptionally dark brown that is just about black. Infrequently, white.

Using Soay Fiber

Dyeing. The wool will take dye but is most often used in its natural colors.

Fiber preparation and spinning tips. Try spinning from the locks after gently opening them. Carding is also an option, because of the fiber length. As we saw with a number of breeds' rooed fleeces, some of our Soay samples had what appeared to be persistent skin flakes at their bases, which we thought was a result of a natural part of the growth cycle in these animals. These would not wash or pull out, and this wool processed best, despite its short staple length, on mini combs. When spinning, bring some experience and a creative approach to Soay. Because of its short fibers, you will probably need to spin fairly fine.

Knitting, crocheting, and weaving. The yarn you get and its use depends on the mix of fibers (including whether you want to, or can, separate the types). Traditionally, residents of St. Kilda used the very finest wools for underwear. Other ranges of fiber were spun and woven into tweedy yardage.

Best known for. Being the most primitive sheep currently in existence. Naturally colored and shedding wool with exceptional fineness (and variability) in the breed.

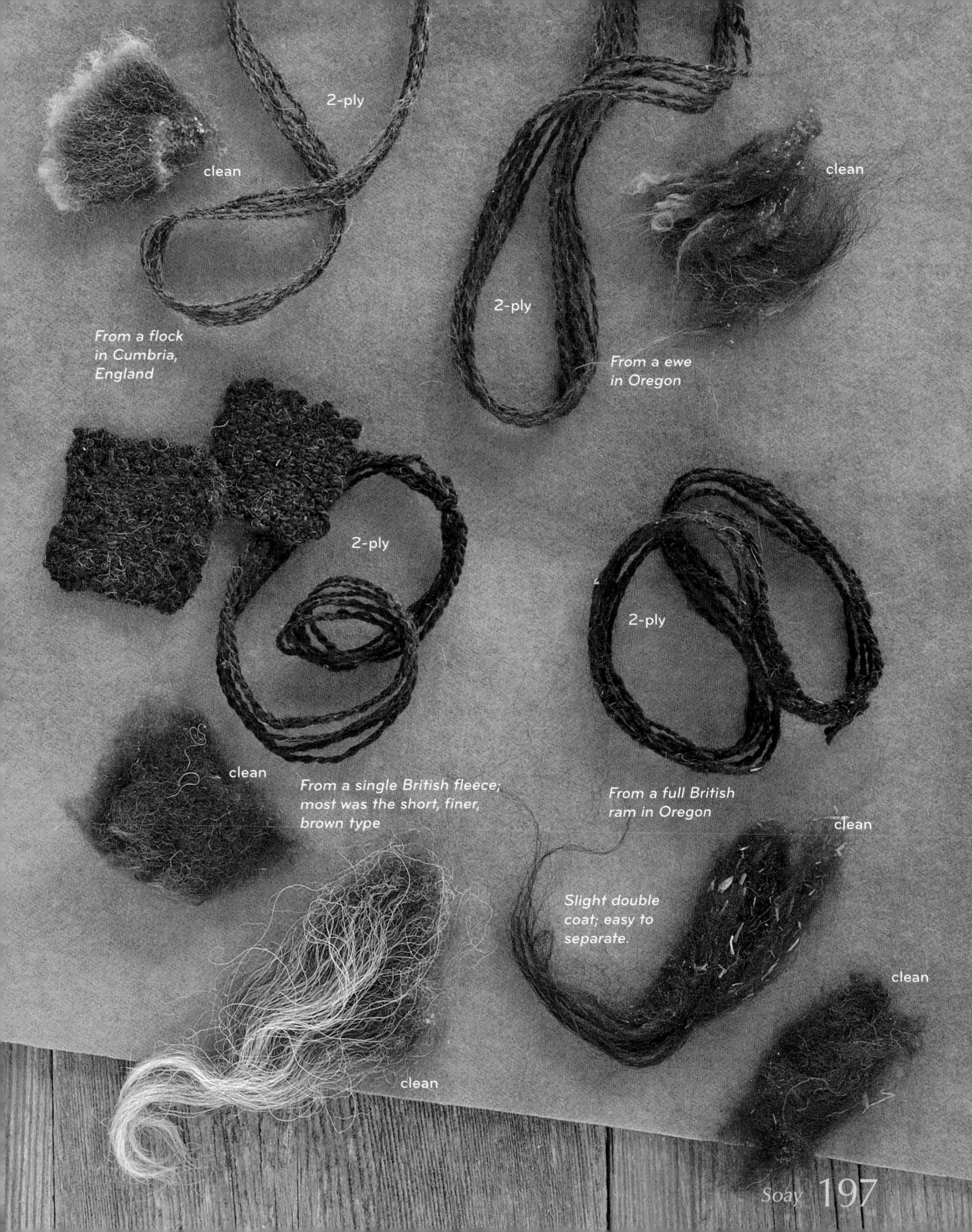
2-ply
clean
2-ply
clean
From a flock
in Cumbria,
England
From a ewe
in Oregon
2-ply
2-ply
clean
From a single British fleece;
most was the short, finer,
brown type
From a full British
ram in Oregon
clean
Slight double
coat; easy to
separate.
clean
clean

Beulah Speckled Face
Badger Face Welsh, Torwen (left) and Torddu (Right)

Welsh Hill and Mountain

FAMILY

Wales, which is a bit smaller than the state of New Jersey, is bordered on the west by the Atlantic and on the east by England. In comparison to England, Wales has heavier rainfall, higher elevation, and poorer-quality soils. Yet 81 percent of the Welsh land area is still used for agricultural production, especially livestock, and sheep remain the prevailing livestock species in this harsh place. Sheep account for about a quarter of the country's agricultural revenue and about 15 percent of the total sheep production for all of Europe. "Since the Neolithic period, sheep have been the dominant livestock of upland Wales," according to Juliet Clutton-Brock and Stephen Hall in *Two Hundred Years of British Farm Livestock.*

Not all sheep breeds in Wales are mountain sheep. We've bundled into this group not only the true mountain breeds, almost all of which have double-coated fleeces, but also their more refined relatives, the Welsh hill sheep such as the Hill Radnor, Kerry Hill, and Lleyn. Thanks to their hardiness, the Welsh Mountain varieties and their close relatives are still the dominant animals today on the

Those Wily Welsh Women

Historically, Welsh farmers sold the meat of their sheep (prized for its sweet taste) to England, while the wool was kept for home use. Welsh women would spin and dye the wool and knit or weave garments for their families. The traditional color women used to dye wool for their cloaks was red.

There is a great story about these "redcoats." During the War of the First Coalition (1792–1798), Napoleon sent four ships to invade southern England. But due to weather, they missed their target, landing instead on the coast of Wales. The French soldiers came ashore late in the evening at the small community of Fishguard.

First thing the next morning, one woman, Jemima Nicholas (now known in Wales as Aunt Jemima), gathered a group of women in their red cloaks on a hillside above the beach. The women walked along the crest of the hill, dropped behind it to get back to the beginning of the line, and walked the crest again, repeating the walk continuously to give the Frenchmen the impression that there were thousands of British soldiers above them. The French quickly surrendered.

hills and mountains, though Cheviots (see page 54), Scottish Blackface (see page 47), and other breeds are also found at lower elevations.

Traditional Welsh Mountain breeds graze the mountains from April through October and spend the winter in the slightly lower and more forgiving valleys, close to the farmsteads. The upland summering grounds, referred to as *sheepwalks*, are still often held as common lands and shared by multiple farmers. Today the common lands of the United Kingdom are seen as a critically important natural resource because they have been less intensively developed than private farmlands. They provide habitat for wildlife, sustain heritage areas, and bring in ecotourism and recreation dollars, as well as continue to provide grazing lands for sheep and, to a lesser extent, cattle and ponies.

Sorting through the Welsh Mountain breeds and variants takes time and concentration. These are old landrace breeds that, with few exceptions, have not seen significant changes over many centuries. They tend to be relatively small, hardy animals. Their wools have strength and body — and may include varying amounts of kemp. The big-bodied South Wales Mountain sheep are known for lots of kemp, while kemp gets closest to negligible in the Black Welsh Mountains, which have been bred for many years for finer, softer, more consistent wool. This has made them the handspinners' darling in this group, though the whole family is worth exploring, and you'll learn a lot about the ranges of wool in the process.

A Cyber Adventure for Welsh Mountain Sheep

As we've researched this book, we've kept our eyes out for unusual stories about the many critters we cover, but one of the craziest we found was the story of a particular Welsh Mountain sheep.

Have you heard of *geocaching* (pronounced "geo-cashing")? Basically, it involves scavenger hunts — coordinated via Internet sites — in which participants use Global Positioning Systems, or GPS devices, to find goodies hidden in waterproof containers by other participants. Anyone with a GPS device can try to locate a geocache; after finding one, the searcher signs the enclosed logbook. In one variation, the finder can swap something of equal or greater value for the goodie in the box. In another variation, the finder moves the box and its contents to another locale and sends the new GPS coordinates into cyberspace to keep the hunt going.

In this story, the goodie is a small statue of a Welsh Mountain sheep. On the geocaching website, the message attributed to the sheep says, "My birthplace is next to the Menai Strait in North Wales. I would like to visit as many different States in the union as possible. Please pick me up and photograph me grazing in places of interest. Ultimately I would quite like to return to the green, green grass of home in Snowdonia, North Wales."

When we stumbled upon this little guy in October 2009, he had made his way from Phoenix to Florida, with stops in California, Connecticut, Indiana, Iowa, Massachusetts, Minnesota, Virginia, and Wyoming. A few months later, he was visiting Tennessee, then Illinois and New York! We hope he continues to travel the country, and we are going to watch for him to come to Colorado, where the two of us might go on a search for the traveling Welsh Mountain sheep!

You can check out the photos and find out where the sheep is now by visiting www.geocaching.com/track/details.aspx?id=78321.

Badger Face Welsh Mountain

Conservation Breed

THE BADGER FACE looks primitive, and we could find no information available on the breed's early history. Sheep of similar description were described in *The Domesday Book*, a two-book compendium commissioned by William the Conqueror in 1086 to assess the extent of the land and resources he could raise taxes on.

There are two types of Badger Face Welsh Mountain sheep. One, called Torddu, which means "black belly" in Welsh, has dark undersides and dark markings on its face, but the rest of the sheep is light colored; the other type, with primarily dark fleece, is called Torwen, which means "white belly." If you plan to learn the meanings of the Welsh terms, remember to reverse: The "black belly" has light wool; the "white belly" has dark wool.

Because breeding of these sheep is meat oriented, the wool is definitely secondary: It is typically a lightweight fleece, classed as carpet or coarse wool, and most often little attention is given to consistency, either among sheep or

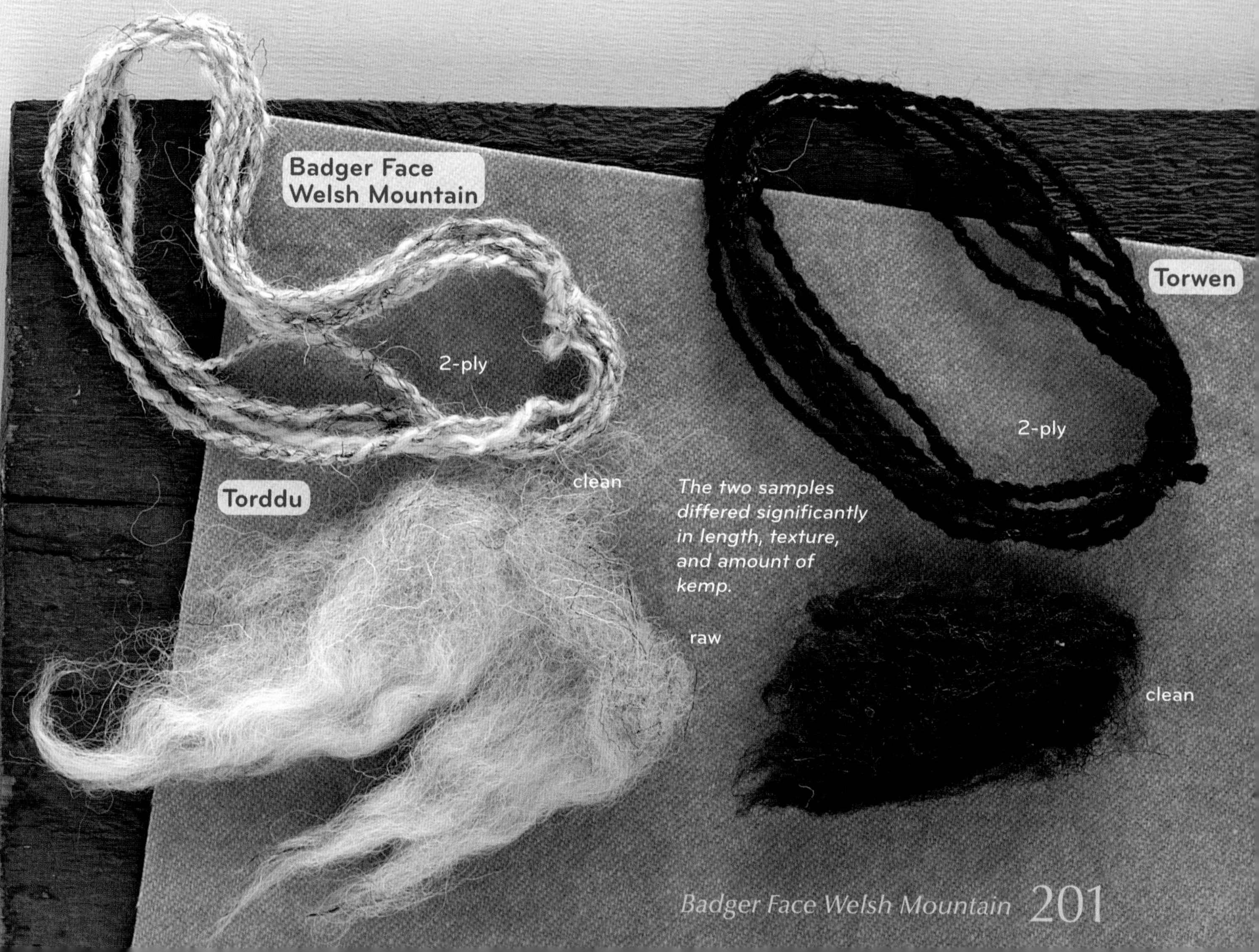

The two samples differed significantly in length, texture, and amount of kemp.

within a given fleece. Although the breed society calls for wool that is "soft, firm and close as for Welsh Mountain sheep," in reality it may be either like that or coarser. Kemp can be present, and some breeders like to see kemp as a theoretical indicator of hardiness in the sheep, while others consider it undesirable.

Our dark (Torwen) sample, while sturdy, was on the border of being okay for a cardigan or other work sweater that didn't rest directly against the back of the neck or on the wrists for an extended time. Our light (Torddu) sample was excitingly highly textured, with a light base and a variety of colors of kemp, mostly white and brown. It would make great rough-wear clothing (lined) or rugs or baskets. The two fleeces also varied dramatically in staple length.

Badger Face Welsh Mountain Facts

▶ **FLEECE WEIGHT**
3½–4½ pounds (1.6–2 kg)

▶ **STAPLE LENGTH**
3–4 inches (7.5–10 cm)

▶ **FIBER DIAMETERS**
On the books, in nearly the same range as Black Welsh Mountain: 26–37 microns (spinning counts 46s–56s); in practice, may be much coarser

▶ **LOCK CHARACTERISTICS**
Rectangular locks with relatively short and tapering tips. Definite but disorganized crimp in both fibers and locks.

▶ **NATURAL COLORS**
Torddu Upper wool is white to light beige or tan, with lighter colors preferred.
Torwen Upper wool is black or brown, with a preference for very dark fiber.

Using Badger Face Welsh Mountain Fiber

Dyeing. The light-colored Torddu fleeces will be interesting to dye, with the varying fiber types taking the color differently and giving a tweedy result. The kemp fibers will retain their original colors.

Fiber preparation and spinning tips. The staples will not be particularly distinct from each other in the mass, although you should be able to separate them out if you locate and pull on the tips. A fairly homogenous fleece can be carded, flicked, combed, or spun from the lock. One with a lot of variety may just need to be picked to open any clumps and spun directly from there. Keep a light hand in drafting, especially if you have a wide range of fiber types, as we did in the light-colored sample, or you will find yourself spinning one or two types, with the others either falling on the floor or staying behind in your hand.

Knitting, crocheting, and weaving. Know you will end up with a firm texture and a solid fabric that is likely to be too rough for next-to-the-skin wear (although a 26- to 30-micron fleece with minimal kemp might work well for everyday garments).

Best known for. Variety in color and texture; mostly for tweeds, upholstery, and other rugged applications; fun (if unpredictable) to work with.

Balwen Welsh Mountain

Conservation Breed

BALWEN WELSH MOUNTAIN sheep originate in a valley called Tywi in southwestern Wales. Although the River Tywi is the longest river that flows entirely within Wales, covering a distance of 75 miles (120 km), its valley is fairly small; due to the breed's limited geographic area in the valley, it almost died out in 1947 — one of the coldest and snowiest winters in the United Kingdom in recent centuries. A single surviving ram was used to redevelop the breed. And it's good the effort to reestablish the sturdy Balwens succeeded, because they are interesting in many ways, starting with the striking coloration of their bodies: Balwens are primarily black, dark brown, or dark gray, and each sheep has a white blaze on its face (Balwen comes from the Welsh word for "white blaze"), four white socks, and a white-dipped tail — dark on the top one-third to one-half of the tail and white on the lower portion.

There is no wool description in the breed society's standard, because this is primarily a meat breed. Nonetheless, an auxiliary page on the site of the Balwen Welsh Mountain Sheep Society suggests soft-to-medium wool. Other resources indicate much coarser fibers. Our sample gave us more robustness than softness. Its fiber length and lock form look like they would be associated with a gentler wool. The wool is a great dark color, with dark hairs or kemp adding texture.

Balwen Welsh Mountain Facts

▶ **FLEECE WEIGHT**
2¼–4½ pounds (1–2 kg)

▶ **STAPLE LENGTH**
2–3½ inches (5–9 cm)

▶ **FIBER DIAMETERS**
The breed society says about 32 microns (spinning count 48s), and other sources we've found say 40–50 microns (spinning counts coarser than 36s); our sample feels like it's a combination moving from 35 microns (spinning count 44s) up, with some hairy bits that are likely in the 40- to 50-micron range

▶ **LOCK CHARACTERISTICS**
Blocky staples with short but picturesquely curly, possibly sun-lightened, tips.

▶ **NATURAL COLORS**
Black (preferred by the breed society), brown, or dark gray.

Balwen Welsh Mountain

2-ply

clean

Spun from flicked locks. On this fleece, the tips were sunburned and fragile, and the bases of the locks needed to be coaxed apart.

Using Balwen Fiber

Dyeing. This might be an option on a dark gray. The other natural colors are too dark for dyes to have an impact.

Fiber preparation and spinning tips. Flick or card, in general, although the longer fleeces can be combed. Because of the coarseness (notable) in relation to the length (fairly short), this wool requires a delicate balance between grist (on the light-weight side) and twist (enough to hold the yarn together but not so much it becomes wiry, unless that's what you want).

Knitting, crocheting, and weaving. Unless you come across an unusually fine specimen without much kemp, Balwen is best used for outerwear, tweeds, and sturdy fabrics, perhaps a woven bag that will be relatively lightweight and can take a lot of abuse.

Best known for. Dark and almost tough wool from truly enchantingly colored sheep.

Balwen yarn from Garthenor

Beulah Speckled Face

ALSO KNOWN AS Eppynt Hill and Beulah Speckled Face sheep (yes, that's all one name), in part for the hilly area in south-central Wales where they presumably originated, these sheep are grown primarily for meat and for crossbreeding. Though fairly common in Wales, they are little known outside the Welsh countryside. Since the 1940s, Mynydd ("mountain" in Welsh) Epynt has been a British military training facility. (Mynydd Epynt is spelled with one *p*; the breed name, according to the breed society, is spelled with two.) The army displaced 219 shepherds and their flocks, and it still controls about 40,000 acres (16,000 hectares) on and around Epynt.

The Beulah's origins remain a complete mystery. It may be a truly native breed, the result of natural and human selection in a small geographic area, as there is no mention of crossings with other breeds in any literature.

Beulah fleeces vary from medium (26–31 microns) to quite coarse (micron counts up to the mid-30s). At the finer end, they can make a nice, springy, midrange, multipurpose yarn for everyday wear. The coarser fleeces, which will also tend to be the larger, heavier ones, are best in tweeds and household textiles that need to stand up to wear.

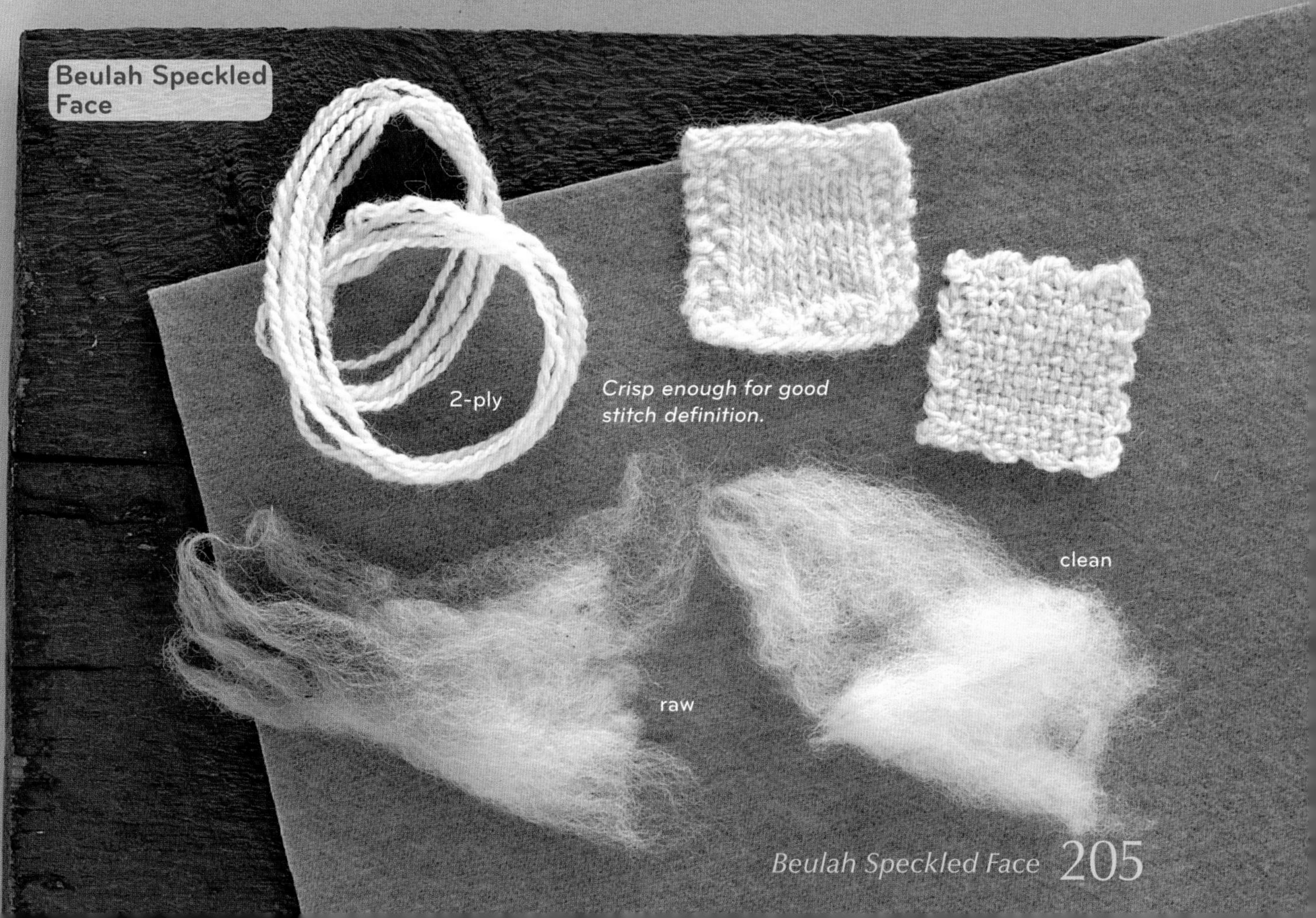

Beulah Speckled Face

2-ply

Crisp enough for good stitch definition.

clean

raw

Beulah Speckled Face Facts

▶ **FLEECE WEIGHT**
3–5 pounds (1.4–2.3 kg)

▶ **STAPLE LENGTH**
3–5 inches (7.5–12.5 cm)

▶ **FIBER DIAMETERS**
25–36 microns (spinning counts 44s–58s), mostly 26–31 microns (spinning counts 50s–56s)

▶ **LOCK CHARACTERISTICS**
Our sample had pretty disorganized staples. There was nice, relatively fine crimp that was not organized in the fibers.

▶ **NATURAL COLORS**
White, with no dark fibers.

Using Beulah Speckled Face Fiber

Dyeing. Produces clean colors, because of consistent whiteness of the wool.

Fiber preparation and spinning tips. Information varies depending on the fineness or coarseness of the individual sample and its length. It can be carded by hand or drum-carded and spun woolen, or combed and spun worsted. The fleece is quite easy to prepare because the locks tend to open nicely.

Knitting, crocheting, and weaving. This fleece is good in the finer ranges for everyday garments and household textiles, and in the coarser ranges for items that need to take a beating, like rugs.

Best known for. Being a nice midrange white wool, not emphasized in husbandry and marketing.

Beulah Speckled Face lamb

Black Welsh Mountain

Conservation Breed

BLACK WELSH MOUNTAIN sheep, which come from the southern mountains of Wales, are the only breed of Welsh Mountain sheep found in North America. All animals on this side of the Atlantic come from a 1972 importation of three rams and thirteen ewes, though breeders have since imported sperm to add genetic diversity to the North American flock.

Black Welsh are now recognized as a unique breed, though they were developed strictly through years of selection for the color — which shows up from time to time in the general, or white, Welsh Mountain breed. According to John Williams-Davies, author of *Welsh Sheep and Their Wool*, the black color has been around since medieval times, when it was called *gwlân cochddu* (red-black wool) and was considered a highly desirable commodity. The Black Welsh Mountain's wool is a true black, not just a very deep brown (although its tips may be sunburned). These sheep don't gray with age, as most dark-fleeced sheep do, so the black sticks around. Yet Black Welsh fleeces also differ in quality from the wools grown by other breeds in the Welsh Mountain family. They are appreciably softer and more kemp free, in part due to their separate breeding for animals used for show in parklands instead of for more pragmatic, meat-producing flocks. While slightly crisp, they remain within the appropriate range for making many garments and household textiles — from sweaters, hats, and mittens to the blanket realm.

Black Welsh Mountain Facts

▶ **FLEECE WEIGHT**
2¼–5½ pounds (1–2.5 kg), usually 3–4 pounds (1.4–1.8 kg)

▶ **STAPLE LENGTH**
2–4 inches (5–10 cm), usually 3–4 inches (7.5–10 cm)

▶ **FIBER DIAMETERS**
Generally 28–36 microns (spinning counts 44s–54s), although members of the North American breed association are aiming for a finer range of 26–32 microns (spinning counts 48s–56s)

▶ **LOCK CHARACTERISTICS**
Dense and firm, not especially long. Almost completely free of kemp and of any fibers other than the nice solid black. Individual fibers have significant crimp that is not organized in the locks, which blend together in a mass except at their slightly pointed tips.

▶ **NATURAL COLORS**
Deeply black, perhaps with a reddish cast, but the intensity of the black is remarkable; likely to have slightly brown tips.

Black Welsh Mountain yarns from Garthenor

Using Black Welsh Mountain Fiber

Dyeing. All applied colors will be lost in deepest space on this black wool.

Fiber preparation and spinning tips. Locks are not especially distinct but can be easily separated from a fleece by locating the tips and pulling outward. You can card any length or comb the longer fleeces. Blend with white wools to produce a spectrum of grays. (Pick a blending wool that is similar in length and fiber diameter.) Spinning from the lock is an option. The fibers tend to maintain their independence nicely in a clean fleece, so drafting is easy, although our sample felt a little dry, and with more time we would have experimented with light oiling.

Knitting, crocheting, and weaving. The wool is relatively soft and extremely durable at the same time. The yarn will have good loft or bulk, making fabrics constructed from it both lightweight and warm.

Best known for. Being uniquely solid black wool in a relatively uniform fleece that falls within the workaday-knitting range of fiber diameters.

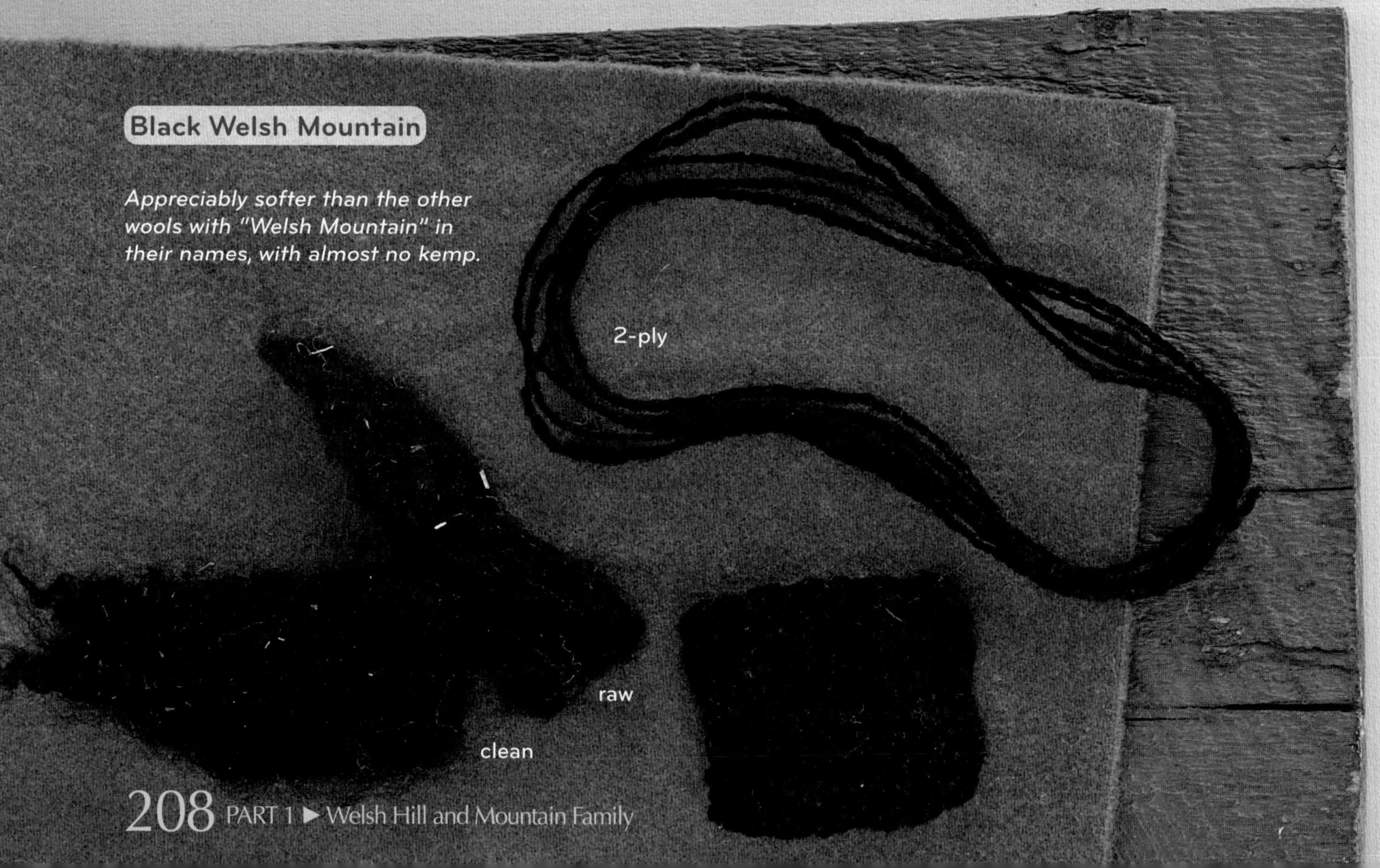

Black Welsh Mountain

Appreciably softer than the other wools with "Welsh Mountain" in their names, with almost no kemp.

Hill Radnor

Conservation Breed

THE MYSTERIOUS HILL RADNOR has been recognized locally for many centuries in central Wales, along the country's border with England, yet there's a real dearth of written references about the breed, even in very old agricultural texts. This was an isolated rural area, even well into the twentieth century, so the breed likely developed primarily from selection within the region, with few or no additions of outside bloodlines. It is similar to other Welsh mountain breeds, yet a bit larger and rangier, with a slightly finer fleece.

Early in its history, the Hill Radnor was often referred to as the Radnor Tanface due to its decidedly tan face. The breed society for the Radnor formed and created a registry in 1955 and settled on the name Hill Radnor at that time.

Although the Hill Radnor's fleece evolved to protect the animal from the elements, it accomplishes this job without being rough. The wool may contain some kemp and colored fibers, which gives it the potential to produce tweedy yarns. The wool is dense and crisp feeling, and it is a good candidate for maintaining the texture of a fabric construction or for color work that remains distinct instead of blurring.

Hill Radnor

2-ply

clean

clean

A little kemp, easy to pull out. Crisp yarn, with good body.

raw

raw

Hill Radnor Facts

▶ **FLEECE WEIGHT**
4½–5½ pounds (2–2.5 kg)

▶ **STAPLE LENGTH**
2–6 inches (5–15 cm), usually 3–5 inches (7.5–12.5 cm)

▶ **FIBER DIAMETERS**
Approximately 27–33 microns (spinning counts 48s–56s)

▶ **LOCK CHARACTERISTICS**
Mostly rectangular staples, indistinct, with short, pointed tips; also some narrow staples with more visible crimp patterns. Overall, medium crimp that is generally disorganized.

▶ **NATURAL COLORS**
White.

Using Hill Radnor Fiber

Dyeing. Produces clear colors with a matte finish.

Fiber preparation and spinning tips. Information varies depending on length; pick and spin from the locks, comb, or card. The fiber is relatively easy to draft, with a little drag (which helps the spinner maintain control) because of the crimp patterns.

Knitting, crocheting, and weaving. The yarn has nice body. Some fleeces will be good for skin-touching garments, like sweaters and hats, while others are a bit too coarse and are better for outerwear. If kemp and dark fibers are present, the yarn and finished items will be subtly and pleasantly tweedy.

Best known for. Not well known. A good all-purpose wool with the potential for tweedy effects.

Kerry Hill

Conservation Breed

The Welsh village of Kerry is well known not only for its sheep, with their charming facial markings, but also for its twelfth-century stone church and for the Kerry Ridgeway walk, once the main route for droving cattle, sheep, and geese from Wales to the markets of London; part of this historic walk is now a 15-mile (24 km) trail offering lovely views to hikers, mountain bikers, and sojourners on horseback. The breed, whose wool is loosely associated with the Down breeds thanks to some infusions of Shropshire bloodlines, is thought to have been developed mainly from crosses of the Kerry Hill area's old landrace breed with the Beulah Speckled Face (see page 205). It was a well-defined breed by the early 1800s.

Kerry Hill fiber is a candidate for long-term relationships: It's a wool to enjoy working with repeatedly, and we suspect that items made from it will have a tendency to become long-lasting, well-worn favorites. It's friendly, a clear white that will take dyed colors beautifully, and it arrives in a relatively narrow range of fiber diameters that fall into the realm of reliable worsted-style knitting yarns, just soft enough and durable enough to be practical for a wide range of everyday garments, like sweaters, hats, socks, and mittens, as well as household textiles like blankets or pillows.

Fleece Art

Collisions between different aspects of reality are often best developed through the arts. Artist Steve Messam cofounded Fold — an artist-led organization in England that promotes access to contemporary art in the rural environment. In 2009, he was commissioned by the Oriel Davies Gallery in Newton, near Kerry Hill, to provide a work for an exhibition on "the cultural meaning of patterns."

Messam's chosen medium? The fleeces of three hundred Kerry Hill sheep, with color accents of Black Welsh Mountain wool, as the covering for a timber-frame building dating from the thirteenth century. The work, called *Clad* (which means "encased" or "covered," as well as "clothed"), was designed to help people think about local agriculture and about architecture's aesthetic impacts on the environment, with the white and black wools echoing the white and black patterns of the buildings that have been constructed in this landscape since the thirteenth century.

Kerry Hill Facts

▶ **FLEECE WEIGHT**
5–6½ pounds (2.3–3 kg)

▶ **STAPLE LENGTH**
2½–5 inches (6.5–12.5 cm), usually around 4 inches (10 cm)

▶ **FIBER DIAMETERS**
26–29 microns (spinning counts 54s–56s)

▶ **LOCK CHARACTERISTICS**
Nicely defined but not exceptionally tight crimp, which will produce a lot of loft in the yarn. Dense but not especially distinct locks that can nonetheless be separated out nicely and kept in good order for combing, flicking, and so on. Look for the slightly pointed tips and pull on them to release individual locks from the mass.

▶ **NATURAL COLORS**
White.

Using Kerry Hill Fiber

Dyeing. This exceptionally white wool will take colors nicely, and the bit of luster will make those colors clear.

Fiber preparation and spinning tips. This fiber plays well any way you want to go. Card the shorter fleeces, comb or flick the longer ones; any staple length can be opened out and spun from the lock, even from the fold on a long fleece. Worsted style will make the most of the subtle shine and will still produce a yarn with some loft and good insulating qualities. Woolen style will result in a matte-appearing yarn with even more air-trapping capacity, increasing the potential warmth.

Knitting, crocheting, and weaving. This fleece makes a nice high-bulk yarn suitable for use at many weights and in multiple techniques. The yarn will be relatively lightweight for its grist, or size.

Best known for. Dense, white, high-quality wool.

Easy to comb and bulky; the finished yarn is lightweight for its size.

Llanwenog

LLANWENOGS came into existence in the western coastal counties of Wales (the lowlands and hills of Wales as opposed to the higher mountainous country of eastern Wales) in the late 1800s, when the now-extinct local sheep, named Llanllwni after the mountain they were kept on, were crossed with Shropshires. Llanwenogs are known as prolific sheep that are primarily raised for meat.

When you think of Llanwenog wool, think of Bluefaced Leicester — with staples that are not quite as long, fleeces that come in heavier, and crimp that is equally pronounced but far more relaxed in frequency. Our samples spun up with a halo reminiscent of mohair and caused us to think of what a nice surface might result if the finished fabric were gently brushed.

Llanwenog

2-ply

singles

Sleek, shiny, and fun to spin.

clean

raw

Llanwenog Facts

▶ **FLEECE WEIGHT**
4½–6 pounds (2–2.7 kg)

▶ **STAPLE LENGTH**
2–4½ inches (5–11.5 cm)

▶ **FIBER DIAMETERS**
25–28 microns (spinning counts 56s–58s), although the sample we sent for lab analysis averaged 35 microns (spinning count about 44s)

▶ **LOCK CHARACTERISTICS**
Splendid luster, wavy crimp, and locks that reflect the fibers' structure. Lovely pointed, distinct locks.

▶ **NATURAL COLORS**
White.

Using Llanwenog Fiber

Dyeing. The fleece is a very clear white; its luster will make colors shine.

Fiber preparation and spinning tips. Shorter fibers can be carded, but the luster strongly suggests combing whenever possible, to maximize the shine and the sleek quality of the yarn. In spinning techniques, take your choice. Worsted will bring out the luster, and Llanwenog can be spun from the opened lock. It's smooth and a little slick, and very pleasant in the hand.

Knitting, crocheting, and weaving. This is a delightful wool for use in any technique. It is shiny, supple, and quite soft. It will retain good definition in knit or crochet stitches or woven patterns. The unspun locks could be used to make pile fabrics in applications where there would not be a lot of wear.

Best known for. It is beginning to develop a well-earned reputation among handspinners as a soft, silky, shiny delight.

Lleyn

THE LLEYN (or Llŷn) breed was developed in northwestern Wales in the early 1800s. Now-extinct Irish longwool Roscommon rams — named for Lord Roscommon, an Irish peer of Bakewell's (see page 86) — may have been crossed in the late 1700s with native Welsh Mountain ewes on the Lleyn peninsula, a spit of land that juts out into the Irish Sea. In 1810, Border Leicester bloodlines were definitely added to the local sheep, yielding the breed we now know as the Lleyns.

The breed became quite popular in northern Wales through the nineteenth and early twentieth centuries, but by the 1960s only 10 flocks existed. The remaining breeders formed a society in 1970 and began promoting their breed with great success. Today, Lleyns are found throughout the United Kingdom.

Primarily raised for meat, Lleyns produce a nice white wool that falls between Bluefaced Leicester and Llanwenog, on its fine side, and Border Leicester, on its strong side. The fiber is far more matte than the wools of these

2-ply, from finest-feeling fibers

2-ply, from longest, strongest fibers

2-ply, from medium-fine, crimpiest fibers

Lleyn

clean

raw

Color, length, and crimp patterns varied; not sure this was all the same fleece. Crisp-feeling throughout.

Lleyn yarn from Garthenor

comparison breeds, and from our sample we separated out three radically different grades of fiber — one remarkably coarse — and then spun them separately. All of the yarns had enough body to hold good stitch or weave-structure definition. The finer ones would be good for everyday warm garments, with nice bounce and good insulating air-space qualities, and the coarsest would be a sturdy contributor to tweedlike fabrics, household textiles that need to endure some rough treatment, or warp yarns.

Royal Sheep

Prince Charles keeps a flock of 450 Lleyn ewes at Duchy Home Farm, which is adjacent to his Highgrove estate in Gloucestershire. The prince shows his sheep at county fairs and considers them to be an essential component for maintaining the land on his organic farm. He also keeps small flocks of Hebridean sheep and recently added a pair of Cotswold sheep, broadening his commitment to helping maintain these rare breeds.

Lleyn Facts

▶ **FLEECE WEIGHT**
4½–6½ pounds (2–3 kg)

▶ **STAPLE LENGTH**
3–5 inches (7.5–12.5 cm), although the coarse portions of our sample were 7 inches (18 cm) long

▶ **FIBER DIAMETERS**
26–31 microns (spinning counts 50s–56s), although parts of our sample were significantly coarser

▶ **LOCK CHARACTERISTICS**
Easily separated, pointed locks, with crimp organized in both fibers and locks. Wavy crimp, sometimes quite pronounced and close (like a Bluefaced Leicester, although not as dramatic) and sometimes more wavy (like a Border Leicester, although these locks were coarser feeling than that breed). Kemp free.

▶ **NATURAL COLORS**
White (a portion of our sample had a distinct warm cast to it).

Using Lleyn Fiber

Dyeing. This fleece takes dyes and shows colors well, with a touch of muted luster.

Fiber preparation and spinning tips. Length permitting (and it usually will), flicking or combing works well. Worsted-style spinning keeps the yarn as smooth as possible and minimizes any hairiness. The coarsest of the fibers from the sample were quite hairy, almost wiry, and produced a yarn with interesting texture.

Knitting, crocheting, and weaving. This fiber is suitable for a range of construction techniques. Lleyn would be especially interesting as a weaving fiber because of its strength.

Best known for. Being another of the wools from a meat breed that can lead to interesting textile experimentation.

Welsh Hill Speckled Face

Conservation Breed

AMONG THE BREEDERS of Welsh Hill Speckled Face sheep, they are affectionately nicknamed simply Speckles. The breed was developed in the mid-twentieth century by crossing Kerry Hill rams (see page 211) and Welsh Mountain ewes (see page 219). The resulting sheep are known for their extreme hardiness, and they graze the higher elevation hilltops in Wales.

Welsh Hill Speckled Face sheep are primarily raised for meat, and the rams are used for breeding mules (see page 328). Like the Beulah Speckled Face (see page 205), the Welsh Hill Speckled Face offers many of the qualities of the Welsh Mountain wools — primarily sturdiness and character — while staying at the finer end of those wools' range, with micron counts in the high 20s and low 30s. This makes it a good all-around wool with more density and stability than similar-diameter wools with fine-wool crimp patterns (compare the crimp in the photo on the next page with the crimp in the Merino photos, page 141).

Welsh Hill Speckled Face Facts

▶ **FLEECE WEIGHT**
3½–4½ pounds (1.6–2 kg)

▶ **STAPLE LENGTH**
3–4½ inches (7.5–11.5 cm)

▶ **FIBER DIAMETERS**
Reports vary, but the range goes from 26.5 to 33 microns (spinning counts 48s–56s)

▶ **LOCK CHARACTERISTICS**
Dense locks in kind of a muddle, but grab a pointed tip and you can pull out a staple that shows wavy and semiorganized crimp in the lock as well as in the fibers.

▶ **NATURAL COLORS**
White.

Using Welsh Hill Speckled Face Fiber

Dyeing. Produces clear colors; a hint of luster in the fiber may give the resulting shades a visual lift.

Fiber preparation and spinning tips. Depending on fiber length, it can be flicked or picked and spun from the lock, carded, or combed. With appropriate preparation, the fleece drafts smoothly for either thick or thin yarns. An easy fiber.

Knitting, crocheting, and weaving. Some fleeces will be soft enough (in the high-20s micron counts) for making sturdy sweaters, hats, and other pieces of clothing that may touch some skin. Others will be better suited to outerwear, bags, rugs, pillows, and the like. There's a lot of body to the yarn, and more stability than elasticity.

Best known for. Like the Beulah Speckled Face, the Welsh Hill Speckled Face is a pleasant and versatile midrange wool.

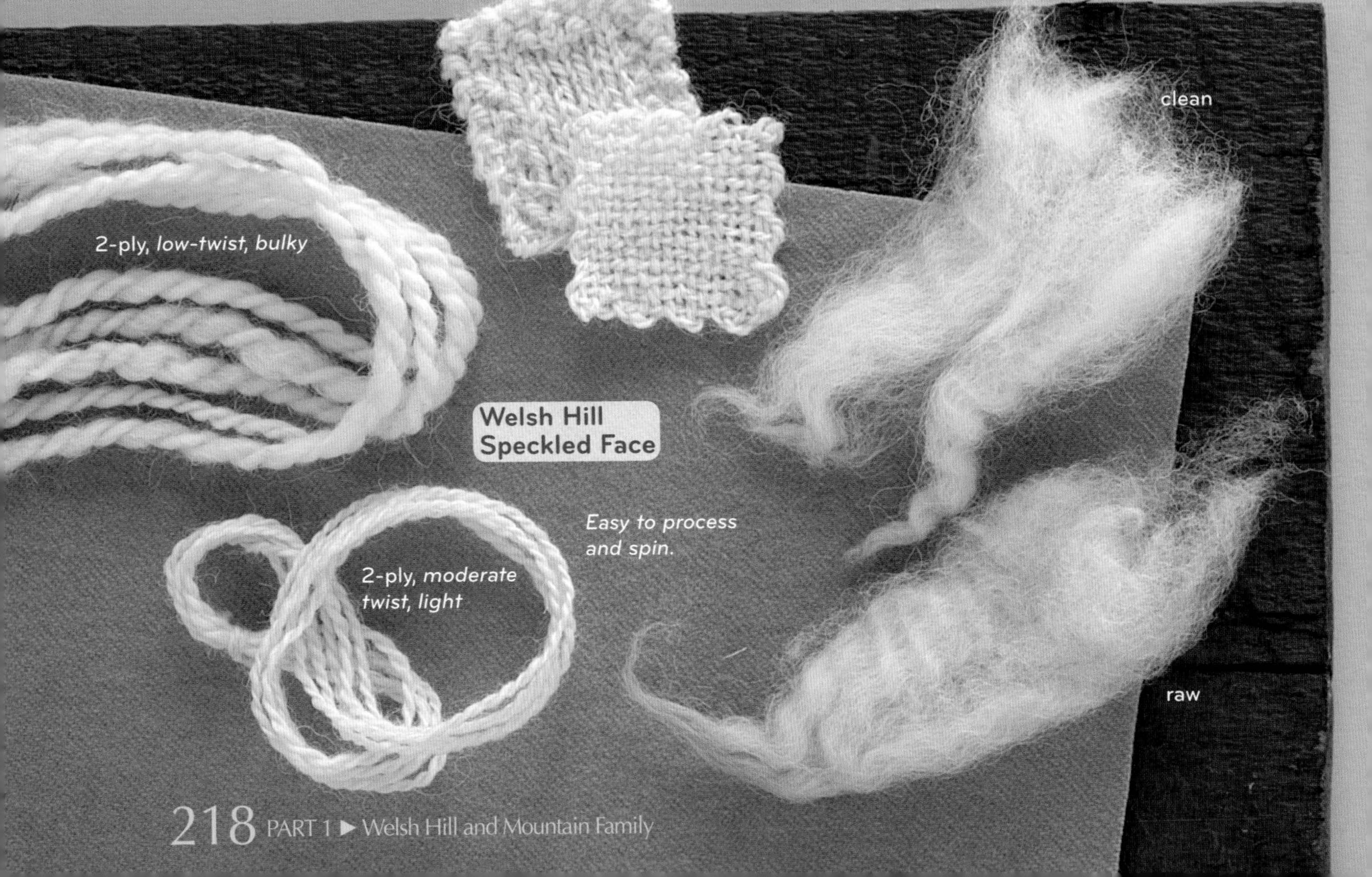

clean

2-ply, *low-twist, bulky*

Welsh Hill Speckled Face

Easy to process and spin.

2-ply, *moderate twist, light*

raw

Welsh Mountain and South Wales Mountain

Conservation Breeds

Welsh Mountain and South Wales Mountain sheep are white, or white with some tan on their faces and legs, and there's some question about whether they are actually different breeds or not. The Welsh Mountain sheep raised in the southern part of Wales are a bit larger than their counterparts in the central and northern mountains, and our research strongly suggests that there are dramatic differences between general Welsh Mountain and South Wales Mountain wools — differences that are profoundly significant for textile workers.

Both Welsh Mountain and South Wales Mountain sheep are primarily bred as meat animals, so there is not much focus on wool production and fiber qualities. Nonetheless, both quality and quantity of wool have been increasing over the past century. There can be a mix of softer wool with both hair and kemp in various colors. Expect the finest wool to be around 30 microns (spinning count 50s) and to see a spectrum moving from there toward the coarse end of the scale, to 38 to 40 microns or more (36s or coarser). Most wool is likely to fall between 32 and 40 microns (spinning counts

Welsh Mountain

36s–48s), although the kemp will be coarser than that, of course.

Our sample of Welsh Mountain contained a small amount of slender kemp throughout — enough to add durability without significantly increasing the harshness factor, although the yarns we spun were unquestionably crisp. We had fantasies of making a nice pair of hiking or work socks. They wouldn't be cushy, but the yarn suggested they'd be comfortable, and they'd definitely be durable. A more traditional use of the yarn would be in woven twill fabrics; our sample encouraged thoughts of tapestries and nice rugs as well, although the wools in our hands were softer than some fibers suggested for those purposes.

On the other hand, the South Wales Mountain sample had abundant kemp distributed throughout the fleece. What we've learned about South Wales Mountain wool suggests great weather resistance, although the scratchiness must be factored in when planning uses for it. Both research and experience indicate that this breed's wool is characterized by white kemp, and also possibly some red, making for a highly textured, tweedy effect in the yarn.

Welsh Mountain and South Wales Mountain fleeces in the 30-micron range without much kemp can become sweaters, mittens, hats, blankets, and other items that may touch but don't need to be able to caress the skin. Up around 40 microns, the wool lends itself to use in rugs and artwork and other places where tough and textured wool is at home. The midrange can become tweedy, sturdy fabrics.

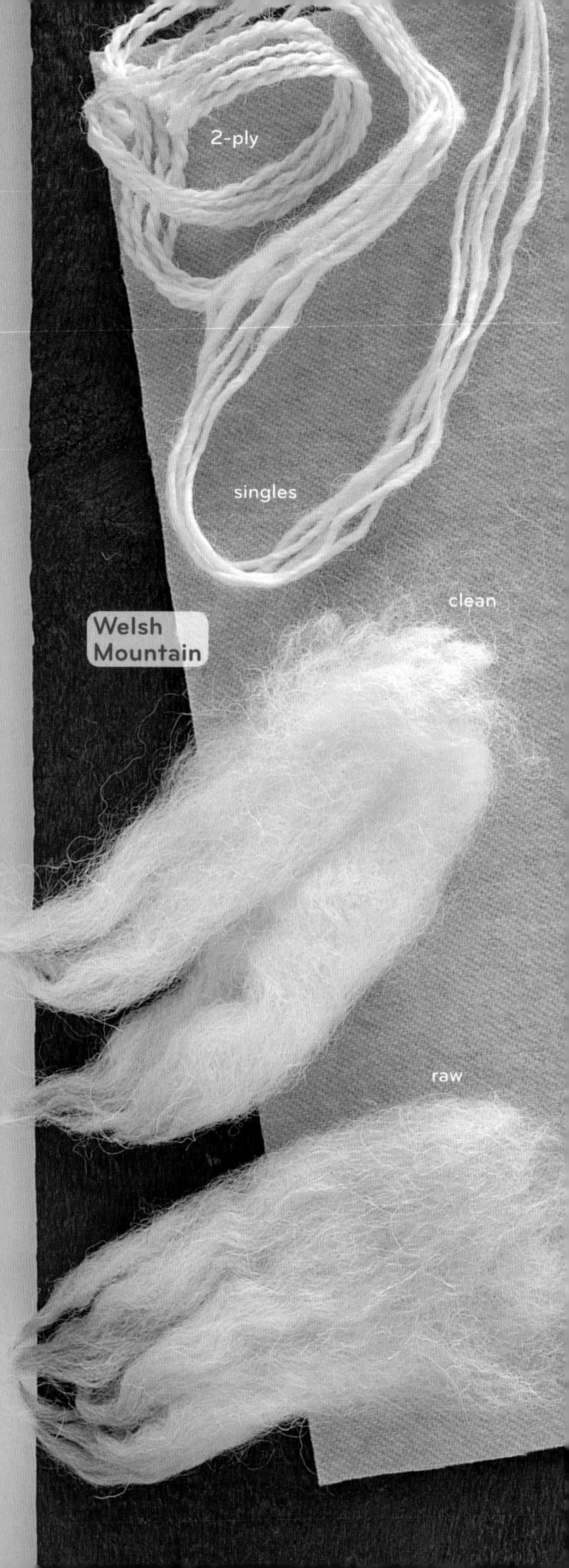

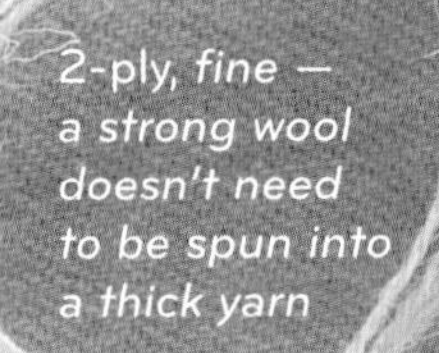

2-ply, *fine* — a strong wool doesn't need to be spun into a thick yarn

2-ply, *medium*

clean

raw

Welsh Mountain sample (at left) contained small amounts of easy-to-remove kemp.

South Wales Mountain sample (at right) boasted abundant kemp that strongly resists twist; trimmed off bases of staples, which were stuck together.

Welsh Mountain and South Wales Mountain Facts

▶ **FLEECE WEIGHT**

2½–4½ pounds (1.1–2 kg), tending toward the lighter end for the Welsh Mountain and the heavier end for the South Wales Mountain. The few references we found on the Sennybridge Welsh (see page 223), which we think may be a strain of Welsh Mountain, indicate that this type may produce significantly larger fleeces, spanning 3–15 pounds (1.4–6.8 kg).

▶ **STAPLE LENGTH**

2–4 inches (5–10 cm), though possibly longer; our sample of South Wales Mountain was a full 7 inches (18 cm), and information on the Sennybridge Welsh type indicates 3–7 inches (7.5–18 cm)

▶ **FIBER DIAMETERS**

See the notes on pages 219–220, but the general range likely spans 30–40 microns (spinning counts 36s–50s). For the Sennybridge Welsh type, we found significantly finer estimates of 25–31 microns (extrapolated from Bradford grades 50s–58s), with possibly even finer wools within some breeding groups.

▶ **LOCK CHARACTERISTICS**

A dense fleece with somewhat blocky staples that have short, tapering tips. Nicely, if irregularly, crimped in the finer fibers; hardly crimped at all in the hairy fibers; no crimp in the kemp. Information on the Sennybridge Welsh type suggests triangular locks, as are typical of the mountain breeds, with lightly curled points to their tips. Crimp well developed but moderate, and disorganized in both individual fibers and staples.

▶ **NATURAL COLORS**

Welsh Mountain White, with an occasional black lamb.

South Wales Mountain White, with an even mixture of white kemp throughout and often a brown "collar" area.

Sennybridge Welsh Said to encompass solid white, black, tan, or brown, and lighter shades of those colors.

Using Welsh Mountain and South Wales Mountain Fiber

Dyeing. The kemp won't show the dyed color with any intensity, but the other fibers in the mix will, resulting in interesting shade and texture contrasts.

Fiber preparation and spinning tips

- The mix of fibers makes preparation either challenging (if you want to comb or card) or a cinch, if you're willing to tease apart the locks and spin them as they come.
- The low-grease and high-kemp content means the locks almost want to fall apart when gently picked.
- Kemp resists not only dye but twist. Use enough twist to control the kemp, and know that if there's a lot of kemp (as in the South Wales Mountain), whenever you let go of your yarn it will immediately untwist, unless you secure the end.
- Use a light touch, endeavor to feed the mix of fibers evenly, and at the end of a kemp-abundant session don't be surprised to find kemp on your lap, on the floor, in your teacup, and in a clump in your hand.
- Tease open the locks and spin directly from them; comb; or card, if you have one of the shorter fiber lengths.

Knitting, crocheting, and weaving. For South Wales Mountain, the kemp needs to be kept in mind, as it does for every other action involving the breed's wool. Kemp prevents smooth spinning and so produces a natural novelty-yarn effect, and it's itchy next to the skin. Plan your approach accordingly. Small amounts placed in nonsensitive areas of garments could be a fantastic design accent. Otherwise it's a good plan to make tapestries, rugs, bags, baskets, and outerwear that is well shielded from the body (for example, lined with something the kemp fibers won't poke through). For the samples we received that were identified as Sennybridge Welsh, the crisp, somewhat dense, strong yarns will show stitch or weave-structure definition nicely and will produce textiles that have a pleasant body to them.

Best known for. Being sturdy sheep with wool to match. When there's kemp, there's probably enough to use to make a statement.

Sennybridge Welsh?

We found sources that refer to a breed called the Sennybridge Welsh, and we received a sample of labeled wool from a very reliable source. But after extensive research, including contacting Welsh agricultural experts, we don't think there is a specific breed by this name. That doesn't mean our source was incorrect or the name is meaningless, and some of our research produced details on wool from Sennybridge Welsh sheep.

Our take is that the Sennybridge Welsh may be a named strain of the regular Welsh Mountain sheep, one that may have a heavier fleece consisting of longer and finer wool than other Welsh Mountains, and that includes a wide array of natural colors.

Sennybridge is the name of a village and a military installation in mid-Wales. The village lies within Brecon Beacons National Park. Just to keep everybody on their toes, Sennybridge Cheviot is an alternate name for the Brecknock Hill Cheviot (see page 56).

We would be delighted to hear from anyone who knows or discovers more information about the Sennybridge Welsh sheep, whether breed or strain! Our study of wools continues, and these critters have presented us with one of many interesting puzzles that we continue to research.

HeRdwick
Jacob
Zwartbles

Other Sheep Breeds

Many sheep breeds don't fall into the family clusters we've used to organize this chapter so far. These breeds may have been developed by combining characteristics from unrelated families, or they may combine independent origins with intriguing history. Despite their eclectic origins and natures, we found a few patterns based on geography, history, or production qualities.

Several breeds came into being to increase productivity in North America. The heat-tolerant California Red has unintentionally good wool. From Western range lands, the Columbia, Panama, and Targhee yield both meat and wool, while the Polypay emphasizes large lamb crops. The Montadale, also a wool-and-meat breed, arose in the Midwest. The Romeldale and associated California Variegated Mutant (CVM) focus on wool quality.

Three breeds that now have uniquely North American identities originated elsewhere. The American Tunis, a meat breed suited to hot, humid climates, traces its ancestry to North Africa. The American Karakul is related to many sheep in Central Asia. At home in the Southwestern deserts, the Navajo Churro carries a lot of Spanish blood.

The mysterious origins of a number of British breeds will likely only be clarified through DNA analysis. One is the Herdwick, strongly tied to the English Lake District. The piebald Jacob displays some primitive qualities (like its horns) along with characteristics of later sheep development, like a single-coated fleece. The Norfolk Horn's obscure parentage may include Soay as well as early Roman sheep. The Portland represents the early tan-faced British sheep. The Ryeland was famous in the Middle Ages for exceptionally fine wool. Associated with the Pennine mountain range, the Whitefaced Woodland may also represent Roman sheep. As a basic type, the Clun Forest may go back more than a thousand years.

The Galway is the only remaining distinctively Irish breed of sheep.

The Devon Closewool and the Exmoor Horn are British breeds known primarily for meat.

This "other" group contains five dairy breeds. From Britain, we get the British Milk Sheep and the Colbred. Continental Europe offers the East Friesian (Germany), the Rouge de l'Ouest (France), and the Zwartbles (the Netherlands), the latter two now often grown for meat.

Meat-focused French breeds include the Bleu du Maine, Charollais, and Île-de-France (now also in dairies). The Netherlands contributes the Texel, with a very strong meat emphasis.

New Zealand and Australia provide a number of breeds that yield large quantities of both wool and meat or that fit specific environments. From Australia we get the Bond, Cormo, Gromark, and Polwarth. New Zealand offers the Borderdale, Coopworth, Corriedale, and Perendale.

Regardless of where these "other" sheep came from or the purposes for which they were originally bred, you will find a fascinating array of sheep (and fibers) in the pages that follow!

American Tunis

Conservation Breed

THE AMERICAN TUNIS (often simply called Tunis in the United States) was developed from Tunisian Barbary sheep, an ancient breed that dates back to biblical times. The original Tunisian Barbary sheep came to North America from North Africa in 1799, as a gift to the U.S. government from His Highness the Bey of Tunis, and George Washington is said to have used a Tunis ram to rebuild his Mount Vernon flock after he retired from the presidency.

The American Tunis, which has had small infusions of Southdown (see page 70) and perhaps Leicester (see page 83) bloodlines over the centuries, is particularly well adapted to hot and humid climates, so early on the breed spread throughout the southeastern and mid-Atlantic states. It was also popular with other presidents, including John Adams and Thomas Jefferson. Unfortunately, large numbers of Tunis flocks were destroyed during the Civil War, and the breed never really came back to its prewar numbers.

The Tunis is a fat-tailed sheep. One characteristic of fat-tailed sheep is that the meat tends to be tender and tasty. The breed is docile and hardy, and although these sheep are usually associated with hot climates, they adapt well to colder areas.

Generally shown in competitions as meat animals, either slick shorn or with an inch (2.5 cm) or less of wool, American Tunis sheep nonetheless grow a very desirable fleece that can be used for many purposes. With a nice warmth to its color — this is not a bright white wool, but off-white to creamy tan (lambs are born all red) — the fleece should be uniform

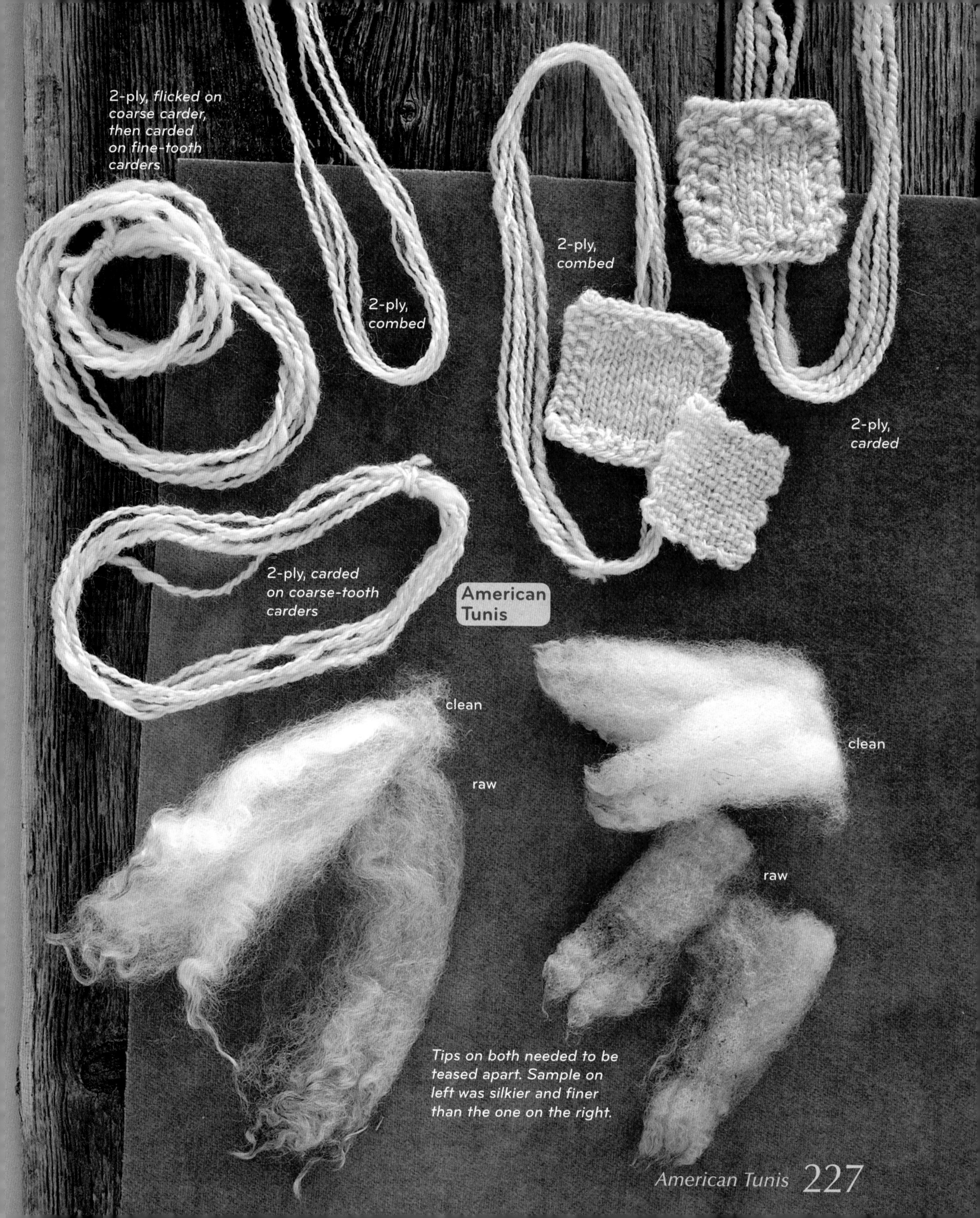
2-ply, flicked on coarse carder, then carded on fine-tooth carders
2-ply, combed
2-ply, combed
2-ply, carded
2-ply, carded on coarse-tooth carders
American Tunis
clean
raw
clean
raw
Tips on both needed to be teased apart. Sample on left was silkier and finer than the one on the right.

throughout and also dense, although less dense in the belly portions. Color within the wool (brown or black spots or red fibers) is discouraged by the breed association. There is a bit of luster, possibly courtesy of some Leicester ancestry, and the wool has a noticeable and even crimp in the fibers, if not in the lock structure; this can make the locks easy to tease open.

See also the notes for California Red (page 233). The fleeces handle similarly, although in general American Tunis sheep grow wool in larger quantities, and California Reds are likely to have coarser red hairs mixed into their fleeces.

A Taste Worth Saving

Slow Food has added the American Tunis to the Ark of Taste, a program designed to help save "delicious foods that are in danger of extinction." According to Slow Food USA, to qualify for the Ark of Taste, food products must be:

- outstanding in terms of taste — as defined in the context of local traditions and uses
- at risk biologically or as a culinary tradition
- sustainably produced
- culturally or historically linked to a specific region, locality, ethnicity, or traditional production practice
- produced in limited quantities, by farms or by small-scale processing companies

Other sheep on the list include the Gulf Coast Native (page 124), Navajo Churro (page 282), and St. Croix (a hair sheep).

American Tunis Facts

▶ **FLEECE WEIGHT**
6–15 pounds (2.7–6.8 kg), averaging 8–12 pounds (3.6–5.4 kg); yield 50–70 percent

▶ **STAPLE LENGTH**
3–6 inches (7.5–15 cm), generally 3½–5 inches (9–12.5 cm)

▶ **FIBER DIAMETERS**
U.S. breed standard is 24.29–29.2 microns (spinning counts 54s–58s); in the field, expect to find 24–31 microns (spinning counts 50s–58s)

▶ **LOCK CHARACTERISTICS**
Relatively open, a bit blocky, sometimes with pointed tips.

▶ **NATURAL COLORS**
Ivory to cream; reddish from first and second lamb shearings before the wool lightens to its adult shade.

Using American Tunis Fiber

Dyeing. The fleece will take colors nicely.

Fiber preparation and spinning tips. A versatile and friendly type of wool, it is willing to please. Shorter staple lengths can be carded, and longer ones can be combed or spun from the fold. Locks should be relatively easy to tease open, so flicking is an appealing approach to preparation. On our samples, the tips of the locks stuck together even after scouring and needed to be teased apart, but the fibers then spun up beautifully.

Knitting, crocheting, and weaving. This is suitable for all fiber techniques. There's enough crispness to hold spaces in openwork, and enough body for good stitch or weave definition.

Best known for. Versatility, warm white color (with red or reddish tan in the lambs), ease of preparation, and appropriateness for a variety of clothing and household uses.

Bleu du Maine

THE BLEU DU MAINE originated in northwestern France. It is named for its bluish-black to bluish-gray head and for the traditional province of Maine, the western part of which is now called the department of Mayenne. Although this area is probably more famous for the Le Mans car race than for its sheep, it was the region in which the Bleus were first developed from the crossing of Leicester Longwools and Wensleydales to the now-extinct Choletais breed in the mid-to-late 1800s. Bleu du Maines were introduced to Britain in 1978, and these sheep are also found widely through northern Europe, including in Germany, Belgium, and the Netherlands.

The Bleu du Maine and Rouge de l'Ouest (see page 302) were developed from similar breedings, yet their fleeces are quite different. Bleu du Maines have staples about twice as long as those of the Rouges, and the Bleus' shearing weights are about twice as heavy per fleece. Bleu du Maine fibers have somewhat larger diameters than those of the Rouge de l'Ouest. Also, at least as far as we can tell from our samples, the wavy crimp pattern visible in the Bleu du Maine locks leans toward the English Longwool side of its lineage, whereas the crimp on the Rouge de l'Ouest is closer to the

Millennium Bleu

This is a new breed, developed in Britain in 2000 by crossing Texel rams (see page 309) with Bleu du Maine ewes. The Millennium Bleu is primarily being used in meat production, and we didn't find any wool for it, but you may hear about it in the future.

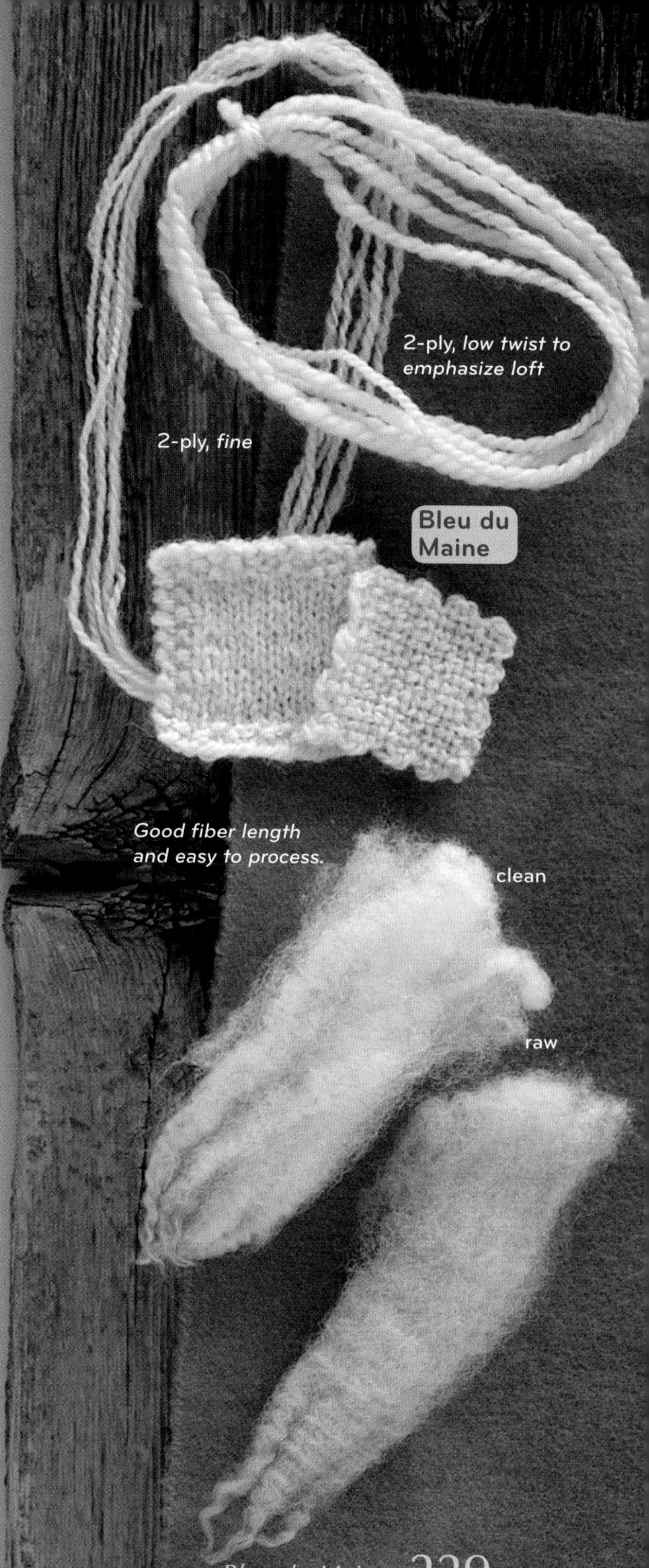

2-ply, *low twist to emphasize loft*

2-ply, *fine*

Bleu du Maine

Good fiber length and easy to process.

clean

raw

fine-wool end of the spectrum. Bleu du Maine also has a hint of luster.

Bleu du Maine can be used to fabricate a variety of versatile, midrange yarns with nice bulk. Some will be soft, while others will be sturdy. In general, they'll be suited for sweaters, hats, mittens, blankets, and other homey textiles.

Bleu du Maine Facts

▶ **FLEECE WEIGHT**
8¾–13¼ pounds (4–6 kg)

▶ **STAPLE LENGTH**
2–4 inches (5–10 cm)

▶ **FIBER DIAMETERS**
Approximately 26–31 microns (extrapolated from Bradford grades of 50s–56s)

▶ **LOCK CHARACTERISTICS**
Long staples with short pointed tips; well-developed crimp relatively disorganized in the individual fibers at the same time that a wavy pattern is visible in the locks.

▶ **NATURAL COLORS**
White.

Using Bleu du Maine Fiber

Dyeing. Produces clear colors, with a hint of shine from just a touch of luster.

Fiber preparation and spinning tips. The Bleu du Maine is primarily a meat breed, so shearing is utilitarian and cutting lines on the fiber may not be smooth and even. However, the Bleu du Maine's staple length means that second cuts should not be problematical. Your choice of preparation methods, depending on staple lengths: flick or tease and spin from the locks; card; or comb. The wool is pleasant to spin, with either woolen or worsted techniques. A woolen approach will emphasize the fiber's loft; yarn spun with a worsted approach will still have a light quality.

Knitting, crocheting, and weaving. Most Bleu du Maine yarns will be suitable for everyday clothing and other textiles. Less crisp feeling than strands spun from their close relative the Rouge de l'Ouest, Bleu du Maine yarns still have a pleasantly full hand.

Best known for. Not well known as a breed-specific fiber for textile work, although it can produce a midrange, versatile series of yarns.

British Milk Sheep

OKAY, YOU GET one guess where these sheep originated and what they're known for, and you can even guess when you're half-asleep. British Milk Sheep were developed in the 1970s from a mixture of other breeds (probably Friesian, and possibly Bluefaced Leicester, Polled Dorset, Lleyn, and some others) for use in sheep dairying with a secondary purpose of meat production. But we're going to let you in on a secret: The British Milk Sheep has a surprisingly pleasant fleece. Some fleeces may even be suitable for some people's next-to-the-skin wear. The long but manageable staple lengths and soft shine of the fibers make this an appealing wool to explore.

British Milk Sheep Facts

▶ **FLEECE WEIGHT**
9–11 pounds (4.1–5 kg) for ewes, and up to about 14 pounds (6.4 kg) for rams

▶ **STAPLE LENGTH**
Medium, 4½–7 inches (11.5–18 cm)

▶ **FIBER DIAMETERS**
Approximately 28–31 microns (spinning counts 50s–54s)

▶ **LOCK CHARACTERISTICS**
Long, fairly narrow staples with pointed tips; pronounced, open, and slightly irregular crimp; and a touch of luster.

▶ **NATURAL COLORS**
White.

2-ply

clean

raw

Fairly open with irregular, pronounced crimp. Easy to process.

British Milk Sheep

Using British Milk Sheep Fiber

Dyeing. The fleece will take colors well and with clarity.

Fiber preparation and spinning tips. Spin the fleece from the locks; pick; or comb. Card it if you're comfortable working long fibers on carders. It is a sweet and cooperative wool, with a lot of options for making fine yarns or, with the longer staple lengths, relatively thick, low-twist yarns.

Knitting, crocheting, and weaving. The fiber is good for general-purpose knitting and crocheting of sweaters, hats, mittens, gloves, socks, and blankets. It is also a nice weaving yarn for similar applications. It has a pleasant amount of loft and enough crispness to work well in lace patterns.

Best known for. Being a surprisingly nice wool from a breed best known for twin and triplet lambs, significant milk production, and meat.

Sheep Dairying

Sheep dairying was virtually unheard of in North America before the late 1980s and early 1990s, although it has a strong tradition in many other parts of the world. In fact, Roquefort, a traditional blue cheese, is always made from sheep's milk, as is Pecorino Romano.

Ewe's milk is higher in fat, calcium, and vitamins C and E than cow's milk. It also has higher proportions of solids and protein, so to make 1 pound (0.5 kg) of cheese you only need 4 to 5 pounds (1.8–2.3 kg) of sheep's milk, whereas you need 10 pounds (4.5 kg) of cow's milk.

Luckily, a number of artisanal producers in the United States and Canada are making yummy cheeses from the milk of sheep.

California Red

Conservation Breed

THE CALIFORNIA RED offers pleasant fiber that wasn't meant to be. The breed was developed in the 1970s by Glenn Spurlock, a sheep Extension specialist with the University of California, Davis. Spurlock, who was working with his personal flock during his off-hours, intended to produce a meat breed that didn't need to be shorn. He failed at the no-shearing part of his plan, but we can be thankful for that!

The 1970s was a period during which many breeders were experimenting with the development of hair breeds suitable for production in the United States. For shepherds specializing in meat, a hair breed reduces labor. Many of the existing hair breeds around the world were known for being quite prolific and for breeding "off-season," or throughout the year, rather than having the limited breeding window that most of the European wool breeds were known for. Many of the hair breeds, including the Barbados Blackbelly, were also known for a high degree of disease- and parasite-resistance — a trait that is less common in wool breeds raised in commercial quantities. Knowing this, Spurlock crossed the American Tunis (see page 226) with the Barbados Blackbelly, a hair sheep whose history on the Caribbean island for which it is named dates back to the island's first settlement by Europeans — the English in 1627. The first sheep to come to the island are thought to have been an African hair type carried on the slave-trade ships that supplied labor to grow the island's main export crops of cotton, tobacco, and sugar. By the 1680s, written descriptions of the island's sheep sounded like the Barbados Blackbelly that is still common there.

A small number of Blackbellies were imported to the United States in 1904 by the United States Department of Agriculture, but these animals died out, leaving no trace. Then in the 1970s, researchers at North Carolina State University imported another group that provided the foundation for the small number of purebred Blackbelly sheep still found in the United States, as well as a starting point for the California Red.

The California Red grows a nice fleece with an appealing, subtle color and texture. What sets it apart is its tan color, often accented by red kemp fibers, although our spinnably sized sample was kemp-free. The lambs are born a dark rust or cinnamon color, and the wool lightens over the first 12 to 18 months, often (but not always) retaining a beige or oatmeal cast. Most of the wool is likely to fall within a fairly narrow range of fiber diameters, all suitable for making general-purpose yarns that will work well for everyday garments and blankets. A fleece is big enough to make a cardigan, mittens, and a hat, or a couple of nice pullovers.

California Red offers some predictability in softness and length, along with subtle variation in color and texture. It's a great wool for textiles you'll use every day.

California Red Facts

▶ **FLEECE WEIGHT**
5–7 pounds (2.3–3.2 kg) for a ewe; yield about 50 percent

▶ **STAPLE LENGTH**
3–6 inches (7.5–15 cm)

▶ **FIBER DIAMETERS**
Generally 28–31 microns (spinning counts 50s–54s), although some fleeces may be as fine as 24 microns (spinning count 60s) or extend on the coarse end into the low 30s (spinning counts 46s–50s)

▶ **LOCK CHARACTERISTICS**
Blocky staples, with tips that may stick together and need to be teased apart. The crimp is well developed but somewhat disorganized.

▶ **NATURAL COLORS**
The fleece may have darker-colored, somewhat coarser hairs mixed in. Our sample was a warm, clear white without additional hairs.

The Caribbean influence. This is a Barbados Blackbelly ram, like the ones Glenn Spurlock used on his American Tunis ewes to develop the California Red.

Using California Red Fiber

Dyeing. The fleece takes color well, with potential tweedy accents from scattered reddish fibers.

Fiber preparation and spinning tips. Fleeces in the shorter range card nicely. Longer ones will handle well on the combs, and combing or flicking will make the most of the fiber's slight luster, although any colored hairs aren't likely to end up in the finished top after this method of preparation. Do take time to open the tips of the locks, if they're sticky. This is an easy fleece to spin.

Knitting, crocheting, and weaving. In knitting and crochet, woolen-spun yarns will have good body for all-around garments; the loft will make them warm for their weight. Worsted-spun yarns will have great pattern definition for texture stitches. The worsteds will weave up like a dream; the woolens will want some breathing room in the sett. California Red offers a terrific balance between durability and softness for light jackets, sweaters, hats, gloves, and other workhorse garments, as well as blankets, afghans, and other home-comfort textiles. It may be fine for garments that touch parts of the skin; make a swatch, and test it against the wrist or the back of the neck.

Best known for. Having an often subtle, unusual color, warmer than a clear white; being versatile.

2-ply, *combed*

2-ply, *carded*

California Red

clean

The combed version looks whiter than the carded yarn! This happens because of the way the fiber reflects light in the smoother (combed) preparation.

Our spinnable samples did not have the red fibers that are often found in the fleece (as the lock at lower right does).

clean

raw

Charollais

CHAROLLAIS SHEEP were developed in Charolles canton (similar to a county) in the Burgundy region of east-central France during the nineteenth century. French shepherds crossed imported British Leicester Longwool sheep with the native landraces. The French government officially recognized and began promoting the resulting fast-growing, muscular breed in 1974, and the first animals were exported from France to Britain in 1976. The first North American imports came to Canada from Britain in 1994; from there the breed spread to the United States.

Because of the meat emphasis, the breed societies do not state wool standards. Charollais grow fine- to medium-quality wools that are dense (lots of fibers grown in close proximity, a characteristic of finer wools) and quite short.

Charollais Facts

▶ **FLEECE WEIGHT**
4½–5½ pounds (2–2.5 kg)

▶ **STAPLE LENGTH**
1½–2½ inches (3.8–6.5 cm)

▶ **FIBER DIAMETERS**
Approximately 23–27 microns (extrapolated from the Bradford count, 56s–60s)

▶ **LOCK CHARACTERISTICS**
Rectangular — possibly, because of the short fibers, close to square.

▶ **NATURAL COLORS**
White.

Using Charollais Fiber

Dyeing. Produces clear colors, likely without luster.

Fiber preparation and spinning tips. Shearing is utilitarian; staples are short, and carding is the most likely preparation method. Because shearing is not fleece oriented, second cuts may make it difficult to produce consistent rolags. Try mini combs, if you're adept with them. Because of the short fibers, the singles will need to be thin and have enough twist to give them integrity. Although a woolen-style draft preserves elasticity, it's easier to control neps with a worsted-style draft.

Knitting, crocheting, and weaving. Like strong Merinos, Charollais wools make soft, general-purpose (but not luxury-level) textiles. Elasticity and bounce depend mostly on spinning technique.

Best known for. Unusually fine fiber for a meat breed, especially a breed not obviously developed from a Merino/Rambouillet base. Short fibers!

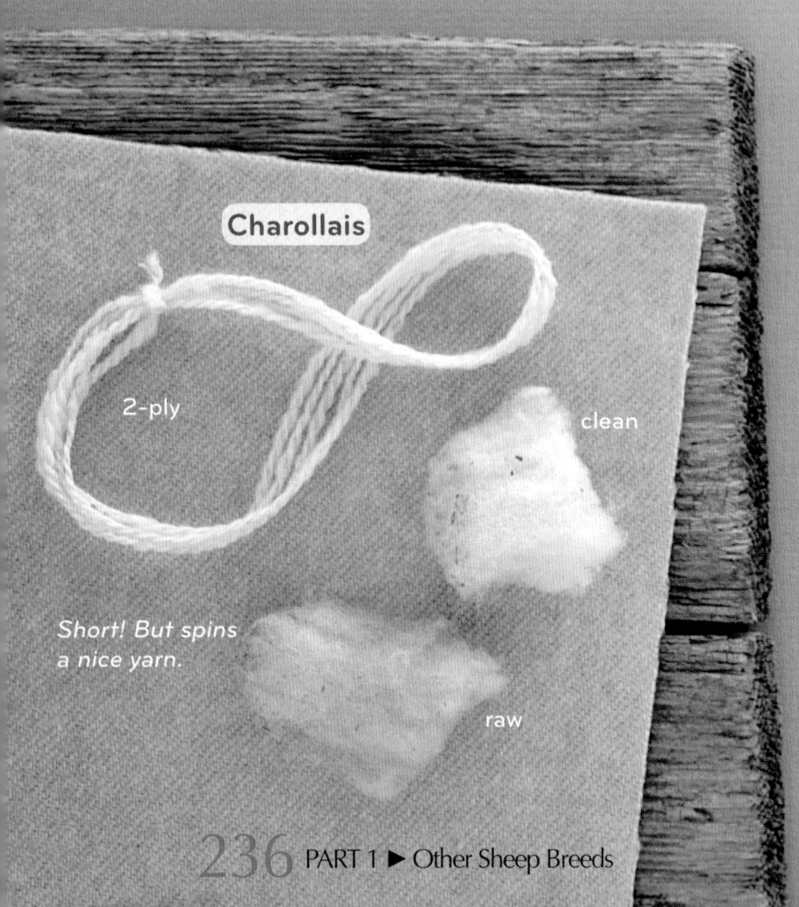

Clun Forest

Conservation Breed

The Clun Forest is an area of open pastures and wooded hills on England's western coast. The area has been home to shepherds for many centuries, and the sheep they kept were hardy and self-sufficient, eking out their sustenance from the rough grasses that were available in the region while still producing lambs and fleece. The Clun Forest breed was developed from these native sheep, with additions of Hill Radnor, Shropshire, and Kerry Hill sheep in the 1860s. The first Clun Forest sheep to arrive in North America were imported from Ireland in 1959, but those animals disappeared. Then in 1970, Nova Scotia shepherd Tony Turner, who had recently emigrated to Canada from Wales, imported 40 ewes and 2 rams. Several additional imports were brought in during the 1970s, providing the foundation for the North American Clun population.

Although emphasis today among Clun Forest breeders tends to be on meat, the show guidelines interestingly (and unusually, for a meat-type breed) permit animals to be entered in competitions with enough wool present to indicate the quality of the fiber — in other words, not slick shorn to show off flesh. And Clun Forest wool is a very appealing fiber that deserves to be better known to textile enthusiasts. Each fleece tends to be of uniform quality, so sorting is relatively simple.

Clun Forest is an ideal fiber for making textiles that are both comfortable and able to stand up to some abuse: perfect for mittens for a snowball fight; or what the Shetlanders call a hap shawl, not the delicate showpiece but the one you throw on every cool day or evening to keep your shoulders warm; or anything else you might dream up that requires similar qualities.

Clun wool is apparently popular for making industrial felts and wool-filled futons. The ordinary filling for futons is cotton; a wool like Clun Forest would retain its loft longer.

Clun Forest Facts

▶ **FLEECE WEIGHT**
4½–9 pounds (2–4.1 kg), generally 4½–6½ pounds (2–3 kg)

▶ **STAPLE LENGTH**
2½–5 inches (6.5–12.5cm), generally 3–4 inches (7.5–10 cm)

▶ **FIBER DIAMETERS**
Most resources indicate 25–28 microns (spinning counts 56s–58s), although the American Sheep Industry Association notes 28–33 microns (spinning counts 46s–54s). This may reflect environmental differences or breeding preferences among different populations.

▶ **LOCK CHARACTERISTICS**
Dense locks and blocky staples with either blunt or somewhat pointed tips. The well-developed crimp shows up well in the fibers and may (or may not) also be obvious in the locks.

▶ **NATURAL COLORS**
White, with no kemp or colored fibers.

Clun Forest yarn from Local Harvest/Touchstone Farm

Using Clun Forest Fiber

Dyeing. The fleece takes and displays colors nicely.

Fiber preparation and spinning tips. Card (with hand carders or a drum carder), flick, comb, or spin from the lock, depending on staple length and personal preferences. Should be very pleasant to spin — lots of air and bounce to the yarn, and as friendly in its own way as Romney.

Knitting, crocheting, and weaving. Clun Forest has a springy quality, contributed by its crimp. Although splendid spun on its own, Clun Forest makes a good addition to blends with less-elastic fibers of similar length, and it contributes resilience and loft to the mix.

Best known for. Not known well enough for such a versatile, elastic, bouncy wool!

The Fairies of Clun Forest

Clun Forest, in the southwest county of Shropshire, is the setting of one of England's oldest fairytales, the story of Eadric Wild and his fairy wife, loosely tied to the life of a real Saxon lord named Eadric, who held off William the Conquerer, the first Norman king of England, from 1067 to 1070.

As the tale goes, Eadric and a page were out hunting and became tired. When darkness fell, they came upon a great manor house in the woods, where neither remembered seeing a house before. It was well lit, and inside they saw many noble ladies, finely clad and taller than other women, dancing with an airy motion and singing in words Eadric could not understand. He took one for a wife. She said he would prosper as long as he did not reproach her about either "the sisters from whom you snatched me, or the place or wood or anything thereabout, from which I come."

Some years later, Eadric came home from a hunt and could not find his wife. When she appeared, he asked if her sisters had kept her. She vanished, and he died of a broken heart.

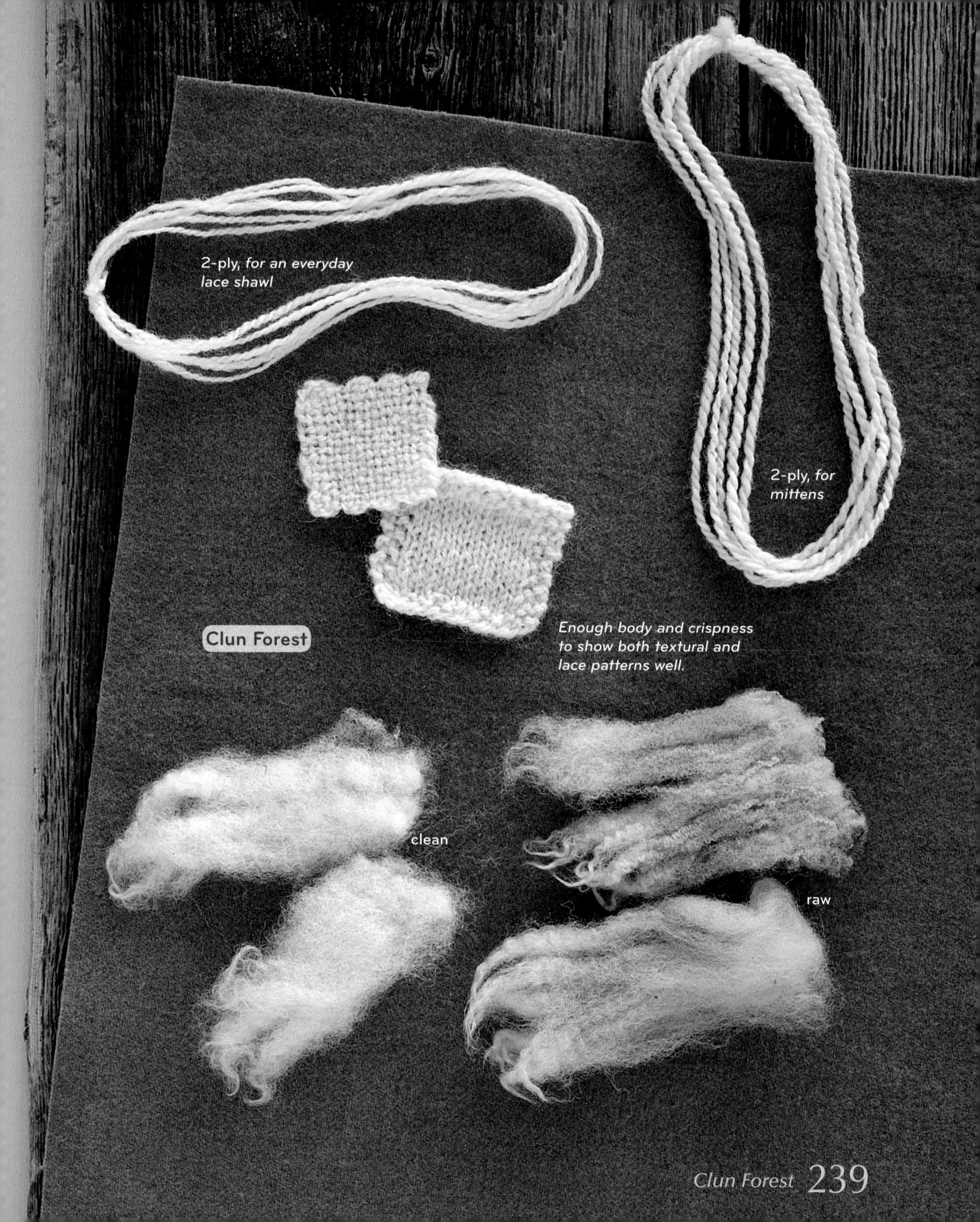

2-ply, for an everyday lace shawl

2-ply, for mittens

Clun Forest

Enough body and crispness to show both textural and lace patterns well.

clean

raw

Colbred

THE COLBRED is a fairly new British sheep that takes its name from the man who first bred it, Oscar Colburn. Beginning in 1956, Colburn crossed East Friesian with Border Leicester, Dorset Horn, and Clun Forest to develop the breed for meat and dairy production. But the Colbred has a versatile, midrange wool. What was unique about it in our experience was how puffy even a combed preparation was. Yarns with lots of twist will emphasize the fiber's crisp qualities, while those with less twist will bring out its bounce and loft.

Colbred Facts

▶ **FLEECE WEIGHT**
5½–8 pounds (2.5–3.6 kg)

▶ **STAPLE LENGTH**
4–6½ inches (10–16.5 cm)

▶ **FIBER DIAMETERS**
26–32 microns (spinning counts 48s–56s)

▶ **LOCK CHARACTERISTICS**
Indistinct, rectangular, puffy locks with short, pointed tips.

▶ **NATURAL COLORS**
White.

Using Colbred Fiber

Dyeing. The fleece takes colors nicely.

Fiber preparation and spinning tips. This probably wants to be combed or, if the fiber length and your preferences align, carded. Could pick and spin locks. Our sample was a little "sticky" to draft, with a lot of cohesion between the fibers. It was still very pleasant to spin.

Knitting, crocheting, and weaving. This is a good multipurpose fiber for durable clothing, possibly next to the skin in the finest part of the range, as well as blankets, bags, and other items you'd make from a standard knitting-worsted–style yarn.

Best known for. Being a reliably moderate fiber with nice loft.

Columbia and Panama

Columbias and Panamas are essentially mirror-image breeds. Both breeds were developed from the same combination of foundation breeds, although the gender contributions were reversed. In 1912, the United States Department of Agriculture began work on developing an improved breed of sheep for "range operators" (see Home on the Range, page 244, for more about range production). They used Lincoln Longwool rams crossed with high-quality Rambouillet ewes to produce large ewes yielding more pounds of wool and more pounds of lamb than their progenitors. At just the same time, James Laidlaw, a Scottish immigrant to Idaho, began developing the Panama by crossing 50 Rambouillet rams on 1,600 Lincoln ewes in order to increase the size and ruggedness of his animals. After five years of breeding his Panama offspring, Laidlaw sold the Lincoln ewes and bred only the Panamas.

The Columbias prevailed over the Panamas. In fact, the Panama breed may be extinct; we were unable to identify any remaining breeders, but that said, if you are in Idaho or Montana you might trip across someone still raising them. (If you do, we would love to hear about it!)

Both breeds produce, or produced, long-stapled, heavy white fleeces with blocky staples. Since we couldn't find Panama breeders, we can't begin to venture opinions about whether there are, or were, notable differences between the wools hiding within the breed similarities in the reported data.

Because of the breed's production-oriented status, Columbia wool has been developed primarily for sale through commercial channels, and white has been the only color encouraged. Nonetheless, the Columbia grows versatile fiber that handspinners can appreciate, and it's a nice, clear white. The wool is lofty and just a touch crisp, without much luster. The finer fleeces qualify for next-to-the-skin use, and even the coarser ones should be fine for most people in sweaters, hats, and the like. Evaluate the fleece in front of you for the best preparation method, but the range of possibilities covers the entire spectrum. Columbia felts without too much effort.

Columbia and Panama Facts

▶ **FLEECE WEIGHT**
Columbia 10–16 pounds (4.5–7.3 kg)
Panama 12–15 pounds (5.4–6.8 kg); yield 45–55 percent

▶ **STAPLE LENGTH**
Columbia 3–6 inches (7.5–15 cm)
Panama 3–5 inches (7.5–12.5 cm)

▶ **FIBER DIAMETERS**
Columbia 23–31 microns (spinning counts 50s–62s)
Panama 25–30 microns (spinning counts 50s–58s)

▶ **LOCK CHARACTERISTICS**
Blocky staples that may have short, recognizable tips. Crimp well developed in the fibers, moderately to significantly disorganized in the locks.

▶ **NATURAL COLORS**
Almost always white, but we've seen a gorgeous brown.

Synthetic Sheep

Breeds like the Columbia, which are developed from the crossing of two other breeds, are often referred to as *synthetic breeds*. However, there is nothing synthetic about the wool of these sheep! You may also hear them referred to as *composite breeds*.

2-ply

Columbia

Very white after washing, and seemed to want a lot of twist.

clean

raw

raw

clean

Using Columbia and Panama Fiber

Dyeing. The fleece takes colors clearly.

Fiber preparation and spinning tips. About half the raw weight of the fleece consists of grease, suint, and the like, so washing before spinning is an excellent idea. Comb, card, flick, or spin from the lock: Columbia is open to possibilities. For spinning approaches, you name it: woolen, worsted, or any combination that's comfortable for you.

Knitting, crocheting, and weaving. Columbia fleeces can become socks, sweaters, pillows, blankets, and other textiles for everyday use. Suitable for any technique you want to use, depending on how you've prepared the yarn.

Best known for. Large quantities of clear white wool that is versatile, especially for clothing, pillows, and the like.

Clothespin sheep made with Columbia by Jacqueline Ericson

Home on the Range

The range is the expansive grassland of the North American West that's memorialized in popular culture as the place of cowboys and Indians, banditos and outlaws, damsels in distress and tender-hearted whores. Most songs and movies glorify the cowboy and his cattle above all the other characters who actually settled the West, but there were plenty of sheep from the earliest days of the region's development. In fact, according to the Bureau of Land Management, in 1870, there were 4.1 million beef cattle and 4.8 million sheep in the 17 western states, and by 1900 those numbers exploded to 19.6 million beef cattle and 25.1 million sheep.

About half of all sheep found in North America today are still part of large-scale range operations in the West. Range flocks, also known as bands or herds, often have thousands of ewes and are moved over large areas by sheepherders and their dogs, all of whom live together for months at a time. Because the sheep spend their time out in wild and open country, some breeds don't do well in range operations, where predators, such as coyotes, can easily pick off solo sheep, even when there are people and dogs nearby. The breeds that do well on the range have a strong tendency to flock tightly together when threatened, and to stay close to their flock mates at all times.

Coopworth

Named for Ian Coop, a college researcher in New Zealand who developed the breed in the 1950s using Border Leicesters and Romneys, Coopworths are known for high productivity. From the start, registration of breeding animals was predicated on production traits — not parentage. Traits evaluated for registration include things such as fleece growth, fertility and lamb survival, and growth rate.

Coopworths can have differing types of fiber, depending on where they're from, although the fiber always has lengthy staples, luster, good crimp, moderate feltability, and overall better suitability for outerwear and utility textiles than next-to-the-skin garments. The Coopworth Society of New Zealand gives only a general description of high-quality wool. The breed standards for the two North American registries for Coopworths specify different wool qualities and use different descriptive systems: The Coopworth Sheep Society of North America calls for finer wool, using a grade designation of 46s to 50s (this means about 30-36 microns), and the American Coopworth Registry standard specifies wool with micron measurements of 35 to 39 (the spinning count equivalent is about 36s-44s). The American Sheep Industry information aligns most closely with Australian Coopworth standards, on the fine end of the overall range (30-35 microns, or spinning counts 44s-50s).

What does it all mean? In short, some Coopworth flocks produce much coarser wool than others. At 30 microns, you might be able to make garments that work well in limited contact with some folks' skin, although you may want to plan instead on sturdy garments and household textiles. At the stronger end (up around 39 microns), you've got significantly more durability than softness going on. The wool is similar in feel to Border Leicester, Perendale, or Romney.

2-ply, *medium twist, medium fine*

2-ply, *more twist*

2-ply, *low twist, bulky*

2-ply, *less twist*

clean

Coopworth

Versatile and sturdy. Really puffed up on the combs! It was easy to overload them.

roving

2-ply

raw

singles

Sold as top, but looks and acts like roving (fiber length)

Coopworth Facts

▶ **FLEECE WEIGHT**
8–18 pounds (3.6–8.2 kg), so think in terms of 12 pounds (5.4 kg)

▶ **STAPLE LENGTH**
5–8 inches (12.5–20.5 cm)

▶ **FIBER DIAMETERS**
30–39 microns (spinning counts 44s–50s), depending on the grower's goals

▶ **LOCK CHARACTERISTICS**
Although the breed produces several styles of wool, they all have these things in common: Any given animal's wool should be consistent; crimp is well defined in both the individual fibers and the lovely pointed locks; good luster; long staples; and not much weight loss from grease during scouring.

▶ **NATURAL COLORS**
White is the traditional color, but natural-colored sheep can be included in some registries. Grays and a coloring called English blue, involving silvery, variegated grays, are more common than browns, which also exist in the breed.

Coopworth yarn, hand-dyed, from Deer Run Sheep Farm

Using Coopworth Fiber

Dyeing. The whites take colors clearly, and grays overdye with subtlety.

Fiber preparation and spinning tips. The fleece can be spun from the lock, from the fold, flicked, or combed. It is usually too long for carding. When we combed our fleece samples, it was easy to overload the combs because the wool really fluffed up! Coopworth is an easy spinner. The locks we obtained in fleece form were 1½ to 2 times as long as the fibers in the commercial roving we tested.

Knitting, crocheting, and weaving. A slightly crisp hand gives definition to stitch or weave structures.

Best known for. High yield; variability across the breed, within predictable boundaries; and consistency in any single fleece. Nice whites, and some color genetics.

Cormo

Cormos were developed in Tasmania, Australia, when shepherd Ian Downie (who raised superfine Saxon Merinos) called on Helen Newton Turner (one of the world's leading sheep geneticists) to help him improve his flock's performance. Downie, who wanted to increase fertility, frame size, and yield of fleece, ended up breeding Corriedale rams to his Merino ewes, and the name of the resulting breed was created as a contraction of the parent names, *Corriedale* and *Merino*. Downie kept the best animals for breeding stock and used computer databases to analyze production traits, at a time when computers were not commonly employed in agriculture. In 1976, Cormos were first brought into the United States (12 ewes and 2 rams), followed by a large importation in 1979 (500 ewes and 25 rams).

Because the Cormo was developed and continues to be bred based on objective, scientific analysis (which includes evaluation of the wool), Cormo fleeces are extraordinarily consistent. Within the breed, fleeces range from 17 to 23 microns, but the tight control shows up most strongly in the requirement that at least 90 percent of a single animal's fleece be within 2 microns of the average count for that fleece. Shepherds who are raising this breed to provide handspinning fiber often coat, or jacket, their sheep to keep the wool as clean as possible. Jacketing requires skill and attention, because jackets need to be changed (for larger sizes) as the fleece grows.

Cormo wool has well-defined crimp, which gives it excellent elasticity. When it's being prepared for spinning, the overall impression is one of fluffiness. Yet there's enough body in the fiber to maintain clarity in lace patterns or in knit/purl textures. Cormo will felt moderately well.

cormo yarn from Spinning Flock Farm

cormo hats from Elsa Wool Company

Cormo Facts

▶ **FLEECE WEIGHT**
5–12 pounds (2.3–5.4 kg) or more; yield 50–65 percent

▶ **STAPLE LENGTH**
3¼–5 inches (8.5–12.5 cm)

▶ **FIBER DIAMETERS**
17–23 microns (in the realm of spinning counts 62s–80s)

▶ **LOCK CHARACTERISTICS**
Dense, rectangular staples with well-defined, regular crimp.

▶ **NATURAL COLORS**
White, because of the commercial orientation of the breeding programs and registration that is limited to white animals; although many colors in Cormo-based wools come from crosses, some breeding programs have yielded purebred colored strains.

Using Cormo Fiber

Dyeing. Nice clear colors.

Fiber preparation and spinning tips. Because of the fine diameter of the fibers, use a delicate touch in preparing it for spinning, but then anything goes. Spin directly from the locks, flick them, hand- or drum-card (using fine carding cloth), or comb. Be careful not to overwork the fiber or you'll tangle it. Make sure to spin fine enough, and with enough twist, to secure and protect the fibers. Check your finished yarn before committing to a technique: You may think you are spinning extrafine and then discover that the strands bloom when they are plied.

Knitting, crocheting, and weaving. Definitely suitable for next-to-the-skin and baby garments; the finest Cormos are as gentle as cashmere. The wool balances nicely between softness and enough body to define stitch and weave patterns.

Best known for. Having a soft fleece that is long stapled and remarkably consistent throughout a single animal's fleece.

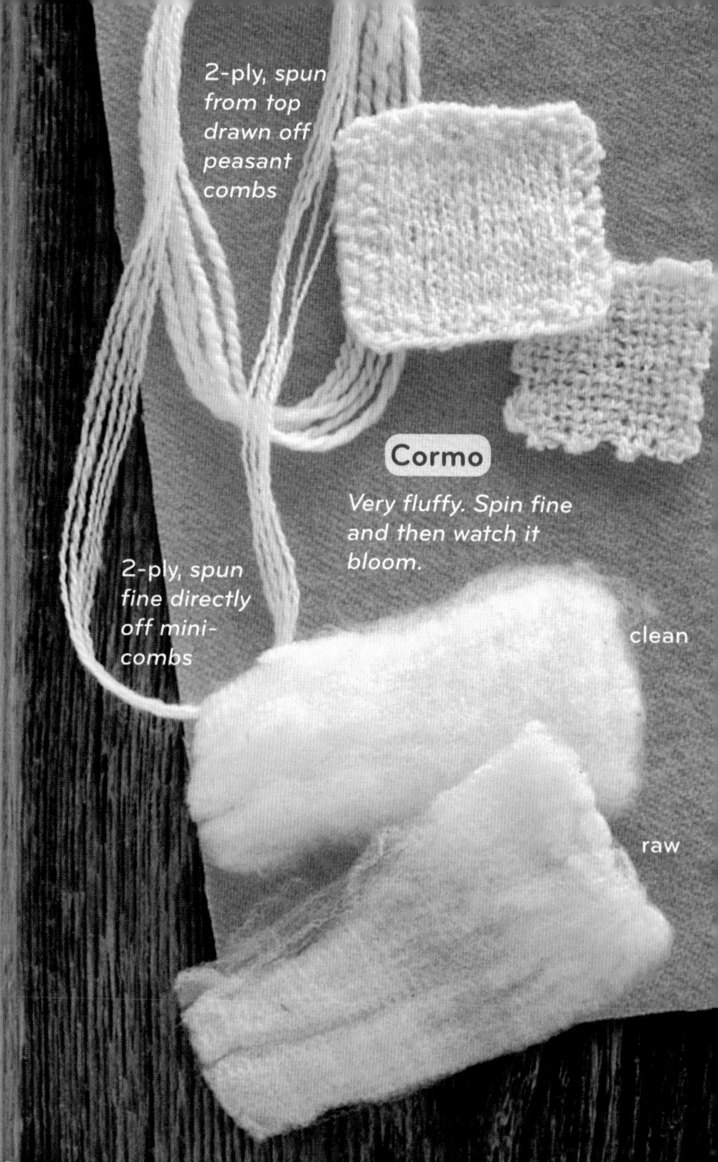

2-ply, spun from top drawn off peasant combs

Cormo

Very fluffy. Spin fine and then watch it bloom.

2-ply, spun fine directly off mini-combs

clean

raw

Natural colored Cormo yarns from Elsa Wool Company

Corriedale and Bond

Development of the Corriedale began in New Zealand in the 1880s, though similar breeding programs began shortly after that in Australia as well. At the time, Merinos, a breed suited to more arid grasslands, and Romneys, which are adapted for lush, rich pastures, were the dominant breeds in both New Zealand and Australia. But both countries had large swaths of intermediate grasslands (areas similar to Kansas or Nebraska in the United States, or Saskatchewan in Canada), so breeders sought a dual-purpose sheep that would perform well in these transition areas. James Little, the manager of a ranch called Corriedale, is credited with being instrumental in the breed's development. Little crossed Lincolns and perhaps some Leicesters with his Merinos until he had a stable breeding population. Corriedales were first imported by the United States Department of Agriculture to the Laramie, Wyoming, research station in 1914.

Corriedale wool has a well-deserved good reputation. It's a reliable, multipurpose fiber that can be a pleasure to spin, knit, or weave. Medium-soft, it has nice long staples, some luster, and well-defined, even crimp, which means it has a lot of loft and elasticity. It's a resilient fiber with enough character to be interesting and not so much that it plays a domineering role in your creative vision.

corriedale yarn from Blacker Designs

Here's where Corriedale's challenge appears: If you are selecting a fleece to spin, you may discover fiber that is very soft, or moderately sturdy and crisp, or anything in between. Corriedale comes in a wide variety of fiber diameters, ranging from the low 20s in micron measurements (spinning counts 62s-64s) for lambs and hoggets (very young sheep) up to about 33 microns (spinning count 46s) for adults, and occasionally even 35 microns (spinning count 44s). Most of the breed's wool will, however, fall in the middle of that span, at 25 to 31 microns (spinning counts 50s-58s), which is mostly soft enough to be used next to the skin for many people, while it also has good durability. It's ideally suited for making clothing, blankets, and the like.

If you're selecting raw Corriedale wool for spinning, you'll find that even though the wool varies from sheep to sheep, within a given fleece the quality tends to be consistent in length, crimp profile, and fineness. While it's hard to determine precisely how soft or crisp a wool is when it's in the grease, you should be able to make a good guess on whether the fleece you're examining will suit the spinning you have in mind. Some shepherds jacket their Corriedales to keep the wool especially clean.

Many of the sheep grown in the Falkland Islands and in South America are Corriedales (see page 16). Wool shipped from Chile through the port of Punta Arenas is likely to be Corriedale and is called *punta* or *PA*.

Corriedale is easier than some breeds to find in yarn form, ready for needles, hook, or loom. For felters, it is definitely a wool to experiment with.

Corriedale Facts

▶ **FLEECE WEIGHT**
10–20 pounds (4.5–9 kg), averaging about 12 pounds (5.4 kg)

▶ **STAPLE LENGTH**
3–6 inches (7.5–15 cm), tending to 3½–6 inches (9–15 cm) in North America and the British Isles, and 3–5 inches (7.5–12.5 cm) in New Zealand

▶ **FIBER DIAMETERS**
25–31 microns (spinning counts 50s–58s) in North America, the British Isles, and Australia (up to 32 microns); in New Zealand, 26–33 microns (spinning counts 48s–58s) for white, and up to 35 microns (spinning count 46s) for colored wool. Lambs' and hoggets' fleeces can have micron counts in the low 20s (spinning counts 60s–70s).

▶ **LOCK CHARACTERISTICS**
Normally rectangular and dense, although often soft-feeling, with flat tips. Crimp clearly defined along full length of fiber.

▶ **NATURAL COLORS**
Most are white, but possibilities in smaller flocks include pale gray through black, and beige through moorit on the brown side. A few sheep are spotted.

Using Corriedale Fiber

Dyeing. The bright whites take color well. The natural grays and browns can be overdyed.

Fiber preparation and spinning tips

- Comb, flick, card, or spin from the locks; if you want to card and the fiber is on the long side, you can cut the staples in half to get a length that works well for woolen preparation.
- Even combed fibers will puff up nicely when the yarn is washed.
- Watch out for, and trim off, weathered tips, which will cause noils in the yarn, especially if you are carding the fiber.
- A freshly shorn Corriedale fleece can be spun in the grease, but most people will want to wash the wool first; select a fleece that is not sticky or full of vegetable matter to make the washing and handling easier.
- Corriedale is a versatile fiber, easier to spin than many wools of similar fineness. Spin a finer strand than you think you want, because washing will cause the yarn to bloom.

Knitting, crocheting, and weaving. Select a softer or crisper fleece or yarn, depending on what you plan to make. Corriedale makes great sweaters, socks, blankets, pillows, and other clothing and household textiles.

Best known for. Dense, medium-fine, long-stapled wool, with well-defined locks and even, clear crimp — an excellent and reliable handspinning wool that is also versatile in knitting, crocheting, and weaving.

clean

2-ply, *top pulled from peasant combs*

raw

2-ply, *top pulled from mini-combs*

2-ply, *spun directly from mini-combs*

2-ply, *top pulled from peasant combs*

clean

clean

raw

raw

clean

2-ply,
carded

raw

Corriedale

2-ply,
top pulled
from mini-
combs

2-ply, spun
directly from
mini-combs

A versatile, handspinner-friendly wool with nice body and crimp. Notice the color and length shifts from raw to clean.

clean

clean

raw

raw

Bond

2-ply, spun directly from mini-combs

2-ply, top pulled from peasant combs

2-ply, top pulled from peasant combs

Even after they have contracted after washing, the staples are 3½ to 5 inches (9–12.5 cm) long. Crimp is well developed and even.

clean

raw

clean

raw

2-ply, *spun directly from mini-combs*

clean

2-ply

raw

Nice combination of length and relatively fine fiber, plus many natural colors.

2-ply, *spun directly from mini-combs*

clean

raw

singles, *top pulled from peasant combs*

raw

clean

Bond Sheep

The cross of Lincoln rams on Merino ewes actually occurred independently in several places in Australia and New Zealand during the late 1800s and early 1900s. You may hear reference to Bond sheep, which is another name for a breed developed in Australia in 1909 by Thomas Bond, who used Lincoln rams on his Saxon/Peppin Merino ewes; he selected more for fineness of wool than other Corriedale breeders. Today in Australia, sheep from Bond's strains are referred to simply as Bond sheep, or as Bond Corriedales, and they have their own breed association.

Although individual shearings from the two breeds can be quite similar, the Bond's wool is, overall, finer (22–28 microns, or spinning counts 56s–62s) than the Corriedale's, with longer staples (4–7 inches, or 10–18 cm) and heavier fleeces (12–16½ pounds, or 5.5–7.5 kg). Bonds are known for softness, good bulk, and elasticity, due to their well-defined and organized crimp pattern. Because of the fiber length, Bond is a better choice than Merino if your goal is soft, thick singles, yet it also makes excellent fine yarns.

In 2000, Joanna Gleason, a Colorado shepherd and fiber lover, imported four moorit-colored Bond sheep from Australia to improve fiber in her own Corriedales. She has been expanding her flock by breeding the purebred Bonds, and upgrading to Bond qualities by crossbreeding to some of her Corriedale ewes. She has developed a lovely array of colors in her long-stapled, fine-wooled flock.

Devon Closewool

Conservation Breed

DEVON CLOSEWOOLS were developed in the mid-1800s from a cross between Devon Longwools (page 104) and Exmoor Horns (page 261). Today they are raised primarily as a meat breed, and they are often crossed with Bluefaced Leicesters to produce Closewool Mules.

The name is a bit misleading: *Closewool* implies a short-stapled fleece, but in reality Devon Closewools have medium-length wool, with staple lengths reaching 4½ inches (11.5 cm). Their wool is crisp and lofty and has lots of body. The breed is similar to the Exmoor Horn, except that it's naturally polled.

Devon Closewool Facts

▶ **FLEECE WEIGHT**
5–8¾ pounds (2.3–4 kg) for ewes, up to 13 pounds (5.9 kg) for rams

▶ **STAPLE LENGTH**
3–4½ inches (7.5–11.5 cm)

▶ **FIBER DIAMETERS**
28–35 microns, extrapolated from Bradford grades (spinning counts 46s–54s, generally 48s–52s)

▶ **LOCK CHARACTERISTICS**
Dense, blocky, and relatively indistinct staples with a tendency to separate from the mass in clumps rather than individual locks; crimp well developed and mostly disorganized. There may be some kemp, but the fleece should not include black fibers of any type.

▶ **NATURAL COLORS**
White.

Using Devon Closewool Fiber

Dyeing. There may be a touch of luster to make colors lively; kemp, if present, will not display dyed colors well.

Fiber preparation and spinning tips. The fleece is easy to process on coarse combs, or shorter fiber lengths can be carded. Picking and spinning directly from the lock are also options for textured results. It is easy and fast to spin.

Knitting, crocheting, and weaving. The fiber is versatile for a variety of techniques; the generally strong nature of the fiber leads in the direction of outerwear, blankets, bags, and the like, although some finer examples of the breed may be suitable for sweaters, hats, mittens, and so on.

Best known for. Bulk, loft, and crispness.

Blocky staples separated from the fleece in clumps; crimp well developed but disorganized.

East Friesian

THE EAST FRIESIAN is a German dairy sheep that's sometimes referred to as the Holstein of the sheep world (after the large and high-producing black-and-white dairy cows) because of the breed's extremely high milk production. There are actually several types of Friesians from the area where Germany, Belgium, and the Netherlands come together. The German type was recognized as the East Friesian breed in 1936 and was first imported into Canada in 1992. The breed made its way from Canada to the United States in 1993, for sheep dairying.

Not surprisingly, with the emphasis on dairy production, this breed's wool is an afterthought, which leads to a broad array of descriptive ranges for its fiber. It may be as fine as 26 microns, but for the most part it is coarser. Whether you'll want to use your East Friesian fiber to make everyday clothing, blankets or rugs, heavy-duty pillows, or bags will depend on the grade that's in your hands. The samples we spun were in the 30-micron neighborhood, give or take a bit, and had nice bounce and body to them.

Both samples were a nice, very dark brown.

Roving was neppy and a little difficult to draft.

East Friesian Facts

▶ **FLEECE WEIGHT**
9–13 pounds (4.1–5.9 kg)

▶ **STAPLE LENGTH**
3–6 inches (7.5–15 cm)

▶ **FIBER DIAMETERS**
May be as fine as 26 microns (spinning count 56s) or as coarse as 37 microns (spinning counts 40s–44s), but for the most part averages are likely to be between 28 and 33 microns (spinning counts 46s–54s)

▶ **LOCK CHARACTERISTICS**
Blocky, dense locks with sun-bleached tips (in the blacks) that don't extend far into the staple length. Our sample had well-developed crimp that was not very organized in either lock or staple.

▶ **NATURAL COLORS**
White, black.

Using East Friesian Fiber

Dyeing. The whites are low luster and take dyes well; the natural blacks will overrule any dyestuff.

Fiber preparation and spinning tips. How to prepare and spin the fleece depends on where in the spectrum of possibilities the particular batch of wool lands. The shorter staple lengths can be carded, by hand or drum carder, and the longer fibers will want to be combed or spun from the fold. For a smooth yarn, spin worsted; for extra bounce, but perhaps more scratchiness, get as much air into the rolags or batts as you can and spin woolen style. Either way, there is likely to be both crispness and lightness (air) to the yarn.

Knitting, crocheting, and weaving. Although some fleeces may be fine enough for standard cardigans, pullovers, hats, and the like, our samples inspired us to think in terms of outerwear with a slight rough edge to it. We envisioned a jacket to wear while walking the dog, or a sturdy blanket to keep in the car and haul out to spread on the ground for impromptu picnics.

Best known for. Easy care, dual-purpose sheep with nice fleece, which are highly productive across a fairly wide swath of North America.

Exmoor Horn

Conservation Breed

THE EXMOOR NATIONAL PARK, located on the southwest coast of England, was previously a royal forest for many centuries. Owned by the British royal family as a hunting preserve, the land was used by local farmers in the summer for grazing their animals. The Exmoor Horn is a native breed that has been documented in the area since at least the 1800s.

Exmoor Horns are a meat breed, so their wool is an interesting, overlooked by-product. Wool data is scarce. They grow white wool that may be in the medium-wool neighborhood, or even coarse to very coarse. Suggested ranges from different sources include 26 to 32 microns (spinning counts 48s–58s) and 32 to 38 microns (spinning counts 40s–48s), and the breed society suggests the fiber hovers around 36 microns.

The sample we obtained felt very crisp, nearly crunchy, with some spring to it. Not a candidate for next-to-the-skin wear, it would make a great walking or barn coat, or perhaps a hearty winter hat, well fulled to be wind- and water-resistant. If you're inclined to make a jacket for your snow-romping dog, Exmoor Horn would be a good choice. There's so much body in our sample that each fiber wants its own space: All the individual fibers are completely willing to walk in the same direction (it's easy to spin), but they insist on keeping some air in between them, instead of nestling down together. While we were spinning, the wool felt wiry, yet the finished yarn did not.

We liked its independent character.

Exmoor Horn Facts

▶ **FLEECE WEIGHT**
4½–6½ pounds (2–3 kg)

▶ **STAPLE LENGTH**
3–5 inches (7.5–12.5 cm)

▶ **FIBER DIAMETERS**
Varies, depending on animal, with reports running from 26 to 38 microns (spinning counts 40s–58s), likely falling on the coarser end of that range

▶ **LOCK CHARACTERISTICS**
Blocky staples with slightly pointed tips. Relatively large number of crimps per inch; semiorganized crimp in the locks. There may be some kemp, but everything is white.

▶ **NATURAL COLORS**
White.

Using Exmoor Fiber

Dyeing. The fleece will take colors clearly.

Fiber preparation and spinning tips. Card shorter fibers or comb longer ones. It can also be spun from the lock nicely. It may feel wiry in the hand — a reason to keep the yarns smooth (worsted preparation and spinning) if the end product is intended for use in clothing.

Knitting, crocheting, and weaving. The body and resilience of the yarn mean it can be cushiony feeling and will also show fabric textures nicely. The stitch or weave structure definition will be clearly defined unless the fabric is felted or fulled.

Best known for. Crisp hand, undisturbed white color, and nice fulling ability.

Has so much body it feels like each fiber wants its own space. Crisp, nearly crunchy.

Galway

Conservation Breed

AT ONE POINT IN HISTORY, Ireland probably had a fair number of native sheep breeds, but today the Galway is the only breed you might consider a native — though it was largely influenced by English Leicesters. The breed was once called the Roscommon, though in actuality there was an earlier Roscommon that was more like the grandparent of the Galway. Lord Roscommon (whom we talked about with respect to the Lleyn; see page 215) imported Dishley Leicesters — lots of them — from Robert Bakewell in England; he heavily crossed them to the old native sheep, yielding the Roscommon breed, which strongly resembled its Leicester roots. By the early 1920s the Roscommon's numbers had fallen significantly, so in 1923 a group of breeders gathered in the city of Galway and selected a small handful of the finest animals to register in a new breed book. The older and less-refined Roscommons became extinct, while the animals registered in the Galway book trucked on and on.

Because it is referred to as an Irish longwool, it's tempting to think that the Galway might be like the English luster longwools — sleek, shiny, exceptionally long stapled, and with robust fiber diameters. It's not, although the shape of the locks and the wavy crimp pattern that can be seen in those locks do resemble the more familiar English Longwools that originated on the other side of the Irish Sea. Galway is more crisp than sleek and has a matte surface, as well as moderate staple lengths and fiber diameters.

What does this mean for yarn spun from Galway? This wool spins up more smoothly than many midrange wools (the Down breeds and others with Down-like, finer crimp), producing yarns that have a pleasant, firm reliable-feeling character. In simple yarns, like two-plies, the individual strands will retain their identities, giving even stockinette or plain-weave fabrics made from them a sense of underlying texture. In novelty yarns, which are spun to have more texture, the designer effects — like loops, feathery bits of loose surface fiber, slubs, or formed knots — have enough body not to be crushed into oblivion. And in yarns spun for smoothness, that fiber-deep character will reveal itself in the fabric; for example, with stitch definition in a textured knitted pattern formed with combinations of knit and purl stitches, or in a nubbly twill weave structure.

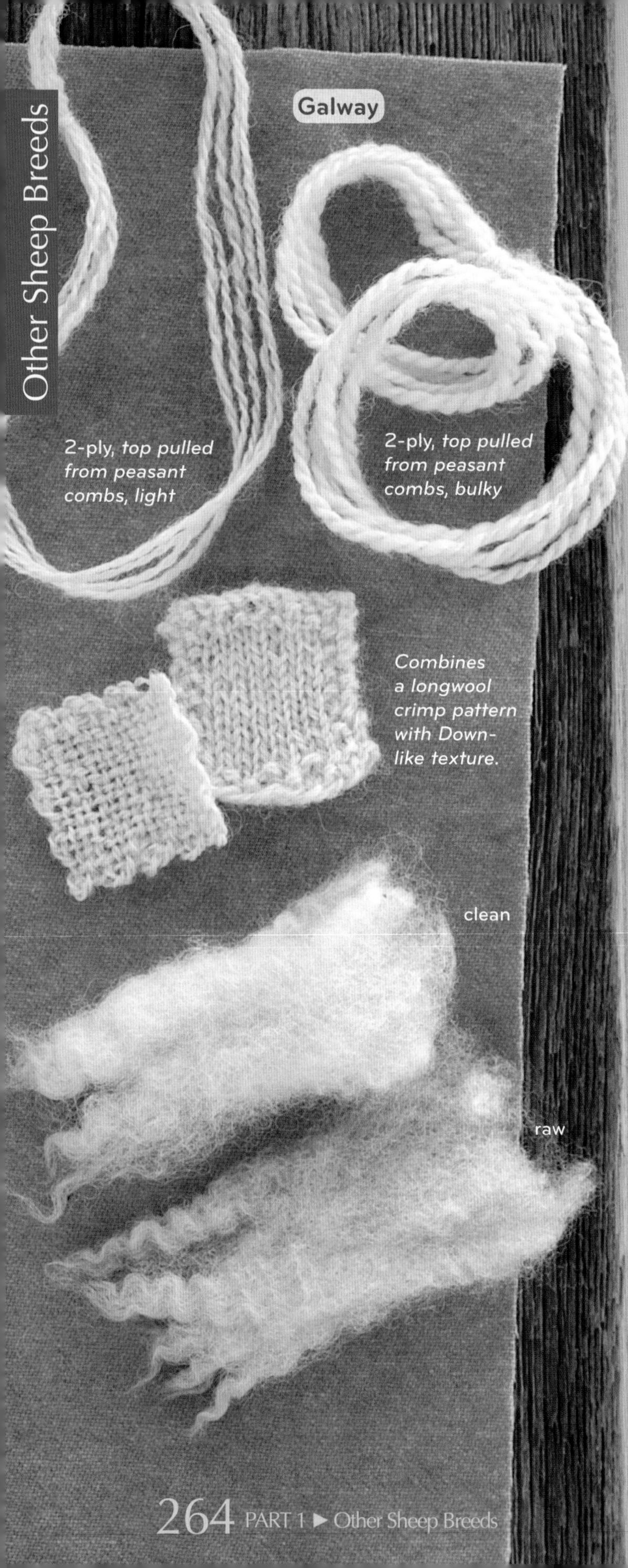

Galway

2-ply, top pulled from peasant combs, light

2-ply, top pulled from peasant combs, bulky

Combines a longwool crimp pattern with Down-like texture.

clean

raw

Galway Facts

▶ **FLEECE WEIGHT**
5½–8 pounds (2.5–3.6 kg)

▶ **STAPLE LENGTH**
4½–7½ inches (11.5–19 cm)

▶ **FIBER DIAMETERS**
26–32 microns (extrapolated from Bradford grades 48s–56s)

▶ **LOCK CHARACTERISTICS**
Locks have the lengthy profile of a longwool, although in miniature, with pointed, slightly curled tips. The crimp pattern is longwool style and wavy.

▶ **NATURAL COLORS**
White.

Using Galway Fiber

Dyeing. Produces clear colors, with apparent depth due to the crimp profile rather than luster.

Fiber preparation and spinning tips. This is a natural for combing, although spinning from loosened locks is an option (they'll need to be teased open), as is carding if the fiber length is in your comfort zone for that technique. The fleece drafts easily to either thick or thin singles, and the amount of twist can vary from relatively low (fiber length permitting) to fairly high (thin, sturdy yarn).

Knitting, crocheting, and weaving. Some will be suitable for next-to-the-skin (if not luxury) wear, and some will be slightly coarser; stitch definition is a forte, whether for textures or openwork and regardless of technique.

Best known for. Not very well known. It's a lovely white with more density and heft than most comparable wools.

Gromark

GROMARKS ORIGINATED in Australia from the same cross that was used in New Zealand to produce the Borderdale — Border Leicester with Corriedale. Although Gromarks were primarily developed for meat production, their breeding also emphasizes good fleece qualities; their two ancestors are about equally represented in the wool, which felts reasonably well. Compared to Borderdale fleeces, Gromark fleeces are, on average, a touch smaller (although still generous), a bit longer stapled, and both finer and more consistent (across the breed) in fiber diameter.

While we were able to locate Gromark locks, our spinnable samples never arrived, so we can't speak from experience about handling the wool.

Gromark Facts

▶ **FLEECE WEIGHT**
9–11 pounds (4.1–5 kg)

▶ **STAPLE LENGTH**
4¾–5¾ inches (12–14.5 cm)

▶ **FIBER DIAMETERS**
28–34 microns (spinning counts 46s–54s)

▶ **LOCK CHARACTERISTICS**
The locks are long, with well-defined, wavy crimp and pointed tips.

▶ **NATURAL COLORS**
White.

Using Gromark Fiber

Dyeing. The fleece should take colors well, since the Border Leicester heritage has likely contributed some luster; this idea is supported by the locks we have seen.

Fiber preparation and spinning tips. The crimp pattern and length suggest that this would be an easy-spinning fleece to work from the lock, flicked, or from Viking combs. Locks at the length of the indicated annual growth could be cut in half and carded, although an experienced carder might want to handle them at full length. Because of the style of crimp, yarns — whether spun by hand or commercially — are likely to be fairly lofty and bouncy, even when spun worsted (this comes from the Corriedale part of the Gromark's ancestry).

Knitting, crocheting, and weaving. The fiber diameter indicates a good wool for sweaters, mittens, blankets, and everyday textiles.

Best known for. Comfortably long locks with pleasant, wavy crimp, resulting in a versatile midrange wool.

Herdwick

Conservation Breed

NATIVE TO THE Lake District of Cumbria in western England, the Herdwick breed is unique in many ways. For starters, the lambs are born jet-black. Their faces and ears soon turn white. By the time they are a year old, their fleece has become dark brown, and with age most of them turn gray.

The breed's origin is an utter mystery. Stories abound. One says Herdwicks came from the wreck of a ship of the Spanish Armada (similar rumors exist about a number of other sheep), but the breed shares few traits with Spanish sheep. Another story suggests they came from Viking stock, though most of the breeds associated directly with the Vikings are naturally short-tailed; the Herdwicks have a long and rather muscular tail. The name, which was first documented in twelfth-century writings, is thought to be derived from the old Norse *herdvyck*, or "sheep pasture," which gives a bit of credibility to a Viking heritage.

When we went looking for Herdwick samples, the shepherd who ultimately got some for us said, "*Why* do you want Herdwick wool?"

Why?? It's truly fascinating and useful stuff. It's just not even remotely like the standard commercial varieties of wool. The animals aren't much like the stereotyped understanding of sheep, either: They aren't white and fluffy and sweet looking. Herdwicks are one of our favorite sheep breeds to have depicted in artwork on our walls, because they look sort of like granite boulders wrapped in shaggy cloaks — and sort of like wise old souls.

Our sample fiber, when we got it, was uncharacteristically short. Herdwick wool is

a lot easier to use when it's long, because it consists of a mix of fiber types that can be separated but pretty much want to hang together. These types include the undercoat of relatively soft wool, the guard hairs, the heterotypic hairs (which change consistency with the season: more wool-like for warmth in winter and more hairlike for shedding rain in the summer), and the kemp. Sometimes there is a lot of kemp, the fiber type that is famously stiff, slippery, and hard to dye. When the staples are long, Herdwick is easier to draft than when it's short; the short stuff is a challenge.

Think sturdiness and texture for Herdwick, along with a fantastic tweedy range of mixed gray colors. A creative commercial use for the fiber involves making it into naturally fire-retardant, completely recyclable insulation. Of course the result is gray. Working with this wool by hand, you can come up with rugs, durable woven fabrics for jackets, or components for baskets and other sculpture. If your fleece does lend itself to separation of the fiber types, that undercoat could make nice sweaters or hats. It'll never be underwear (unless you have a very tough hide), but Herdwick is an uncommon type of fiber that will help you develop a creative take on life.

Herdwick fleeces lighten up with each successive year. Some individuals reach near-white, although most end up a mottled gray. As an aside, while on the sheep, Herdwick wool sheds water more efficiently than other fleeces and dries out more quickly than more typical wools. It will probably do the same for you.

A number of ready-spun Herdwick and Herdwick-blend yarns are available for curious and adventurous textile aficionados to experiment with.

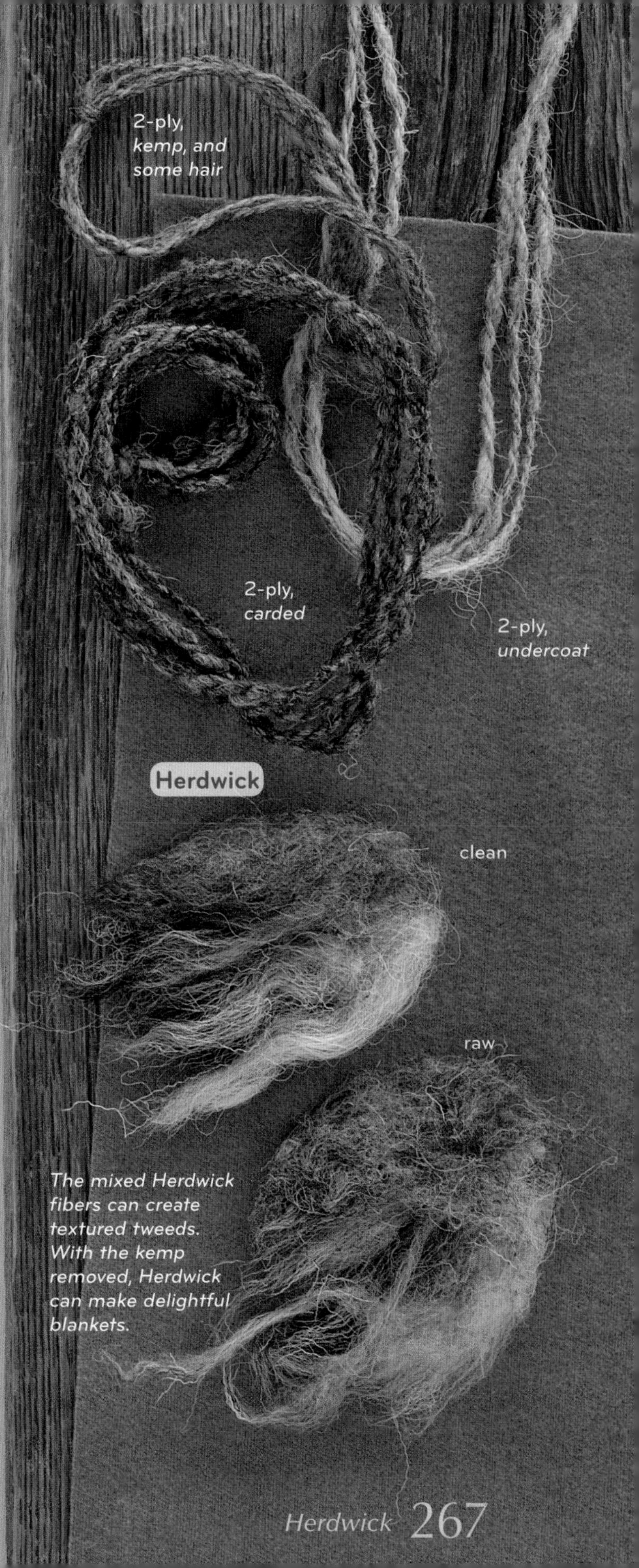

The mixed Herdwick fibers can create textured tweeds. With the kemp removed, Herdwick can make delightful blankets.

Herdwick Facts

▶ **FLEECE WEIGHT**
3–4½ pounds (1.4–2 kg)

▶ **STAPLE LENGTH**
3–10 inches (7.5–25.5 cm), generally 4–8 inches (10–20.5 cm)

▶ **FIBER DIAMETERS**
Variable, depending on which of the fiber components is being examined. Generally described as 36 microns or more (spinning counts 40s and coarser), although the undercoat may be finer, and the kemp fibers can be significantly coarser.

▶ **LOCK CHARACTERISTICS**
Clumpy, dense, composed of mixed fiber types: undercoat, hair, heterotypic hair, kemp. Some parts of the fleece are far more prone to containing the coarse fibers than others — including a mane around the neck and shoulders.

▶ **NATURAL COLORS**
Mostly gray (although the lambs are black), sometimes going into brown and tan ranges or aging to near white; the fleece overall looks like a "frosted" or tweedy mix; spots aren't part of the picture.

Using Herdwick Fiber

Dyeing. The kemp fibers will not show the dye colors with much, if any, intensity; the darkest grays also won't display much applied color; the lighter grays may overdye effectively. The results of a trip through the dyebath will be, as the wool is, tweedy.

Fiber preparation and spinning tips. Experiment. Our suggestion is to open out the locks to facilitate drafting, and go for it. Combing may work to separate some of the fiber types. Carding can work on fleeces with shorter staple lengths, but use your coarsest carding cloth and a light touch. Unless you have been able to separate out the undercoat (which can be spun as any medium wool), keep a very light touch throughout, so the fiber types remain blended as well as possible. While you'll want a yarn with structural integrity, don't fight the natural texture that comes from the fiber mix: Make it a positive contribution to your finished yarn. You'll need enough twist to hold the yarn securely together, but not so much that the strand becomes harsh.

Knitting, crocheting, and weaving. With its mixed fibers, Herdwick lends itself to hard use and outerwear functions. The undercoat may be more suitable to knitting and crocheting and the mixed yarns to weaving, but that's just a generalization.

Best known for. Truly unique sheep and wool, with an unusual mix of natural colors and textures in a single fleece.

Hand-felted Herdwick sheep by Ellie Rowell, of The Wool Clip co-operative in Cumbria, England

Herdwick yarn from eBay/UK

Île-de-France

DEVELOPED IN FRANCE beginning in 1832 when shepherds crossed Dishley Leicesters with French Rambouillet Merinos, the Île-de-France was intended to improve meat production, because wool prices of the day had plummeted. The development was supervised by a professor at the French veterinary college, and from early in the breed's history, testing for specific traits became the norm. In fact, in 1933, when the breed association formed, the organizers established a testing station, and all animals had to be performance-evaluated before they could be registered. The breed was brought into North America starting in 1995 via embryo transplant, starting out at Ian and Deb Clark's Medicine Ridge Farm in Alberta, Canada; from there it expanded into the United States.

The Île-de-France has moved away from its Rambouillet Merino ancestors, and its wool reflects that shift, only slightly overlapping the normal Merino range and landing in what is, by Merino definitions, strong territory. But like Merinos, Île-de-France is a ready felter. It's still suitable for use in next-to-the-skin wear — for general sweaters at the coarser end of its range and allover cuddling at the finer end — while the durability goes up with an increase in average fiber diameter.

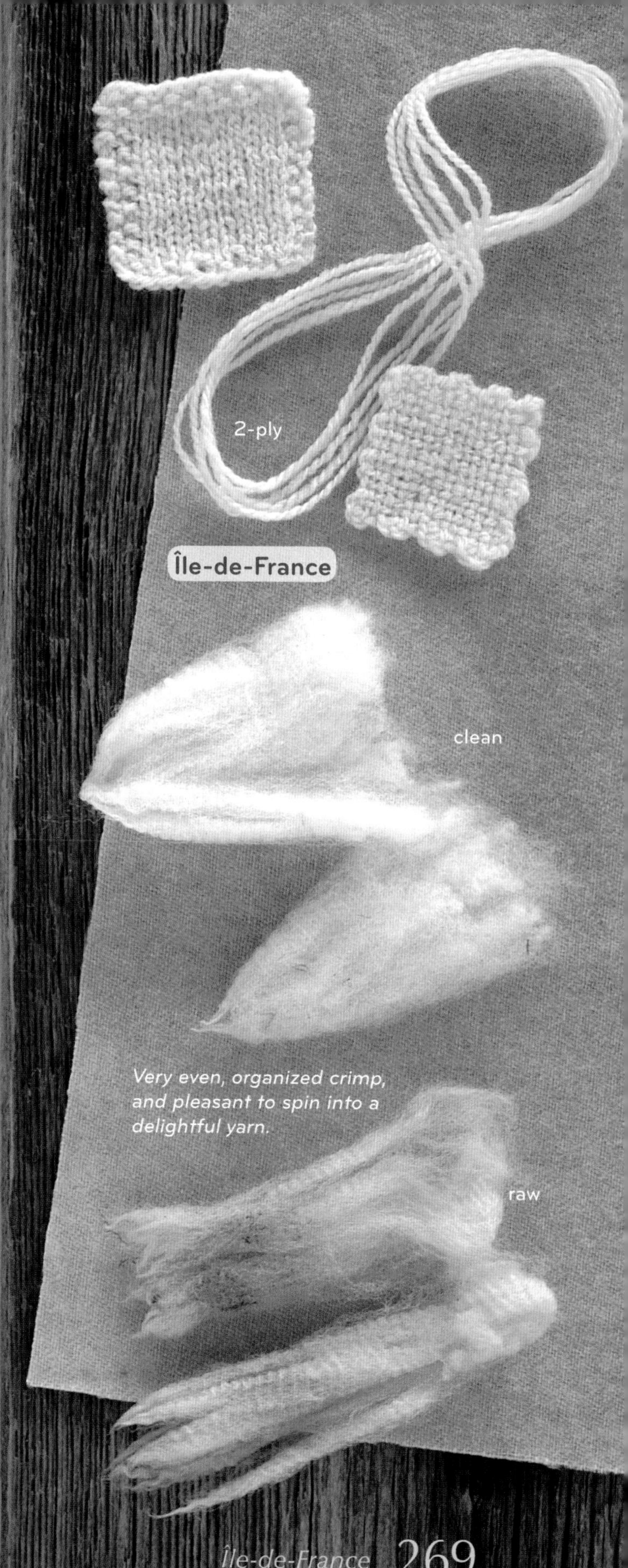

2-ply

Île-de-France

clean

Very even, organized crimp, and pleasant to spin into a delightful yarn.

raw

Île-de-France Facts

► **FLEECE WEIGHT**
6½–10 pounds (3–4.5 kg), up to 13 pounds (5.9 kg) for rams, with a significant amount of grease

► **STAPLE LENGTH**
2¾–3½ inches (7–9 cm)

► **FIBER DIAMETERS**
23–30 microns (spinning counts 50s–60s)

► **LOCK CHARACTERISTICS**
Nice lock definition, easy to separate out, with regular crimp organized in both the fibers and the staples, which are less blocky than Rambouillet and have slightly pointed tips.

► **NATURAL COLORS**
White's what you're likely to find.

Using Île-de-France Fiber

Dyeing. Produces clear colors — our sample was an especially delightful white without any yellow overtones.

Fiber preparation and spinning tips. Depending on staple length, card, flick, or comb. The carded version will have nice loft and insulating qualities; combed, this wool still offers a nice amount of spring. Île-de-France is a good choice if you're new to, and want to experiment with, fine wools. It's not as demanding as Merino or Rambouillet and capable of teaching the spinner ways of approaching those wools. To make the most of the elasticity, straighten but don't stretch the fibers while spinning. The wool will take more twist than you might think, but don't crank in so much that it gets wiry (and if you are plying, judge the twist amount after checking a plied sample).

Knitting, crocheting, and weaving. The fiber combines a soft hand with enough presence for good stitch or weave definition, along with great bounce. Cables or patterns formed with combinations of knit and purl stitches seem to be calling, as does a woven shawl that would conform to the body snuggled in it.

Best known for. A breed largely raised for meat, but one that shows Merino heritage in its wool.

Jacob

Conservation Breed

NAMED FOR JACOB, the son-in-law of Laban from the book of Genesis in the Bible, these sheep have been bred in England for several centuries. In spite of their name, there is no evidence linking these sheep to the biblical flock by direct genetics.

Jacobs are known for their multicolored coats and unusually picturesque horns; they can have two, four, or six horns — a trait known as *polycerate*. Thanks to their spots, they are sometimes called piebald sheep. They never had great commercial value, but they were historically kept as ornamental animals by the landed gentry of England. Today, however, there are two major types of Jacob sheep. Those in Great Britain have undergone breeding shifts to make them more productive (primarily as meat animals), so they are now significantly larger, while those in North America are still smaller and closer in appearance and other characteristics to the old-style sheep.

If you've seen one Jacob fleece — or even a bunch of them — keep your mind open about the breed and its fiber. Jacobs have one of the widest ranges of acceptable fiber qualities of any kind of sheep. Over time, you may learn what to expect from specific suppliers, depending on where a retailer is getting its fiber, what a breeder is emphasizing in a flock, and so on. For example, the fibers shown (see pages 272–273) include two commercially prepared tops (both from Great Britain) and several samples from a single American Jacob fleece. They're radically different.

In Great Britain, Jacobs are grown more commercially, and most of the wool ends up being industrially processed, resulting in batches of yarn or spinning fiber that are uniformly blended to white, browns, grays, and possibly blacks. There tend to be more browns in British supplies than in North American. The fibers in the two top forms, while quite different in color and texture, are crisp feeling, immediately reminiscent of tweeds. Both contain a bit of kemp: The medium brown has enough kemp to add a roughness to its overall texture, and the dark brown has just a sprinkling.

Breeders in North America tend to have smaller flocks and to select for fleece quality, keeping the body of the sheep much like it was in the past. In North America, there are several breed associations with slightly varying criteria for color proportions and fleece qualities.

The colors in the raw fleece we started with did, as is usual, vary slightly from each other in texture and length, so, for example, the whites when sorted out were visibly different from the blacks or grays. Overall, this fiber was drastically softer than the top, with finer crimp. It felt almost buttery smooth, especially the gray. There was not even a hint of coarser fiber, let alone kemp.

While Jacob is not one of the Down breeds of sheep, the wool has many Down-like qualities, especially in the middle of its exceptionally wide fine-to-coarse range. Kemp, if present, may give the wool a tweedy effect, and most Jacob has a springy quality. The sheep are often predominantly white (one standard says 60 percent white) with spots of another color — black, brownish, or what is called lilac, which is not a single color and is not yet entirely understood from the genetic perspective. It may be a gray or light brown. The spots should be clear, with the white fiber growing from white or pink skin and the colored wool growing from darker skin.

The Genesis of Selective Breeding

In the book of Genesis, Jacob, a lowly shepherd, asked his wealthy father-in-law, Laban, if he could keep all the spotted sheep as payment for his work. His father-in-law said yes, and in a dream, God told Jacob to use only spotted rams for breeding. He soon had all spotted sheep.

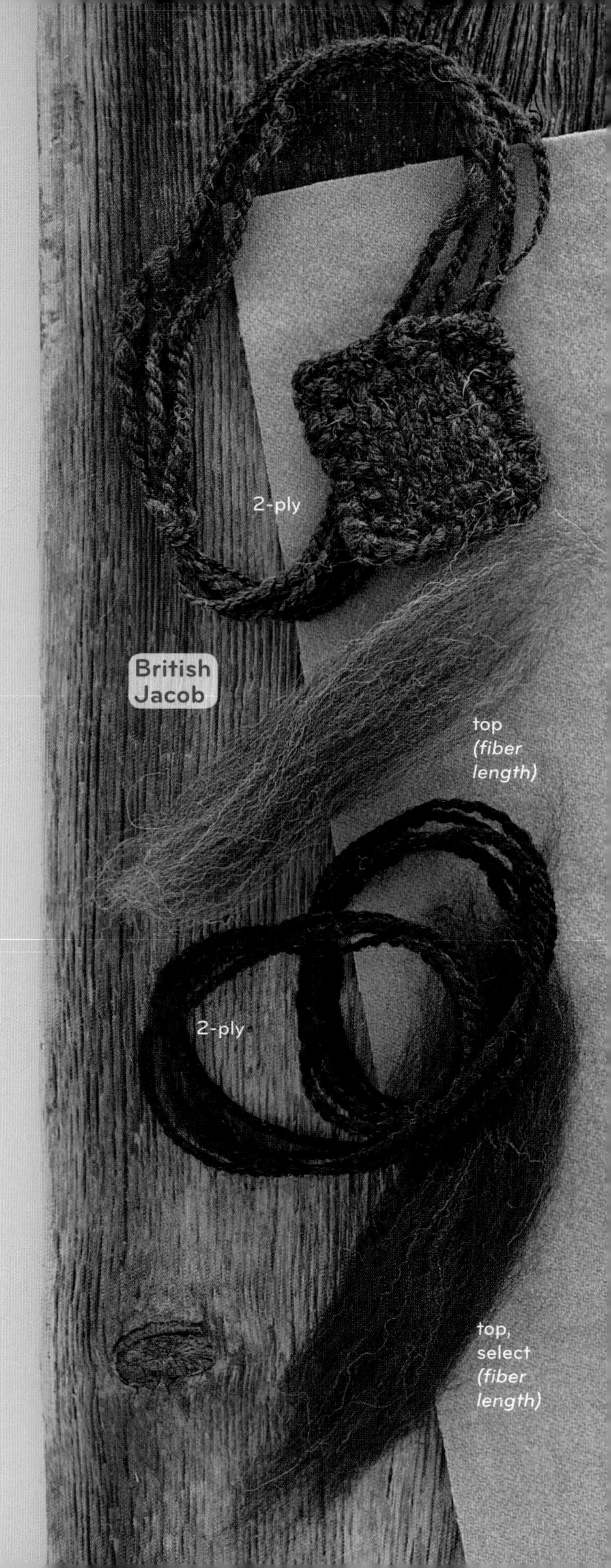

Within a single fleece, the colors vary in length, texture, and crimp. When spinning, you can either sort or mix colors. The American Jacob yarn samples are all from the same fleece, as are the locks.

Jacob Facts

▶ **FLEECE WEIGHT**
3–6 pounds (1.4–2.7 kg), or a bit more; yield 50–65 percent; British Jacobs will, overall, have heavier fleeces than American Jacobs (they're bigger sheep and have been bred to produce more wool and meat)

▶ **STAPLE LENGTH**
3–7 inches (7.5–18 cm), usually in the 3- to 6-inch (7.5–15 cm) range

▶ **FIBER DIAMETERS**
Mostly 25–35 microns (spinning counts 44s–56s), and even though this is already a wide range, some may be a bit finer and some even coarser

▶ **LOCK CHARACTERISTICS**
Single coated. Slightly pointed tips on generally jumbled locks with moderate crimp and some luster. The different colors in a single fleece are likely to have different lengths and texture patterns, although pronounced variation is called a *quilted fleece* and is not considered a good thing. There may be some kemp in coarser Jacob fleeces.

▶ **NATURAL COLORS**
White, black, brown-black, and a color called lilac that is a soft gray or brown.

Using Jacob Fiber

Dyeing. Well, you could dye it, and the results might be very intriguing. Most people consider the existing colors enough to deal with.

Fiber preparation and spinning tips. As Jacobs' fiber varies greatly, so do preparation methods. Shorter fibers will need to be carded, possibly as a random blend, to either preserve or obscure the distinct colors (the more you card, the more blurring of colors will occur). Longer fibers take nicely to combing, with the type of combs matched to the fibers' fineness. Colors may be separated, spun as they come, or blended to make a variety of shades.

Knitting, crocheting, and weaving. The colors can be a real boon to designing, or an overwhelming opportunity that turns into a jumble. Get to know Jacob with small projects and ease into larger efforts when you're ready. Keep in mind that the texture and color can vary dramatically; buy enough of your fiber supply — yarn, top/roving, or fleece — for your whole project right at the start.

Best known for. Multicolored fleece. Always. Of unusually widely varying qualities, depending on lineage and breeder's preferences.

Jacob yarn from eBay/UK and felt from Local Harvest/Kenleigh Acres

Gene Therapy Potential: Tay-Sachs Disease and Jacob Sheep

In 1999, Fred Horak and his wife, Joan, of Lucas, Texas, took two dying Jacob lambs to Texas A&M University, suspecting an acute case of a skeletal congenital defect called occipital condylar dysplasia; but that wasn't the problem. The Texas A&M pathologists said the lambs had a lysosomal storage disease, and they wanted to draw blood and take tissue samples from the Horaks' other sheep. The Horaks agreed. They soon learned that their sheep carried a defective gene.

Most often when breeders learn of a genetic disorder of this type they simply quit breeding the carriers, but the Horaks decided to work with the pathologists in order to better understand the disease. They continued breeding the affected lines, even though at the time both parties considered the project to be of limited interest to the outside world. They thought that for a breed whose numbers were so low, it could be helpful for the long-term protection of the breed to understand what was happening genetically to produce the disease.

In 2008, the researchers called the Horaks with startling news: The lysosomal disease in their Jacobs was the same form as found in Tay-Sachs disease, a human hereditary disorder that is fatal to affected children, just as it was to the Horaks' lambs. There is currently no cure for Tay-Sachs disease. Human children begin to show symptoms in their first year and usually die from the disease before they reach the age of four. Researchers suddenly wanted to use the carrier sheep to test a new gene therapy that could save not only lambs, but also human infants and toddlers!

This is yet another case where conserving a rare livestock breed may have benefits beyond the obvious feel-good value. In an e-mail exchange with Fred Horak, he said, "You can help make a difference for fighting Tay-Sachs by spreading the story among your circles: The Jacob may be a key to improving the human condition. Look for Jacobs that may be affected or carriers of the lysosomal disease because more numbers of carriers are needed for this research." If you learn of Jacobs who have this disorder (more have recently been identified), get in touch with the Horaks and the researchers — or get in touch with us through our publisher, and we'll help you make the connection.

Karakul

Conservation Breed

THE KARAKUL BREED is very, very old. It's native to the steppes and deserts of central Asia, concentrated in the area of the "stans" (particularly Uzbekistan and Turkmenistan, but also extending into parts of Tajikistan, Kyrgyzstan, Kazakhstan, Afghanistan, and Pakistan, as well as Iran). This region of the world is one of the cradles of the domestication of livestock, so the Karakul is considered one of the oldest sheep breeds.

Like Tunis, Karakuls are classified as fat-tailed sheep, because they deposit fat down their backs, in their rumps, and in their tails. This adaptation is also found in several breeds of sheep that are not found in North America but are common to the seasonally barren landscapes of Persia and Africa. During the rainy season, when feed is abundant, the sheep build up these fat and nutrient stores, and then during the long dry season, the animals basically live off these reserves. Historically, in the Karakul's native range, prize-winning ewes would be fed until their tails reached such great weight that they required wheeled carts strapped to their haunches to carry them around!

When it comes to the wool featured here, we're talking not about Central Asian Karakuls but about American Karakuls. This breed was developed from several imports to the United States, starting in 1909 and continuing through the first half of the twentieth century. Regional variations have occurred within the breed since its arrival in North America, including some

Faux Furs

Today fur is less popular than it used to be, but as recently as 2000 designers were featuring Persian lamb furs in their collections. Historically very popular, these "furs" (also known as Astrakhan, broadtail fur, or even Kara-Cool fur) actually come from the skins of newborn Karakul lambs that are slaughtered within hours of birth, or even fetal lambs whose mothers are killed to obtain the lamb's pelt about 10 to 15 days before the lamb would naturally be born.

Why has this practice been followed with Karakul lambs? Because the lambs are born with tightly curled, exceptionally lustrous black (or occasionally red) coats. About ten days after the lambs are born, the curls open and the luster dims to what is normal for the breed. Personally, we'd rather let the lambs wear their own fleeces and grow up to produce sturdy, gorgeous, colorful, mature Karakul wool.

infusions of Lincoln, Tunis, Navajo Churro, Cotswold, and other breeds' bloodlines. Starting in the 1970s, many so-called spinners' flocks (small flocks of sheep kept by spinners to generate wool for the new practitioners of a recently rediscovered craft) incorporated Karakul, partly to contribute a spectrum of color genetics and partly to produce wools that were open, relatively light in grease content, and particularly easy to spin by hand. Because the hand-me-down line of traditional spinning knowledge had been broken in the years following the Industrial Revolution, many of these handspinners lacked the necessary knowledge to handle fine wools; also they were, as spinners continue to be, intrigued by naturally colored wool, which was not readily available at the time. Shetlands and Icelandic sheep had not yet been brought into North America, and most wool was grown for the wool pool (see page 68), where colors other than white were anathema.

Into this historic moment Karakuls brought dominant genes for color, although spinners frequently wanted less-coarse wool, so they crossbred to get the fiber they wanted. The American Karakul came out of this melee with its individual character intact, even though the spinners' breeding choices still influence many flocks. Karakuls are double coated, with a soft, fine, shorter inner coat and a coarse, sturdy, longer outercoat, although sometimes the distinctions between the two are quite narrow. The coats can be the same color, but in some intriguing fleeces they are differently colored. Colors can also vary along the staple from butt to tip. The wool has a nice luster.

When relatively free of dust and vegetable matter, Karakul lends itself to being spun in what little grease it has. It also washes up beautifully. In preparation, you'll want to decide whether to emphasize the texture (and possibly color) variations or to even them out by blending.

Elasticity is not Karakul's strong point. It excels in durability, willingness to felt, and its array of colors. While Karakul is a standby for spinners who venture outside the "soft and softer" fiber realms, Karakul yarns are also available for exploration by knitters, crocheters, and weavers who want to discover its unusual qualities. A bag, pillow, rug (knitted, crocheted, woven, or braided), or pair of boots made of Karakul will endure hard wear. Karakul is one of the quintessential felting fibers.

Karakul Facts

▶ FLEECE WEIGHT

5–10 pounds (2.3–4.5 kg); yield 80–85 percent

▶ STAPLE LENGTH

6–12 inches (15–30.5 cm)

▶ FIBER DIAMETERS

The quality of Karakul fleece is more important than its fiber diameter, which is variable and often listed as 29 microns, either as an average or a "greater than" orientation point. The American Sheep Industry Association lists it as 25–36 microns (spinning counts 44s–58s), which is the appropriate ballpark. The inner coat is likely to have micron counts in the 20s and the outercoat in the 30s, or possibly stronger.

▶ LOCK CHARACTERISTICS

Open, lustrous, with wide bases (a characteristic of double coatedness) gently tapering to the tips. Locks separate out fairly easily.

▶ NATURAL COLORS

A remarkable range, with black predominant, although the spectrum includes a variety of grays and browns (some tending into unusual gold and reddish tones), and some whites. The wool can be solid colored or display shades of color within the staples. Traditional names are associated with some of the colors: *arabi* (black), *guligas* (rosy), *kambar* (reddish brown), *shirazi* (blue-gray, from a mix of black and white), and *sur* (an unusual brown base with lighter tips). The black fleece of Karakuls, like the black hair of people, gets grayer with the years.

Using Karakul Fiber

Dyeing. The fleece accepts dye well, although when planning colors the underlying natural shades need to be taken into account.

Fiber preparation and spinning tips. In the more pronounced double-coated fleeces, combing or flicking can be expected to separate out the undercoat, but these techniques generally work well to prepare the fiber, despite the higher levels of loss. Those who are comfortable carding long fibers may find they enjoy preparing Karakul on regular or coarse carding cloth, with hand carders or a drum carder. It's a relatively easy-spinning wool, willing to make either fine or thick yarns, and you can draft to maximize texture or go for smoothness instead.

Knitting, crocheting, and weaving. Remember the strength of this fiber, evaluate the unique yarn in front of you, and experiment. In the Middle East, socks are made of fiber from the local Karakul-type sheep, although the results are coarser than most Western knitters prefer (but also very long lasting in the face of extreme wear). Karakul is exceptional as the weft in weft-faced textiles, including exquisite, hard-wearing rugs. It also makes a good warp yarn.

Best known for. Durability, luster, range of colors (with variegation within many of those colors), and stability (often called lack of elasticity, but that makes it sound like a negative, which it's not).

roving
2-ply
2-ply
roving
2-ply
roving

Note the lamb tips on the locks on this page.

Montadale

THE YEAR WAS 1932. A lamb buyer from Missouri, E. H. Mattingly, brought the first Columbia sheep (purchased in Kalispell, Montana) east of the Mississippi. The animal was a ram, and Mattingly began crossing him with Cheviot ewes. Mattingly's goals: to increase fertility, hardiness, and fleece quality, all while maintaining excellent meat production and quality. His experiment was one of those close-but-no-cigar trials: It didn't quite produce the critter he hoped for, so in 1933 he turned the breeding scheme on end, using a Cheviot ram on some Columbia ewes he had also acquired in Montana. That combo worked as he had hoped and provided the foundation for the Montadale.

Because the Montadale was developed as a dual-purpose breed — or a breed with reasonably good and equally valued meat and fiber traits — plenty of attention is paid to the fleece quality. Breeders are looking for a heavy clip of wool that is focused toward the next-to-the-skin realm of softness and won't cause problems for mechanical processing systems. The breed standards discuss the wool in the most old-fashioned and nonspecific terms, using the blood system (see page 11) of defining fiber qualities. They reflect the goal of ⅜- and ¼-blood fleeces; low ¼-blood is undesirable, and sheep with common or braid wool are denied registration. The standards call for a minimum 12-month growth of 4 inches (10 cm).

Montadale wool can be a lovely fiber for hand processing, with a very white color, good staple length, and fine but not fussy amount of crimp. A friend who was nearby when we were spinning the samples, and who had access to demonstrations of a number of breeds, got starry-eyed when he touched the Montadale skein and began dreaming out loud about a hat. The fleeces of black Montadales (which produce a range of grays) are intended only for handspinners and small-scale artisan yarn production, because they are useless to the mass-industrial system. Lucky us!

Montadale Facts

▶ **FLEECE WEIGHT**
7–12 pounds (3.2–5.4 kg); yield 45–60 percent

▶ **STAPLE LENGTH**
3–5 inches (7.5–12.5 cm); breed standard calls for a minimum of 4 inches (10 cm) for 12 months' growth

▶ **FIBER DIAMETERS**
25–31 microns (spinning counts 48s–58s), sometimes coarser, but an average greater than 34 microns (spinning count 46s) will result in disqualification from registration

▶ **LOCK CHARACTERISTICS**
Dense and uniform, with nice crimp in both fibers and locks.

▶ **NATURAL COLORS**
The vast majority are a clean, bright white, but there are black Montadales as well.

Using Montadale Fiber

Dyeing. The clear white wool will give clean colors.

Fiber preparation and spinning tips. It is nice to card (if short enough), comb, or spin from the lock. Because this breed is mostly grown in large flocks for a meat as well as a wool crop, the fleece may need careful cleaning before it can be prepared by hand, although it's dense enough that most of the dirt should be in the tips. Well-cleaned and prepared fiber can be spun with any technique you prefer.

Knitting, crocheting, and weaving. This is a lofty wool with a crisp quality that shows textures nicely. Not noted for luster, but the clarity of the white color gives the fleece a light-catching quality anyway.

Best known for. Very white wool of good staple length.

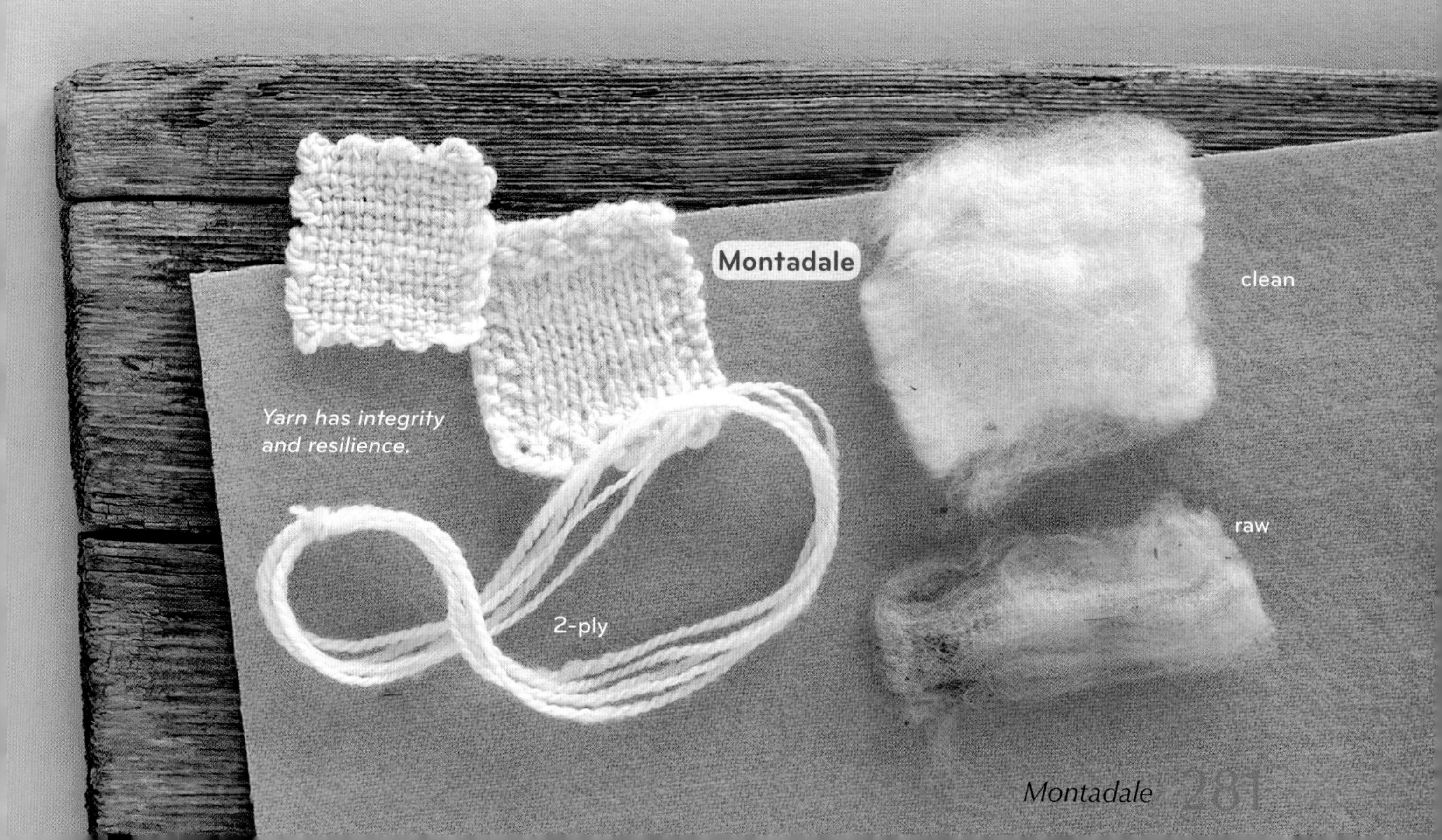

Yarn has integrity and resilience.

Navajo Churro

Conservation Breed

NAVAJO CHURRO SHEEP are the oldest breed in North America. They were developed in the southwestern United States from the Churra sheep — an old native breed that Spanish explorers brought to the New World from the Iberian peninsula in the fifteenth and sixteenth centuries. Native Americans in the Southwest acquired animals through trade, raids, or luck (when the Spanish lost their animals). The Navajo in particular took to shepherding and soon relied on the Churra sheep for weaving, which provided the basis of their economy.

As later movements of European settlers entered the Southwest, the name was changed from Churra to Navajo Churro, although the sheep are also known as Churros or Navajo sheep. In the late 1800s, the U.S. Army killed tens of thousands of Churro sheep in an attempt to control the Navajo tribe, and later still the federal agencies pushed crossbreeding of Churros in an attempt to "improve" the sheep Navajos kept. That "improvement" did not take the environment into consideration. The larger crossbred sheep placed an unsupportable grazing load on the landscape's meager vegetation and were unable to survive periodic droughts as capably as their tough little predecessors, who had been almost completely exterminated.

Yet small pockets of original Churros remained hidden in valleys and isolated areas of the Navajo nation, and these animals, which livestock-breed conservationists began taking an interest in during the early 1970s, are still remarkably similar to the sheep brought to the New World by the Spanish Conquistadors. The breed's wool has contributed substantially, and in ways that cannot be replicated by other breeds' fibers, to the development of the textile traditions of the Navajo, Hispanic, and Puebloan communities in the southwestern United States.

Wool characteristics. Navajo Churro wool takes some time to get your mind around. It's double coated, but it often gives the impression of being single coated. It can feel soft and fine but has a well-earned reputation for durability. It's described as having luster, and it does — but unlike the English luster longwools, Navajo Churro seems to have an inner gleaming quality. Like many of the so-called primitive wools (meaning here the ones that have not been bred into narrow specification channels to accommodate mechanical processing), there's a lot of variety, and any generalization we make will by no means apply uniformly to every fleece the breed produces.

That said, the breed standards of the Navajo Churro Sheep Association give us a good way to understand the apparent contradictions in the description above. There can be three types of fiber in a Navajo Churro fleece. The *undercoat*,

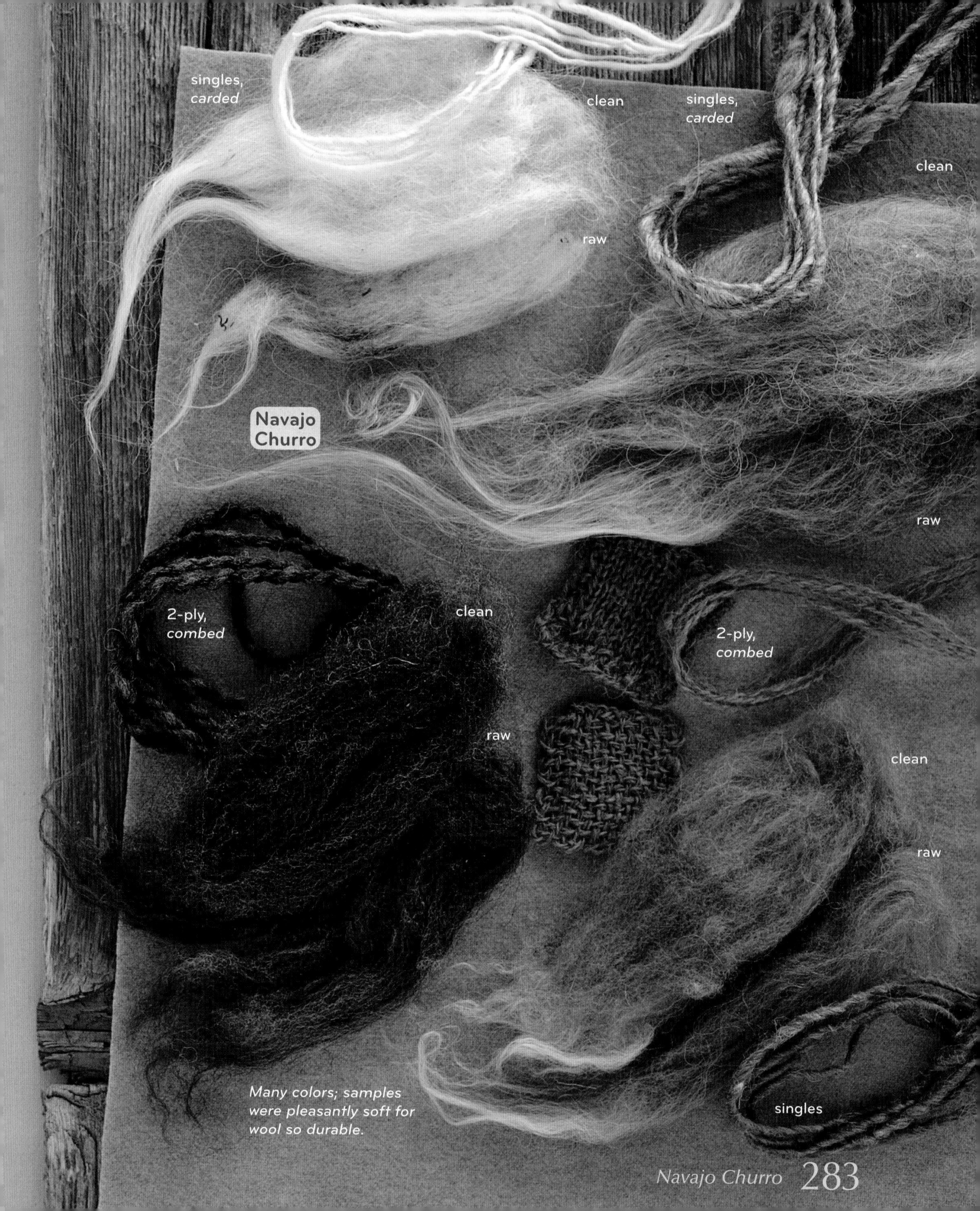

Many colors; samples were pleasantly soft for wool so durable.

which makes up about 80 percent of the fiber, is about 3 to 5 inches (7.5–12.5 cm) long and measures between 10 and 35 microns — a remarkably wide range (spinning counts from 44s to much finer than 80s). Some of that is as fine as wool gets at this point in history; some is relatively coarse. Since most of the fleece is undercoat, these fine fibers are dominant in what you touch.

The *outercoat*, sometimes called guard hairs, contributes 10 to 20 percent of the fiber. This part is about 6 to 12 inches (15–30.5 cm) long and measures from 35 microns on up (spinning counts 44s and coarser). These two types of fiber may meet each other at the 35-micron point, so there may or may not be an obvious separation between the two coats. The outercoat may, in fact, appear as a relatively small number of longer, coarser fibers that are nearly invisible within the base of the lock and extend out to form the lock's narrow, slightly curling tip.

The third type of fiber in the fleece, *kemp*, measures 65 microns or more (off the scale for spinning counts) and is short and prickly. It contributes up to 5 percent of the total. While kemp fibers may exist, they're not a large part of what's going on. Churro wool does felt.

Working with Navajo Churro. Most Navajo rugs today are woven with commercially prepared, dyed yarns of a generic type. There are exceptions. For example, the Two Grey Hills style of rug and tapestry weaving is known for displaying the natural colors of Navajo Churro wool — gray, brown, black, and white — that is frequently handspun.

Commercially spun Navajo Churro yarns, in both natural and dyed colors, are available for use by knitters, crocheters, and weavers. Because of the constraints of processing fiber at the necessary scale, these yarns are best for use in textiles where durability, body, and luster are all needed. The softest of fleeces will be reserved for handspinning. The most alluring Navajo Churro fleeces we have ever seen were in the back room of a trading post and were not for sale. They had been reserved for the use of nearby Navajo weavers, whose amazing skill would turn them into breathtakingly exquisite tapestries. It was enough to caress the wool and leave it to its amazing destiny, and to be haunted years later by how it looked and felt.

"Sheep Is Life"

The Navajo Sheep Project was established in 1977 to bring back the almost-extinct Navajo Churro breed and reintroduce it into Navajo and Hispanic communities with the help of Lyle McNeal, a Utah State University animal science professor. McNeal — a true breed-preservation hero — recognized the genetic and cultural significance of the Navajo Churro to the Navajo, or Diné ("the people"), and set out to bring the breed back from the brink of extinction. He gathered some animals from the few remnant flocks, bred them to increase their numbers, and then facilitated their return to their original people. He helped the shepherds to form a nonprofit, Diné bé Iiná, which began sponsoring the Sheep Is Life celebration as a way to educate people from the local communities and the public from a wider area about the breed and about traditional weaving. It also celebrates the importance of Navajo Churro sheep to the Native American and Hispanic cultures of the Southwest. The event is now held annually in June in either New Mexico or Arizona, so if your travel plans take you in that direction, it's worth going out of your way for! Learn more at www.navajolifeway.org.

Navajo Churro Facts

▶ **FLEECE WEIGHT**
4–8 pounds (1.8–3.6 kg), possibly as light as 2 pounds (0.9 kg); yield 60–65 percent (most loss due to dust, rather than grease)

▶ **STAPLE LENGTH**
Undercoat generally 3–5 inches (7.5–12.5 cm), although it can be as short as 2 inches (5 cm) or as long as 6 inches (15 cm); outercoat generally 6–12 inches (15–30.5 cm), although it can be as short as 4 inches (10 cm) or as long as 14 inches (35.5 cm)

▶ **FIBER DIAMETERS**
Inner coat: 10–35 microns (spinning counts 44s–much finer than 80s), most likely in the low 20s (spinning counts 60s–62s); outercoat: 35 (or more) microns (spinning counts 44s and coarser); kemp fibers: 65 (or more) microns

▶ **LOCK CHARACTERISTICS**
Wide base tapering to a narrow tip. Low in grease and open, which means the fibers can easily be separated, but this wool has an interesting cohesive quality as well: the fibers seem to have an affinity for each other, not joined, as other fleeces' fibers are, by lanolin or strict similarity.

Navajo Churro yarns, natural dyed, from NearSea Naturals

▶ **NATURAL COLORS**
Many are white, although the breed is also well known for its variety of light to dark browns, some of them with reddish undertones, as well as its grays and blacks. The outercoat and undercoat can be different colors. Some sheep have spots.

Using Navajo Churro Fiber

Dyeing. The whites take color nicely; the natural colors can be overdyed.

Fiber preparation and spinning tips

- Washing is optional, because of the low grease quantities.
- Can be spun from the picked lock, carded, or combed (if the fleece is not the sort where combing results in too much waste).
- Despite the fiber length, the customary Navajo preparation method is carding, which keeps the orginal fiber blend together. Regular carding cloth is in order.
- Traditionally, Navajo Churro yarn is spun from rolags on a Navajo hand spindle, in a two-step process where the rolags are first attenuated and joined together, with a small amount of twist added, and then are fully drafted out and turned into yarn. You can use any spinning techniques or tools that you find comfortable.
- The wool can be a bit "draggy" in the drafting, and you'll want to be sure you keep the different types of fibers feeding together evenly.

Knitting, crocheting, and weaving. Navajo Churro is an ideal yarn for making weft-faced rugs and tapestries. The results look and feel soft, but stand up astonishingly well to wear. Finer fleeces also lend themselves to use in garments; how close you want to have them to your skin will depend on exactly how fine the specific wool is. Some would make excellent, soft camisoles, although most will be better suited for cardigans, hats, or mittens.

Best known for. Long heritage, natural colors, sweet luster, and durability.

Norfolk Horn

Conservation Breed

THE NORFOLK HORN became extinct after contributing (along with the Southdown) to the development of the Suffolk (see page 80). Suffolks were intended to be better meat producers and more docile than the feisty Norfolks. Later the Suffolk became the core of a backbreeding effort that produced the New Norfolk Horn, now just called the Norfolk Horn again. There's no telling how much the new breed genetically resembles its progenitor.

Today's Norfolk Horn fleece is closer to the original breed's, according to measurements taken by sheep historian Michael Ryder, than it is to the Suffolk. It is lighter than Suffolk (as befits an agile animal able to survive in marginal conditions), longer on average, and appears

to be more consistent in fiber diameters, staying below 30 microns. Lambs are born with mottled fleeces, although the adults are white.

Norfolk Horn Facts

▶ **FLEECE WEIGHT**
3–4½ pounds (1.4–2 kg)

▶ **STAPLE LENGTH**
3–4 inches (7.5–10 cm)

▶ **FIBER DIAMETERS**
26–29 microns (spinning counts 54s–56s)

▶ **LOCK CHARACTERISTICS**
Blocky staples with square tips, and fibers with fine and well-developed crimp that is not organized. Our sample has a touch of luster.

▶ **NATURAL COLORS**
White.

Using Norfolk Horn Fiber

Dyeing. This fleece gives clean colors.

Fiber preparation and spinning tips. Card, or comb, or pick the locks apart and spin them directly. It's an easy and versatile fleece to spin.

Knitting, crocheting, and weaving. This fiber balances durability and softness, for use in sweaters, hats, mittens, socks, blankets, and similar comforting textiles. The crimp means yarns will be lofty.

Best known for. Perhaps for being extinct and then redeveloped, although with current science we can't evaluate how much the new Norfolk Horn genetically resembles the original. Not well known enough, considering how nice the breed's wool is.

Perendale

THE PERENDALE was developed in the 1950s as a dual-purpose meat and fiber breed by Geoffrey Peren at Massey University in New Zealand. Peren crossed Cheviot rams on Romney ewes to produce this breed. Although Perendales were first imported into the United States in 1977, they are still rather rare in North America. They are longwool animals that do well in cold and wet climates.

Perendale is a bouncy wool, which will spin up with a spring to it, as opposed to the compact sleekness of other (mostly English) longwools. This lofty quality can add warmth to sweaters (for the finer ranges) or cushioning qualities to rugs (for the stronger wools).

There are both finer and coarser ranges within the breed, and New Zealand standards have moved toward the coarse end of the scale lately in response to market demands and husbandry realities. Thus some sheep are producing wools for general knitting yarns, while others grow fleeces best suited for harder-wearing textiles, like rugs, bags, and upholstery.

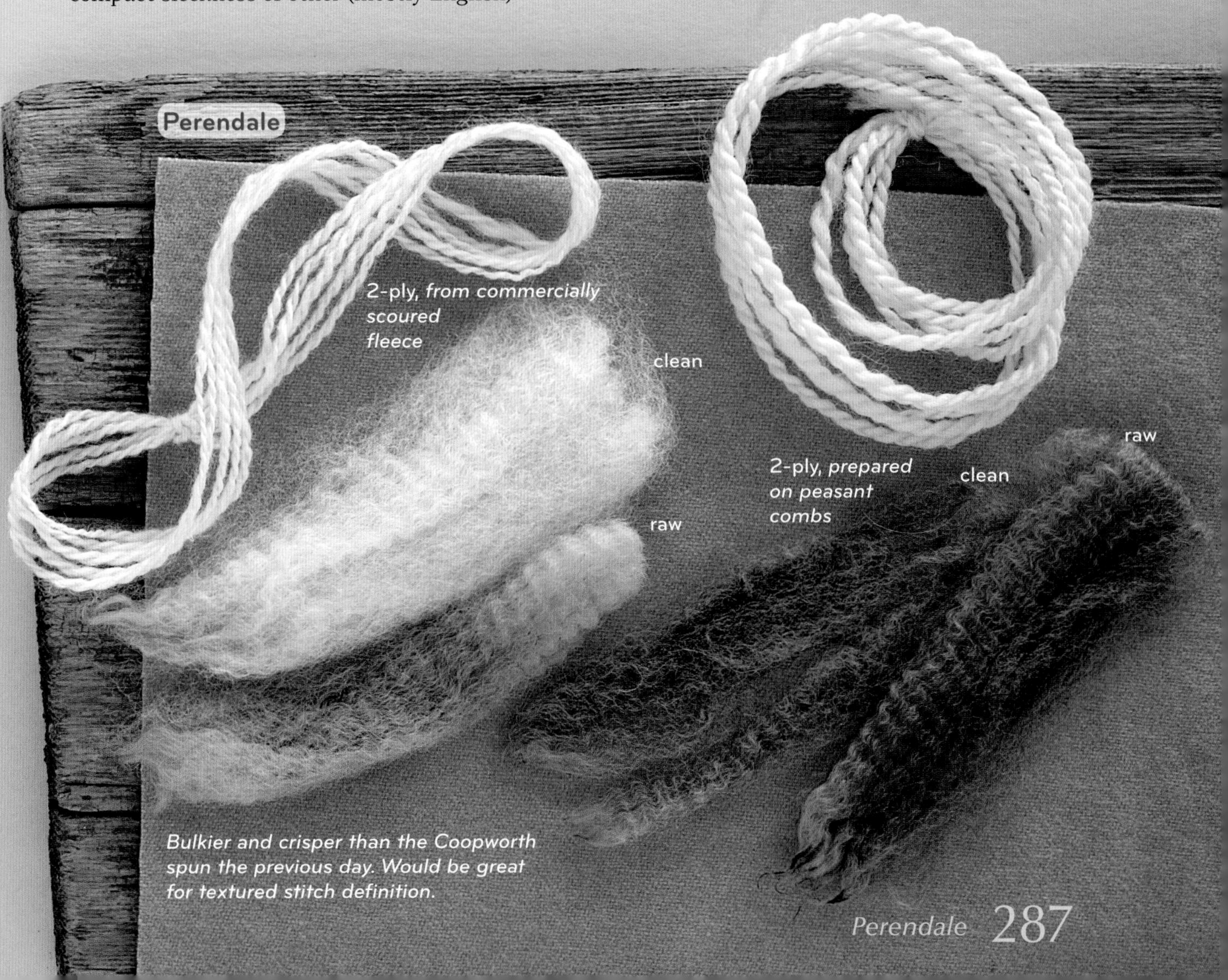

Perendale

2-ply, from commercially scoured fleece

clean

raw

2-ply, prepared on peasant combs

clean

raw

Bulkier and crisper than the Coopworth spun the previous day. Would be great for textured stitch definition.

Perendale Facts

▶ **FLEECE WEIGHT**
7½–11 pounds (3.4–5 kg)

▶ **STAPLE LENGTH**
4–6 inches (10–15 cm)

▶ **FIBER DIAMETERS**
Generally 28–35 microns (spinning counts 44s–56s), with New Zealand breeding aiming for 30–37 microns (spinning counts 40s–50s)

▶ **LOCK CHARACTERISTICS**
The fibers show clear, even crimp, and the staples usually do as well. Low luster, with no kemp or black fibers.

▶ **NATURAL COLORS**
Generally white, although there ar[illegible] colored flocks.

Perendale yarn from Local Harvest/Currow Hill Farm

Using Perendale Fiber

Dyeing. This fleece will take colors well, although without the shiny clarity of other longwools.

Fiber preparation and spinning tips. The long staples immediately suggest spinning from the lock, flicking, or combing, although with shorter staples (naturally short or cut in half), carding is an option. Perendale is easy and pleasant to spin.

Knitting, crocheting, and weaving. The fiber tends to capture air and bounce, so even full worsted spinning will give you loft in the finished yarn. The yarn's slightly crisp quality should help texture patterns show up nicely in knitted or woven textiles.

Best known for. Nice, even crimp; open staples; yarns with light weight and warmth for their size, because they are high bulk.

Polwarth

Conservation Breed

NAMED FOR A COUNTY in southwestern Australia where the breed was developed, Polwarths are not found in North America. They are still common in Australia and New Zealand and have found their way to South America, where they are known as Ideals. Along with Corriedales (see page 250) and Romneys (see page 110), they are common in the Falkland Islands. Polwarths were developed by breeding Merino rams to Merino/Lincoln ewes, with a goal of improving meat production while maintaining high-quality fleeces, until the cross bred true to itself.

Polwarth fleeces are very even, soft-feeling, and lovely to work with, no matter where you encounter them: as raw wool, clean fiber ready for spinning, or prepared yarns. They're not "beginner" wools, because they are so fine, yet they can be an easier introduction to the fine wools than Merino (see page 135) or Rambouillet (see page 148) because of their generous lengths. Note that Polwarth puffs up immediately upon washing. Some fleeces become less organized in the staple in the transition from raw to clean (when the grease is removed, the fibers spring apart), but all the crimp is still there.

These are superb next-to-the-skin fibers, appropriate for camisoles, neck snugglers, and other items with lots of body contact. Treat them as delicates and enjoy. With enough twist in the yarn, the relatively coarser Polwarths (not really coarse at all on the all-wools scale) can be reasonably durable for nice everyday wear — fine for walking the dog or staying warm while snowshoeing, but not so much for yard work.

Polwarth Facts

▶ **FLEECE WEIGHT**
Usually 9–13 pounds (4.1–5.9 kg), although recent breeding efforts are moving some fleece weights toward (and perhaps beyond) 15 pounds (6.8 kg); yield about 75 percent, which is very high for a wool this fine

▶ **STAPLE LE[illegible]GTH**
3–7 inches (7.[illegible]–18 cm), mostly 4–6 inches (10–15 cm)

▶ **FIBER DIA[illegible]ET[illegible]RS**
21–26 micr[illegible]s[illegible]inn[illegible]ing counts 58s–64s), mostly in the 23–[illegible]icron range (spinning counts 58s–62s)

▶ **LO[illegible]RACTERISTICS**
De[illegible]ular staples with flat or very slightly p[illegible]nd well-defined crimp. The crimp may [illegible]lightly more open.

▶ **[illegible]RAL COLORS**
[illegible]lver grays to blacks; tans to dark browns.

Using Polwarth Fiber

Dyeing. The clear whites take dyed shades well, and the other natural colors can be overdyed for more subtle effects.

Fiber preparation and spinning tips. Removing the grease from Polwarth, as from other fine wools, requires hot water and a good cleansing solution; don't let the washing solutions cool off, because the dissolved grease will then be redeposited on the wool. Spin from the lock, flick, comb (a great choice), or card if you have the shorter fiber lengths. In the winter, a spritz of water can help control static during prep. You'll be balancing softness (less twist) and durability (more twist) when you fit your spinning method to your desired end result. For durability, spin fine and use enough twist. In winter, smooth hands help: Greasy hand cream won't be a plus, so use a light lotion, apply it frequently, and make sure it's well absorbed by your skin before you start to spin.

Knitting, crocheting, and weaving. This fleece is definitely suitable for next-to-the-skin and baby garments, although with a bit more heartiness than Cormo. This fine, soft, long-stapled wool is a delight to spin, knit, or weave. The fleece is elastic, resilient, and lofty even when spun worsted. It is known for good draping qualities.

Best known for. Softness, fiber length, bounce, and drape.

Polwarth
2-ply
2-ply
clean
clean
raw
raw
The brown was especially soft.
The crimp on the black was not as fine as on the other samples.
clean
raw
2-ply
The white samples all came from the Falkland Islands. All felt very fine and were evenly crimped. The yarns bloomed after washing.

Polypay

Two brothers, Reed Hulet, an Idaho shepherd, and Charles Hulet, a researcher at the USDA Sheep Experimentation Station in Dubois, Idaho, set out to develop a new breed of sheep in the early 1970s. Their goal was to come up with a breed that would be prolific, regardless of whether it was raised on the range or on farm pasture. They crossed Targhees to Dorsets and Rambouillets to Finnsheep. The crossbred offspring were then recrossed, yielding an animal with 25 percent of each parent breed. The Finnsheep contributed high prolificacy, early puberty, and short gestation; the Rambouillets contributed hardiness and adaptability to range and improved pasture; the Targhees contributed superior fleece quality, large body size, and long breeding season; and the Dorsets contributed superior mothering ability, carcass quality, early puberty, and long breeding season. The four-way cross did perform exceptionally well, so the Hulets coined the name Polypay in 1975, from *poly*, which means "many," and *pay*, referring to the potential for the breed to increase a shepherd's return on investment and labor.

The breed has been developed with significant attention to wool quality for industrial processing, which means the aim is for relatively fine, high-yielding fleeces with staple lengths that can be easily accommodated by mechanical equipment — and those goals are nicely met. But when you find fleeces for handspinning there's still some variability, especially in the general feel, which can be anywhere between sleek and crisp. People who work with commercially prepared Polypay yarns will find a product that hovers around standard knitting-worsted-style yarns, with a nudging toward the fine end of fibers used for that purpose. In mechanically processed rovings and other spinners' preparations, the fine wool of the breed may have a tendency to form neps. Hand-washed and prepared spinning fiber may be noticeably livelier and smoother.

Polpay Facts

▶ **FLEECE WEIGHT**
7–11 pounds (3.2–5 kg); yield 50–60 percent

▶ **STAPLE LENGTH**
3–5 inches (7.5–12.5 cm)

▶ **FIBER DIAMETERS**
The American Polypay Sheep Association emphasizes qualitative wool standards, while specifying spinning counts of 54s–62s (translates to approximately 22–29 microns)

▶ **LOCK CHARACTERISTICS**
Locks can be separated out easily for combing, flicking, or other preparation methods by grasping the slightly tapered tips and gently pulling. The crimp is well developed and readily evident, although it may be partially disorganized.

▶ **NATURAL COLORS**
White.

Using Polypay Fiber

Dyeing. The bright or very slightly creamy white takes colors cleanly.

Fiber preparation and spinning tips. The fiber is fine enough to require a light and somewhat experienced hand, especially in carding. Combing will be the quickest route to consistent, nep-free preparation. It can also be spun from the lock. Keep the handling very light when you spin; otherwise the fiber tends to alternately clump and thin out.

Knitting, crocheting, and weaving. The yarn is great for blankets, sweaters, hats, mittens — a super combination of softness and durability. It has enough body to show texture well, while remaining supple enough for remarkable wearing comfort. It is also good for color patterning.

Best known for. Excellent wool in a breed equally maintained for meat.

Polypay

2-ply

clean

raw

2-ply

roving

Raw fiber and roving came from the same flock. Good combination of body and softness.

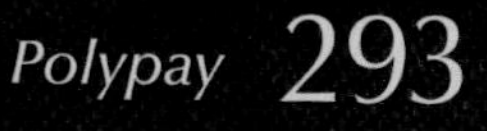

Portland

Conservation Breed

Named for the Isle of Portland in the English Channel, these sheep are a small and very old breed, thought to be closely related to, and representative of, the ancestral sheep that gave rise to the Dorset Horn. The Isle of Portland was once a true island, and its sheep retained genetic integrity by being separated from the mainland, though today this bit of land is tied to the southern coast of England by a spit of gravel known as Chesil Bank or Chesil Beach. In the 1970s, Portlands were almost extinct, with fewer than 100 breeding animals remaining in a half-dozen flocks, but today the breed has rebounded nicely, with over 1,500 breeding animals in more than 100 flocks.

Portland lambs are born with reddish brown wool, and some red, coarse kemp hairs may be found in the upper leg area of a shorn adult fleece. The lambs' wool lightens quickly to a warm, clear white called *creamy*, which is one of the consistent characteristics of the wool; the animals, however, retain tannish red faces with lighter patches around their eyes that give them a bespectacled appearance, and tan to brownish red legs.

Portland wool is described by different people as fine, medium, or coarse. In terms of lock formation, the U.K. Portland Sheep Breeders Group describes the wool as being close, fine, and short stapled, and other sources call the locks blocky with square tips. Our first sample, from a top-notch flock, was very long stapled (twice the expected length), and its locks were quite open, with pointed tips. A second sample, obtained from a different source, was similar in quality and at the generous end of the predicted length. What gives? Our guesstimate: The Portland is an old breed, one that has not been as influenced by breeding adjustments for modern production values in terms of either wool or meat, so there is a good bit of natural variability. The remarkable length of our samples may be the result of unusually good nutrition, or time elapsed since the previous shearing.

The general micron counts assigned to the wool are in the 26 to 31 range, which means some of the fleeces will be fine for use next to the skin, and the wool is also described as suitable for hosiery (socks, that is) and hand-knitting yarns: Think of it as good for making all-purpose worsted-weight yarns, the versatile standby of many knitters over the past 50-plus years.

In spinning, both of our samples were a joy, with locks that were easy to tease open, to comb, and to spin. They showed excellent length, with distinct but irregular crimp in the fibers, and a structure to the staples that varied between wavy and somewhat disorganized.

Portland Facts

▶ **FLEECE WEIGHT**
4½–6½ pounds (2–3 kg)

▶ **STAPLE LENGTH**
2½–4 inches (6.5–10 cm)

▶ **FIBER DIAMETERS**
26–31 microns (spinning counts 50s–56s)

▶ **LOCK CHARACTERISTICS**
Variable. There can be some red kemp in the britch area (upper legs and buttocks).

▶ **NATURAL COLORS**
Creamy white; lambs are born red, and their coats shift to white (or perhaps gray) within a few months.

Using Portland Fiber

Dyeing. The warm tone of the white will affect overlying dye colors and should do so in a pleasant way.

Fiber preparation and spinning tips. Our samples were more like American Tunis (see page 226) and California Red (see page 233) than we had any reason to expect. They were easy and pleasant to spin.

Knitting, crocheting, and weaving. This is a versatile, multipurpose fiber, suited for making sweaters, socks, caps, blankets, and other textiles that are in everyday use.

Best known for. Being a historic breed with a mild-tasting meat that was favored by British royalty, and a fleece that is interesting to work with.

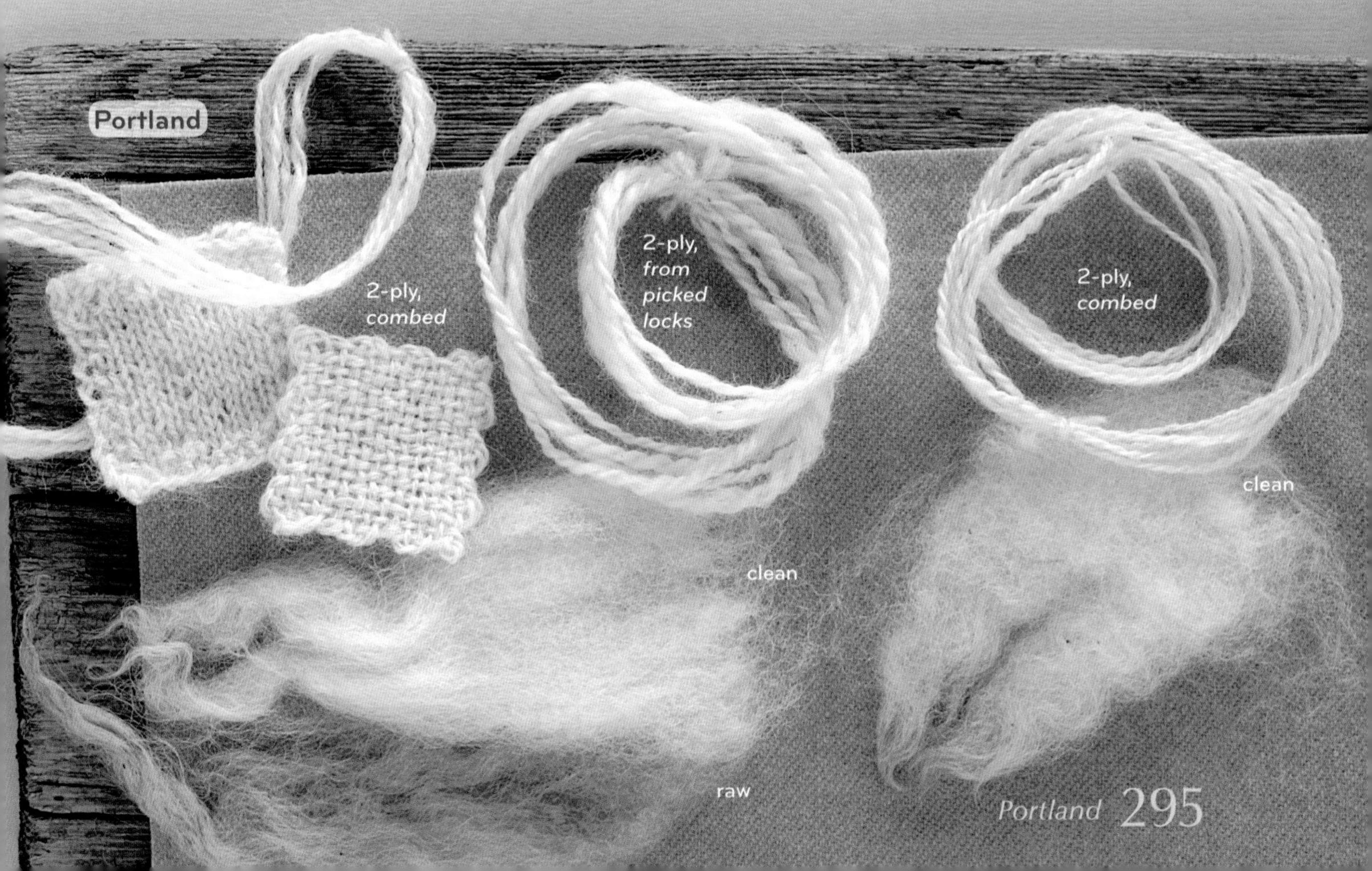

Romeldale and CVM

Critical Conservation Breed

THE ROMELDALES — a cross of Romney with Rambouillet — were developed starting in 1915, when A. T. Spencer, a California sheep man, purchased New Zealand Romney rams being exhibited at the Pan-American Exhibition in San Francisco to breed to his Rambouillet ewes. Only a handful of breeders in California were ever active with the breed, and the main population for many decades was the J. K. Sexton family's Stone Valley Ranch flock in Glenn County, California, the entire clip from which was sold to Pendleton Woolen Mills. Pendleton, like all large-scale manufacturers, frowned upon even single fibers of random colors showing up in their wool, so the occasional colored lamb was instantly culled. The Sextons partnered in developing the breed with a neighbor, shepherd Glen Eidman. In the 1960s, two badger-faced colored lambs — one ewe and one ram — appeared in the Romeldale flock. Eidman, intrigued, bred these sheep through a number of generations, focusing on both color and fleece quality, and called the resulting animals California Variegated Mutants, or CVMs.

Since selling colored fleece to the Pendleton mill was out of the question, Eidman began marketing his CVMs to handspinners, who welcomed the colors, as well as the fineness, length, and crimp of the fiber (also present in the white Romeldales). As a result, additional colored Romeldale lambs were not automatically culled, and owners of Romeldale flocks began to sell more frequently to handspinners and to cultivate a full range of natural brown, gray, and black fleeces, in addition to white. Interestingly, there are sixth-generation Sexton family members still raising Romeldales, now in eastern Oregon.

California Variegated Mutant (CVM)

Some breeders considered the CVMs to simply be colored strains of Romeldales, while others considered CVMs to be a distinct breed. There are some pros and cons to either view, but both the Romeldale and the CVM populations were dwindling, and neither had a really active breed society promoting the animals, so they continued to lose ground until both were classified as critically rare by the American Livestock Breeds Conservancy (ALBC). The staff at ALBC helped form a registry and association that combines both colored and white Romeldales. Under the registry's rules, an animal can be registered as a white Romeldale, a naturally colored Romeldale, or a CVM Romeldale, and flocks can consist of a mix of these types. To be a CVM, a light-colored sheep must have at least five of the dark markings associated with badger-faced sheep, including markings on the face (around the eyes, on the muzzle, and/or stripes down the sides), and on the underbelly, legs, chest (from chin to underbelly), or the area under the tail. (Within the natural-colored Romeldale group, there's also a reverse badger-faced pattern, where the animal is predominantly dark and the marking areas are light.)

Between them, the Romeldale and the CVM offer a multitude of solid natural colors and of variegated fleeces, sometimes spotted or striped. In some animals, the colors change along the length of the wool staple as well as throughout the fleece. This variety of color within a single fleece is one of the unusual qualities of the wool. One more point about the color: Almost all other breeds' lambs have their darkest wool when they're born, lightening up with time, and sheep's wool also ordinarily becomes coarser as the years pass. Romeldales and CVMs are unique in that their wool becomes darker (for the colored strains) and finer with age.

The wool is consistently soft, long stapled, and uniform as to fiber quality within a single fleece, and it does felt. These wools are good selections for next-to-the-skin garments and for knitters who want soft, lofty yarns. They provide a good introduction to fine wools for spinners who may be tentative about working with fine fibers.

This is what we think of as classic Romeldale: soft, white, and with nice crimp producing resilience.

Romeldale

varied gray

soft brown

2-ply

2-ply

clean

clean

Surprise! Romeldale also comes in natural colors. Because the wool was initially sold to high-quality mills, colors were not cultivated, except in CVM flocks.

raw

raw

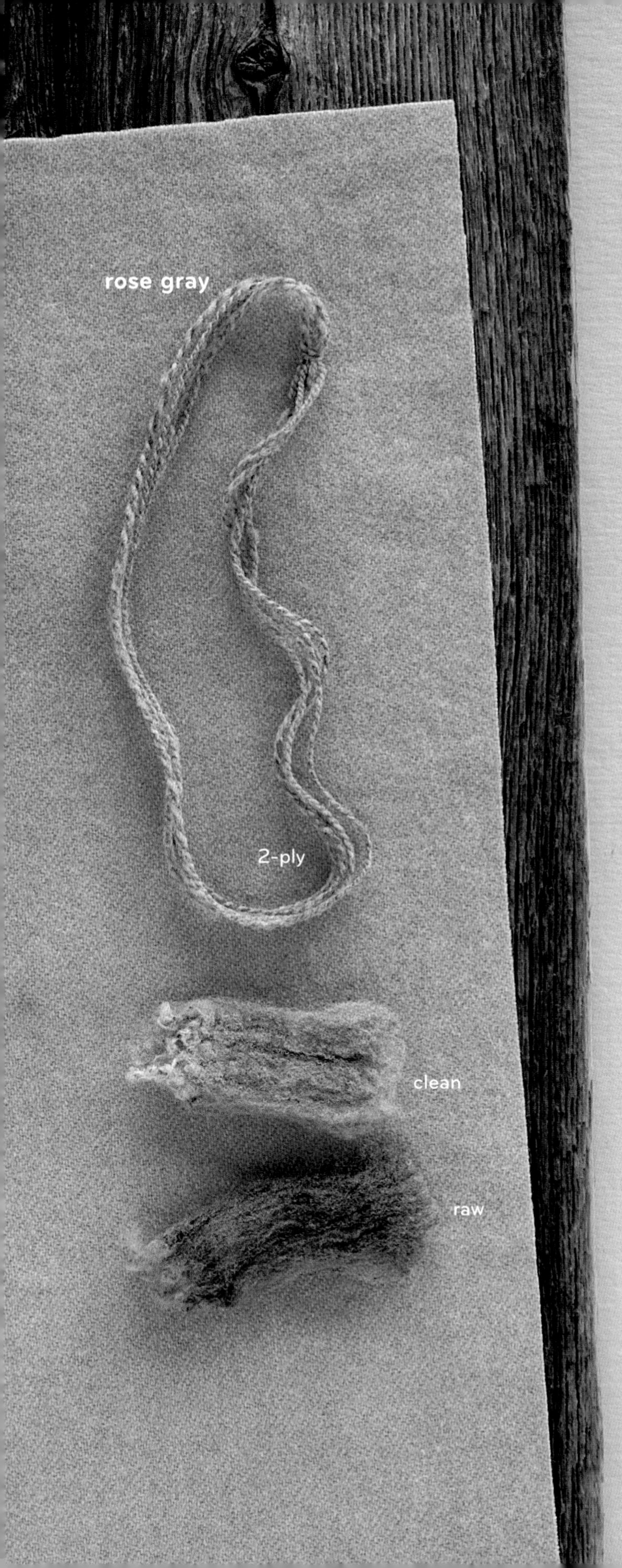

Romeldale and CVM Facts

▶ FLEECE WEIGHT

6–15 pounds (2.7–6.8 kg); the American Romeldale/CVM Association breed standard calls for 6–12 pounds (2.7–5.4 kg), but other sources suggest shearing weights of 10–15 pounds (4.5–6.8 kg); yield 60–65 percent, the latter by breed standard

▶ STAPLE LENGTH

3–6 inches (7.5–15 cm)

▶ FIBER DIAMETERS

The breed standard calls for spinning counts of 60s–64s, which roughly translates to 21–25 microns

▶ LOCK CHARACTERISTICS

Dense, soft, nicely crimped from base to tip of each fiber. Locks have flat or minimally tapered tips. No kemp or hair.

▶ NATURAL COLORS

Romeldales can be white, or shades of reddish to clear brown (dark to light, including both reddish moorit shades and other browns), and a full span of grays through blacks, either mixed or solid. CVMs are multicolored with specific color patterning (described on pages 296–297); the base color is often cream or gray.

2-ply
2-ply
2-ply
2-ply
clean
CVM (California Variegated Mutant)
Different fleece colors may vary in length and texture (as at left)
clean
raw
lamb
raw

Using Romeldale and CVM Fiber

Dyeing. The whites will dye well (with a matte finish), and the other colors, although normally used in their natural state, can be overdyed for more subtle effects.

Fiber preparation and spinning tips. Depending on length, locks can be spun without preparation, flicked, combed, or carded. Because of the crimp and lively quality of the fiber, even combed preparations will have a lot of loft and elasticity. Because of the relatively long average fiber lengths for a fine fiber, in some cases (the longer staples) lower-twist yarns will have enough integrity to function well (although durability will decrease with lower twist levels). Keep a light touch, and remember that the finished yarn will have more loft than the freshly spun yarn promises, because of the crimp and elasticity.

Knitting, crocheting, and weaving. Romeldale and CVM are soft fibers, with fiber diameters approximately equivalent to the upper ends of the Merino and Rambouillet ranges. They are suitable for next-to-the-skin clothing, light-duty outerwear, blankets, and other cuddly items. The colors — both the broad array and the way they mix in the fleeces — offer many of the aesthetic pleasures of working with Jacob (see page 271) fleece, translated to a range of applications that do best with a softer wool.

Best known for. Fine, soft wool of a good length for many preparation methods in an array of natural colors. Both Romeldale and CVM are rewarding and comfortable to spin.

Rouge de l'Ouest

THE ROUGE DE l'Ouest ("Red of the West" in English) developed in the Loire province of western France when English Longwools (particularly Wensleydales and Bluefaced Leicesters) were bred to the native landrace sheep in order to produce a dairy breed; today, however, these sheep are mainly raised for their meat. In Britain the breed is simply called the British Rouge.

The wool of the Rouge is described in terms of its ability to protect the sheep, not its value for the shepherd or textile artisan. It's plenty fine, but very short (although sometimes described as medium, and it's not clear whether this refers to fiber diameter, length, or both) — hard to understand how this breed managed to end up with such short staples when there are English Longwools in its heritage!

Although the Rouge and the Bleu du Maine (see page 229) come from similar breeding and geographic origins, the Rouge fleeces are about half as long in the staple, half as large in the weight shorn per animal, and finer fibered. Although its fiber diameter is within the range of strong Merinos, Rouge de l'Ouest wool has a crisp hand.

Rouge de l'Ouest Facts

▶ **FLEECE WEIGHT**
3½–6½ pounds (1.6–3 kg)

▶ **STAPLE LENGTH**
1½–2 inches (3.8–5 cm)

▶ **FIBER DIAMETERS**
Approximately 23–26 microns (extrapolated from Bradford grades of 58s–60s)

▶ **LOCK CHARACTERISTICS**
Dense and rectangular.

▶ **NATURAL COLORS**
White.

Using Rouge de l'Ouest Fiber

Dyeing. The fleece gives clear colors with a matte finish.

Fiber preparation and spinning tips. Like the Charollais, the Rouge de l'Ouest is shorn for practical reasons, not for spinners' purposes, and second cuts may cause neps — already a risk in a fine wool that is also very short. The shearer doesn't have much maneuvering room, and it's better to nick the wool than the sheep. Carding is the most reasonable preparation method, but resign yourself to a slightly imperfect job because of the pragmatic shearing. If you've got enough fiber length and a pair of mini combs, you can clear out the second cuts that way. Plan on spinning fine, with enough twist to hold the short fibers together and enhance the durability of the yarn.

Knitting, crocheting, and weaving. The fleece is suitable for next-to-the-skin garments, but with a crisp, rather than soft, feel.

Best known for. Short-stapled, high-crimp, fine fibers that aren't, if our samples are representative, cushy.

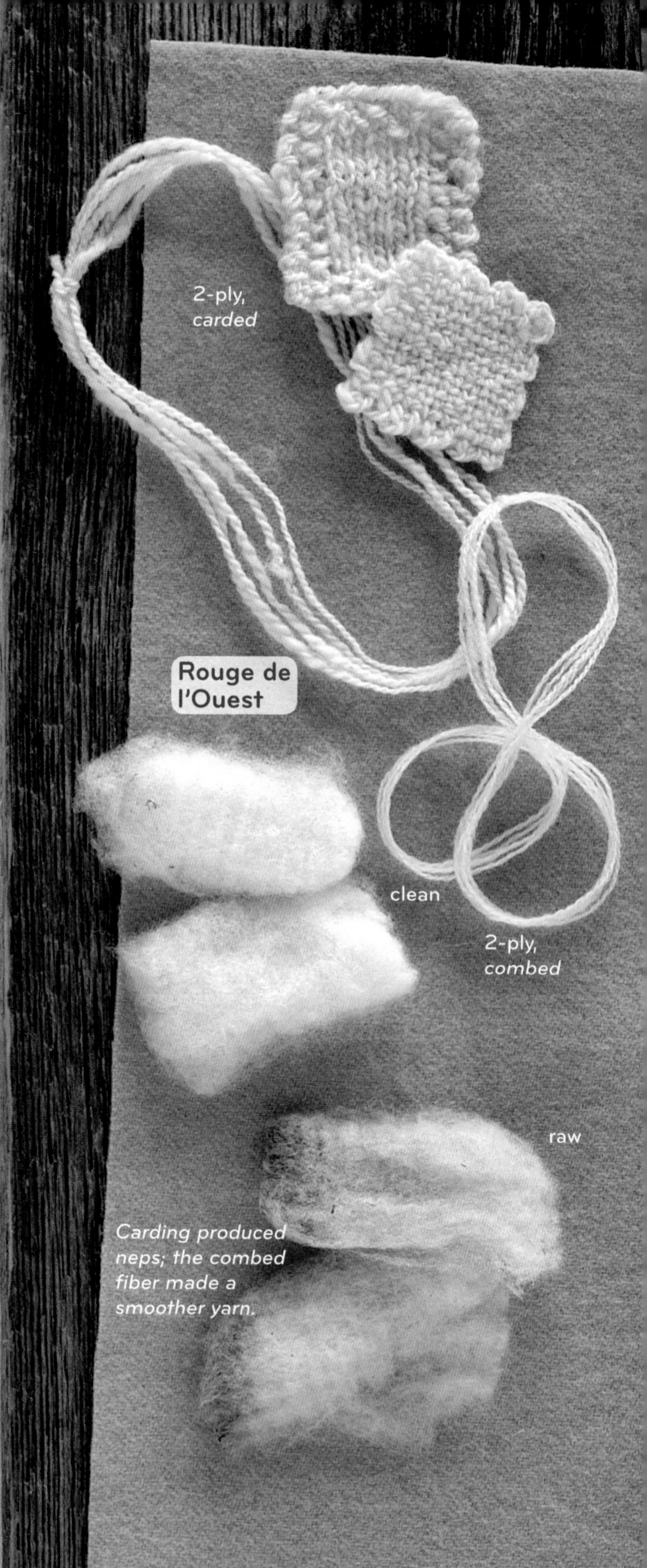

2-ply, *carded*

Rouge de l'Ouest

clean

2-ply, *combed*

raw

Carding produced neps; the combed fiber made a smoother yarn.

Ryeland

Conservation Breed

THE RYELAND IS SAID to be named for the ryegrass pasture it grazed upon in Herefordshire, where documentation indicates the breed was raised by monks as early as the 1300s. As one of the oldest of British sheep breeds, the Ryeland originally had one of the finest wools in the British Isles — said to be so fine that it rivaled that of the Spanish Merinos. In fact some historians suspect that the Ryeland descended from the same ancestral stock, brought to Britain by the Romans, that led to the development of the Spanish Merino on the continent; but as with many sheep breeds, future DNA analysis may be the only way to answer our questions about the breed's origin.

In the sixteenth century, Queen Elizabeth I was given some Ryeland stockings, which spoiled her for stockings made from other wools. By the late eighteenth century, Ryelands lost some of their original exceptionally fine fleece quality when they were crossed with other breeds in order to increase their meat value. Modern Ryeland breeding seems to have branched in two directions. Ryelands in the British Isles and Australia more closely resemble the original source breed, while those in New Zealand produce heavier, and in some cases coarser, fleeces.

Contemporary Ryelands are still relatively small sheep, with ewes averaging 141 pounds (64 kg) and rams 192 pounds (87 kg). Despite the loss in wool quality from their legendary early days, they grow an exceptionally fine, soft, and fluffy fleece, perhaps the quintessential candidate for woolen spinning, which will emphasize the fiber's unusual loft and lightness. Ryeland makes a very springy yarn with good elasticity. Woolen-spun Ryeland yarn brings forth dreams of nurturing and cuddling: cozy sweaters you never want to take off, light blankets that make a bed into a nest, hats that surround a head like a halo. Ryeland can be spun worsted style, too, and yarns formed that way evince the same qualities in subtler forms. It does not felt readily, a fact to keep in mind as a plus for garments that will be worn and washed frequently.

Ryeland Facts

▶ **FLEECE WEIGHT**

Great Britain and Australia 4½–6½ pounds (2–3 kg)

New Zealand 6½–9 pounds (3–4.1 kg)

▶ **STAPLE LENGTH**

2–5 inches (5–12.5 cm), mostly 3–4 inches (7.5–10 cm)

▶ **FIBER DIAMETERS**

Great Britain and Australia 25–28 microns (spinning counts 56s–58s)

New Zealand 26–32 microns (spinning counts variously described as 50s–56s and 56s–58s, but equivalent is roughly high 48s–low 58s)

▶ **LOCK CHARACTERISTICS**

Dense, blocky staples. May have very short, pointed tips. Abundant fine crimp, but disorganized, unlike Merino.

▶ **NATURAL COLORS**

White. British Ryelands have a colored gene (for black or shades of gray and also browns) as well; Australian Ryelands do not.

clean
raw
clean
2-ply,
combed
Ryeland
Both white samples were much shorter than expected for the breed, but the one on the upper right combed up nicely and made a soft yarn with presence: classic for warm winter woolies.
2-ply,
carded
Combed to remove as much vegetable matter as possible, then carded to make a light, insulating yarn.
clean
raw

Using Ryeland Fiber

Dyeing. The fleece colors nicely, with a matte finish.

Fiber preparation and spinning tips. Carding makes the most of the fiber's lightness, loft, and elasticity. Fleeces with longer staples can be combed effectively. Ryeland can also be spun directly from the washed, then picked or teased, locks. The tips may cling together and need a bit of extra attention. Woolen-style preparation and spinning seem to be made for Ryeland. However, worsted-style processing will yield a lightweight yarn with a smoother surface, more compact feel, and increased durability.

Knitting, crocheting, and weaving. Ryeland offers an abundance of lightweight warmth. When choosing a gauge or sett, be sure to give the yarn a little room to breathe.

Best known for. Lightness and softness.

Ryeland yarns from Garthenor

Targhee

IN 1926, RESEARCHERS at the USDA Sheep Experiment Station in Dubois, Idaho, began breeding Rambouillet rams to Corriedale and Lincoln/Rambouillet ewes, quickly backcrossing the offspring to each other. Their goal was to produce an all-around, dual-purpose sheep that would thrive on range and farm operations in the states of the West and the High Plains. They named their new breed Targhee, after the national forest where the station's sheep grazed in the summer months. A large-framed animal, the Targhee indeed has superb production traits for both meat and fleece.

From its fine-wool ancestors, the Targhee gets fine fiber diameters, which means its wool is very soft, yet its handling qualities feel like they come more from its Longwool forebears. The fiber diameters overlap those of medium to so-called strong Merinos and of many Rambouillets. Targhee wool has loft and good elasticity, although of the sort that makes it lively and supple rather than springy, and it lends itself to the sorts of fabrics you'll want to wrap yourself in for both softness and a bit of elegance.

On the sheep, Targhee is a dense, uniform wool; in the hand, it feels a good deal airier than that description makes it sound. It's quite consistent throughout a single fleece and within the breed. It also felts.

Targhee Facts

▶ **FLEECE WEIGHT**
10–14 pounds (4.5–6.4 kg) for ewes, 16–22 pounds (7.3–10 kg) for rams; yield 45–55 percent

▶ **STAPLE LENGTH**
3–5 inches (7.5–12.5 cm)

▶ **FIBER DIAMETERS**
According to the breed standard, 22–25 microns (spinning counts 60s–62s), with occasional forays into slightly finer territory (say, 21 microns, spinning count 64s) and up to a maximum of just under 28 microns (spinning count low 54s or high 56s) as the coarsest fiber on the britch

▶ **LOCK CHARACTERISTICS**
Dense, uniform, matte locks, blocky but sometimes with slightly pointed tips. A whole lot of crimp, which looks disorganized in the individual fibers and semiorganized in the locks.

▶ **NATURAL COLORS**
Predominantly white, because of the commercial nature of the breed's economic positioning.

Targhee yarns from Sweet Grass Wool

Targhee socks from Sweet Grass Wool

Using Targhee Fiber

Dyeing. Colors take well, while the fiber's matte surface gives them a soft quality.

Fiber preparation and spinning tips. Depending on length, the fleece can be combed or carded. Spin from the locks, flicked or not, although the fineness of the fiber may make these approaches harder than other preparation techniques. Use fine carding cloth on hand carders or a drum carder, and handle gently to avoid producing neps. Use a light touch in spinning to keep the fibers flowing smoothly, and put in enough twist to give the yarn stability and durability.

Knitting, crocheting, and weaving. This sweet, soft wool has the potential to produce "everyday luxuries," textiles that feel like an indulgence and yet have enough stamina to keep up with many people's normal activities (unless you spend your days mucking out barns, in which case you'll want a tougher wool for that activity). The bounce of the yarn means fabrics will be cozy and resilient.

Best known for. Heavy fleeces of fine, nicely soft, white wool, with excellent loft and elasticity.

Chief Targhee

The Targhee sheep are named for the national forest where they grazed in the summer, but how did the national forest, in turn, get its name? The Caribou-Targhee National Forest, located primarily in eastern Idaho, western Wyoming, and northern Utah, is the gateway to Yellowstone and Grand Teton National Parks. Established by President Theodore Roosevelt in 1908, the Targhee National Forest was named in honor of Chief Targhee, a Bannock Indian warrior who negotiated one of the earliest treaties with the United States government. The Shoshone-Bannock Tribe still maintains ancestral treaty rights to use of the Caribou-Targhee National Forest.

Texel

THESE SHEEP were named for the island of Texel, off the coast of the Netherlands, where the native sheep were crossed with Lincoln and Leicester Longwools in the mid-1800s to develop the modern Texel. The goal was to create a high-quality meat breed. Texels were introduced to the United States in 1985 by the U.S. Department of Agriculture Meat Animal Research Center (MARC), and some of the sheep were released to breeders after a five-year quarantine period. The private breeders have since arranged for additional importations of live animals, embryos, and semen.

Because Texels are still thought of as primarily meat animals, the wool is definitely considered a secondary crop. The breeders' societies offer little information on wool quality except to mention "high bulk" and a rough description based on the fairly outdated Bradford system. As with all wool clips destined for industrial processing, any black fibers are unacceptable.

Nonetheless, Texels produce a nice, matte, white fleece that can be used for everyday textiles — including socks, sweaters, outerwear, and blankets — at the finer end of the breed's range, or rugs at its coarser end. Medium is the name of the game here. Shorter-stapled fleeces can be carded, and longer ones can be combed, although the lofty quality means that even worsted-processed yarns will have better-than-average insulating qualities (they'll trap air and retain warmth) and will not look anywhere as sleek as a worsted made from one of the long, lustrous wools.

Blue Texels, which have steel-gray to almost black fleeces and badger-patterned faces, are recognized as a separate breed in the Netherlands; a few Blue Texels (and many white ones) are found in the British Isles. Staple lengths and fleece weights are similar to those of regular Texels. Fiber diameters are given as 27 to 28 microns (spinning counts 54s–56s).

Texel Facts

▶ **FLEECE WEIGHT**
7–12 pounds (3.2–5.4 kg)

▶ **STAPLE LENGTH**
3–6 inches (7.5–15 cm)

▶ **FIBER DIAMETERS**
28–33 microns (spinning counts 46s–54s); some may be as fine as 26 microns (spinning count 56s) and Australian Texels may be coarser, in the range of 30–36 microns (spinning counts 44s–56s)

▶ **LOCK CHARACTERISTICS**
Not much luster; there may be some kemp fibers. Locks are springy and a little crisp feeling. The crimp is partially organized in both fiber and lock.

▶ **NATURAL COLORS**
White, for the benefit of mass processing; see note on previous page on Blue Texels, a separate breed.

Using Texel Fiber

Dyeing. The fleece will take colors clearly, but the low luster means they won't glisten.

Fiber preparation and spinning tips. This is a medium wool willing to be prepared any way you want to use it. If you've obtained your wool from a flock that is grown primarily for meat, you may need to put extra attention into washing and prepare the wool with a method that encourages vegetable matter to drop out (combing works well for this). It is easy enough to handle in spinning, as long as you remember that there will be some loft to the yarn even if you work it strictly worsted style.

Knitting, crocheting, and weaving. Texel is flexible in a variety of techniques, depending on the staple length and where on the fineness/coarseness scale a given batch of fiber or yarn falls. The high-bulk characteristic is the factor to keep in mind when designing: Consider color work or stockinette sweaters, cozy and very sturdy blankets, or mats and pillows that offer nice cushioning.

Best known for. A nice, solid wool from a meat-focused breed.

Whitefaced Woodland

Conservation Breed

THE WHITEFACED WOODLAND comes from the southern end of England's Pennine mountains, an area that is also home to a number of the Blackfaced Mountain breeds, such as the Derbyshire Gritstone (see page 40) and the Lonk (see page 42). Whitefaced Woodlands have been in this area for centuries. They were known earlier as Penistone sheep, the name having come from the town in South Yorkshire where general sheep sales began in 1699, when the market received its Royal Charter. In the twentieth century, the numbers of Whitefaced Woodlands plummeted, and by the 1970s the breed was nearly extinct. Following the breed's listing as endangered by the Rare Breeds Survival Trust, it has seen a good comeback but is still rare.

Having been raised primarily for meat, Whitefaced Woodlands lack consistency in their wool, although they are noted as growing one of the finest and softest fleeces from a hill breed. For textile purposes, you'll need to judge each yarn or fiber source by what you feel. Many references list the fineness as 28 to 31 microns (or their Bradford equivalent, 50s-54s). The Whitefaced Woodland Sheep Society gives a Bradford range of 44s to 50s; well-known British spinning author Mabel Ross's table of wool characteristics for handspinners lists 40s to 46s; and the two samples we sent for optical scanning came back with average fiber diameters of 34.7 (spinning count 44s) and 46.7 (spinning count "coarser than 36s," which means "coarser than almost everything in the wool world"). Michael Ryder, in *Sheep and Man*, confirms this broad range within the breed and calls the fleece from Whitefaced Woodlands a "hairy medium wool."

How to pick your way through the confusion? The finer and possibly predominant type of wool from Whitefaced Woodland is the sort you'll find in classic knitting-worsted-style yarns: serviceable, durable, and suited for use in socks, gloves, mittens, blankets, pillows, and sweaters to be worn over shirts. The stronger wools will work best for tough outerwear, rugs, and other textiles that need to stand up to serious abuse. The hairy sample we tested (and spun) that was off the scale for coarseness would make a fantastic, strong rug warp or an interesting addition of texture and color in textile artwork — or for floor pillows. Whitefaced Woodland wool does felt.

Whitefaced Woodland Facts

▶ **FLEECE WEIGHT**
4½–6½ pounds (2–3 kg)

▶ **STAPLE LENGTH**
3–8 inches (7.5–20.5 cm)

▶ **FIBER DIAMETERS**
28–38 microns, or coarser (spinning counts 40s–54s)

▶ **LOCK CHARACTERISTICS**
Somewhat open, with crimp patterns present and noticeable but varying according to fineness or coarseness, relatively organized in the fiber and only semicoordinated in the locks.

▶ **NATURAL COLORS**
White.

Using Whitefaced Woodland Fiber

Dyeing. The fleece will take colors nicely, except perhaps for some kemplike fibers at the coarsest end of the range.

Fiber preparation and spinning tips. Preparation and spinning approaches depend entirely on where in the fiber range your specific fleece falls.

Knitting, crocheting, and weaving. Depending on the fiber profile, you can produce anywhere from serviceable everyday garments to rugs, ropes, and indestructible bags.

Best known for. Varied range of wools, all white, all especially good for something in the hands of a creative person.

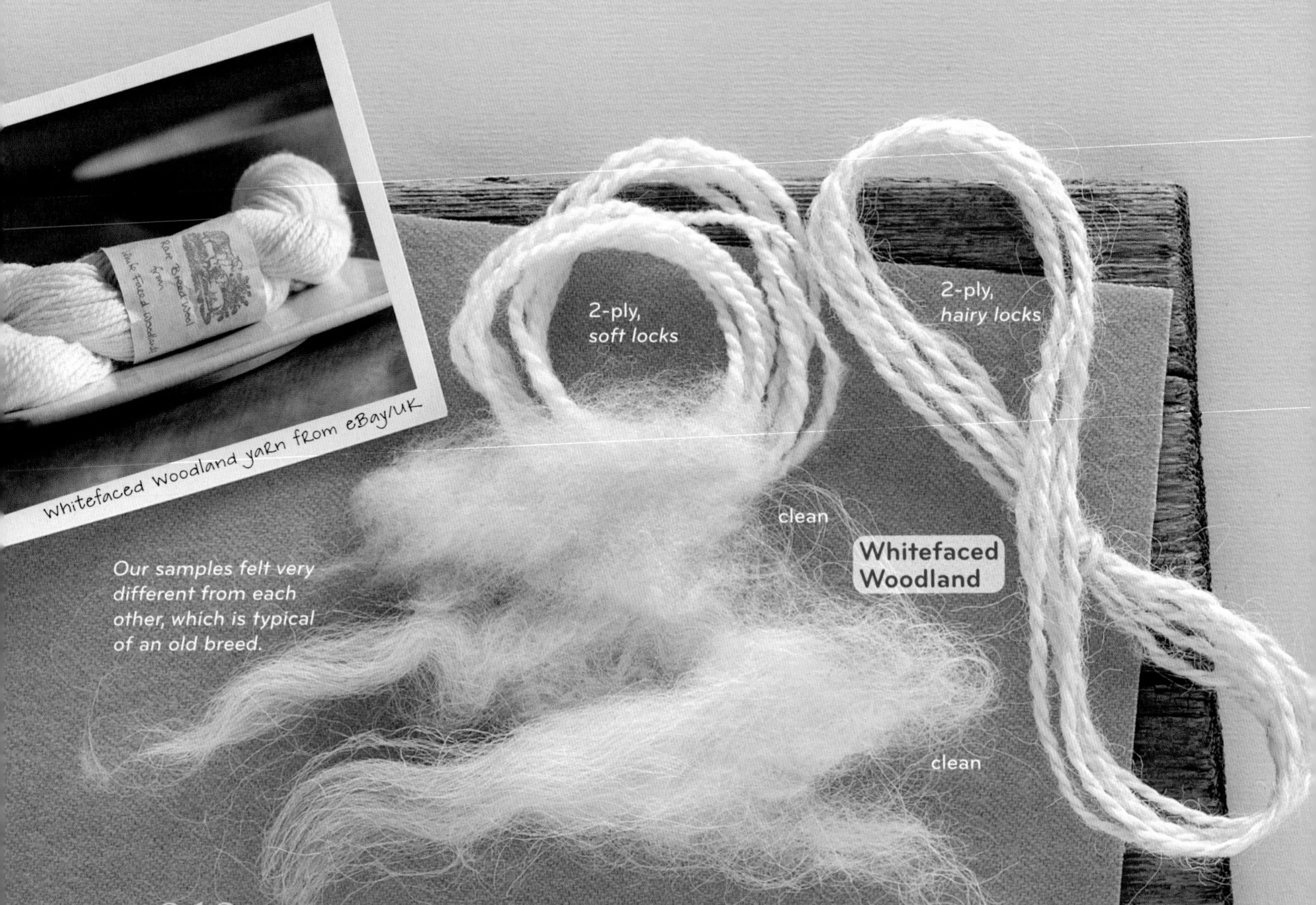

Whitefaced Woodland yarn from eBay/UK

Whitefaced Woodland

2-ply, *soft locks*

2-ply, *hairy locks*

clean

clean

Our samples felt very different from each other, which is typical of an old breed.

Zwartbles

The Zwartbles (now there's a mouthful) were developed from a cross between the two main types of native sheep in the Netherlands, the hornless and short-tailed Friesian milk sheep (see page 259) and the horned and hairy-fleeced Drenthe (outside the scope of this volume). See page 224 for a Zwartbles photo. Zwartbles sheep are traditionally used for milk and meat rather than wool. In the 1990s, British shepherds imported the breed, so now Zwartbles fiber is becoming more available.

Zwartbles wool is medium to fine with excellent crimp. It's unarguably springy, with an unusually dark quality to its black color.

For spinning and knitting, consider it a workaday wool, one that will make durable hats, mittens, gloves, and cardigans, as well as household textiles that need to combine strength with reasonable tactile appeal.

Definitely outerwear material.

Zwartbles Facts

▶ **FLEECE WEIGHT**
6½–10 pounds (3–4.5 kg)

▶ **STAPLE LENGTH**
4–5 inches (10–12.5 cm)

▶ **FIBER DIAMETERS**
Micron counts perhaps for the most part in the high 20s and low 30s (spinning counts 50s–56s). The one data source we located suggested an average of 27 microns (and spinning counts of 54s–56s). The sample we sent to the lab came back with an average fiber diameter of 35.4 microns, with sufficient consistency to qualify for a USDA grade of 44s.

▶ **LOCK CHARACTERISTICS**
Blocky, with well-developed, slightly jumbled crimp in both fiber and lock. Tips may be sun bleached, but because of the density of the fleece, the bleaching is likely not to extend far into the staple.

▶ **NATURAL COLORS**
Black, shading to brown, especially on the sun-bleached tips of the locks, with some silvery or white fibers, although the Zwartbles Sheep Association in Great Britain says that any white in the wool is undesirable.

Using Zwartbles Fiber

Dyeing. Technically, you can dye it. It is pretty pointless, though, because of the natural dark color.

Fiber preparation and spinning tips. The moderate length, relatively open lock formation, and medium fiber diameter mean you can select your favorite preparation technique. It is quite easy to handle when spinning. The biggest challenge (and it isn't much of one) is the bounce factor from the type of crimp.

Knitting, crocheting, and weaving. Zwartbles is suitable for durable, everyday garments and household textiles. The crisp hand will emphasize stitch and weave structure definition, and the springiness will make the fabric resilient.

Best known for. Intensely black wool.

Welsh Mule
Racka
Gute

Wider Circles of Sheep

We'd like to include all the fiber sheep in the world within these pages. That's hardly practical, considering the number of breeds in countries and continents we have either barely touched on or not had time (or space) to consider. Yet with increasing access through the Internet and travel, people interested in textiles can obtain and work with wool from an ever-widening number of sheep types. What's fascinating to us is that even with the immense variety of fibers we've experienced, fleeces with entirely new-to-us combinations of characteristics keep appearing in our lives.

This section hints at a few of the possibilities within that wider circle, still staying (for the most part) with the concept of breeds, rather than crossbreds or regional types that have not been formally recognized as breeds. Some of these wools will already be familiar to particular segments of the textiles-by-hand population, like the Spelsau and Pellsau, regularly used for many years by tapestry and rug weavers as well as felters. Others came to us while we were working on this project, through a combination of generosity and serendipity.

What we'd like to offer here is a quick glimpse of some sheep in territories adjacent to our primary area. These include breeds that English-speaking fiber people can — and do — regularly get their hands on, but that we found more challenging to obtain during our two years of gathering. In part, we're amazed, and we want to share with you that amazement, at how, after handling literally hundreds of samples of fleece, we can receive a box of wools at the last minute from another part of the world, and the fibers it contains open our eyes yet again. They're familiar in their overall qualities but totally unique in their specifics. It's almost magical.

So let's look at a few Canadian-developed breeds, some Scandinavian wools that in most cases still can be found only in their original countries (unlike the Finn, which has been moved out into colonies on other continents), one Hungarian treat, and a type of crossbred sheep that's so much a part of the British landscape that it warrants special attention. Our research into these sheep has of necessity been cursory, but we hope to give you at least a quick idea of where in the magnificent diversity of fibers these additional animals fit. In a couple of months, we're looking forward to seeing wool from some Finnish conservation breeds; we want to know more about the Lithuanian and Estonian and Latvian sheep; and we wish we could study other European breeds (many fascinating options), as well as the sheep of the Indian subcontinent and China. Basically, the list is long. We offer the following as a way of hinting at the nature of this ongoing journey of discovery.

The Arcotts

Canada. The Canadian government developed three breeds of sheep in the 1970s and '80s, each with the acronym Arcott (from Agricultural Research Centre of Ottawa) in its name:

- **The Canadian Arcott** was developed primarily from Île-de-France and Suffolk parents, though there was some Cheviot, Leicester, and Romnelet (a Romney/Rambouillet cross) blood added to the mix.
- **The Outaouais Arcott** was produced by crossing a number of breeds, predominantly Suffolk, Finnsheep, and Shropshire.
- **The Rideau Arcott** mainly used Finnsheep, Suffolk, Shropshire, Dorset, and East Friesian sheep.

In the late 1980s, the Canadian government opted to eliminate its sheep-breeding program. The above breeds, considered the three most successful results of the effort, were named and dispersed to Canadian shepherds in 1988.

The Arcotts are all raised primarily for meat production. Wool is definitely an afterthought in these breeds, but that certainly doesn't mean it can't or shouldn't be used. Fleeces on the Canadian and Outaouais Arcotts are called variable. The Rideau's fleece is described as medium (most likely referring to both length and fiber diameter) and runs about 8 to 10 pounds (3.6–4.5 kg). There's no other information available in the breed standards or the other sources we located.

Dala

Norway. Developed from a crossing of the Spelsau with British breeds, including Cheviot and Leicester, Dala sheep were established in the late nineteenth and early twentieth centuries. Most are white, with about 10 percent black. Considered a production animal, this breed is not currently endangered, but additional crossbreeding means that purebred Dala sheep are becoming less common. The medium wool has an unusual amount of luster.

Faroese

Faroe Islands. The Faroese is another ancient double-coated breed with well over a thousand years of history on a set of islands in the Atlantic Ocean. Faroese sheep are closely related to old-style Spelsau, Icelandic, and Shetland sheep. They are pretty evenly divided in color between white, black, and gray, with some light through dark browns appearing in the mix now and then. Their fleeces average about 2¼ pounds (1 kg). The outer coat may be up to 16 inches (40.5 cm) long; the inner coat is a good deal shorter, finer, and softer. To show off the best qualities of each type of fiber, the coats need to be separated and processed individually, but they can also be spun together.

Faroese

Gute

Conservation Breed

Sweden. Gute (Gutefår) sheep are sometimes called Gotlands, which can be confusing because the modern hornless Gotland (see page 162) is one of this ancient horned breed's descendants. The Gute also goes by the name Gotland Outdoors sheep, reflecting its sturdy ability to survive without human-provided shelter. These old-style sheep nearly disappeared by the middle of the twentieth century, but conservation efforts have helped them survive.

Most Gute sheep, like their descendants the Gotlands, are gray, which may result from a mix of black and white fibers or may be an actual gray. A tan-marked white also exists, and browns occur occasionally. A Gute fleece consists of layers of both fine wool (as fine as 17 microns, which is cashmere range) and coarse hair (around 40 microns), along with both black and white kemp. The fleece sheds in the late spring or early summer. Annual fleece size is in the range of 4½ to 5½ pounds (2-2.5 kg), with a yield of 85 to 90 percent. Because the two coats are about the same length, they are often spun together, although the fiber types can be separated into different qualities and used for a variety of purposes; Gute wool is also very useful for felting.

Pellsau

Norway. An excellent source of felting wools, the Pellsau breed was developed in the 1960s through crossbreeding Swedish Gotlands (see page 162) with gray Spelsaus (see page 323). Most Pellsaus are solid gray, although some are white.

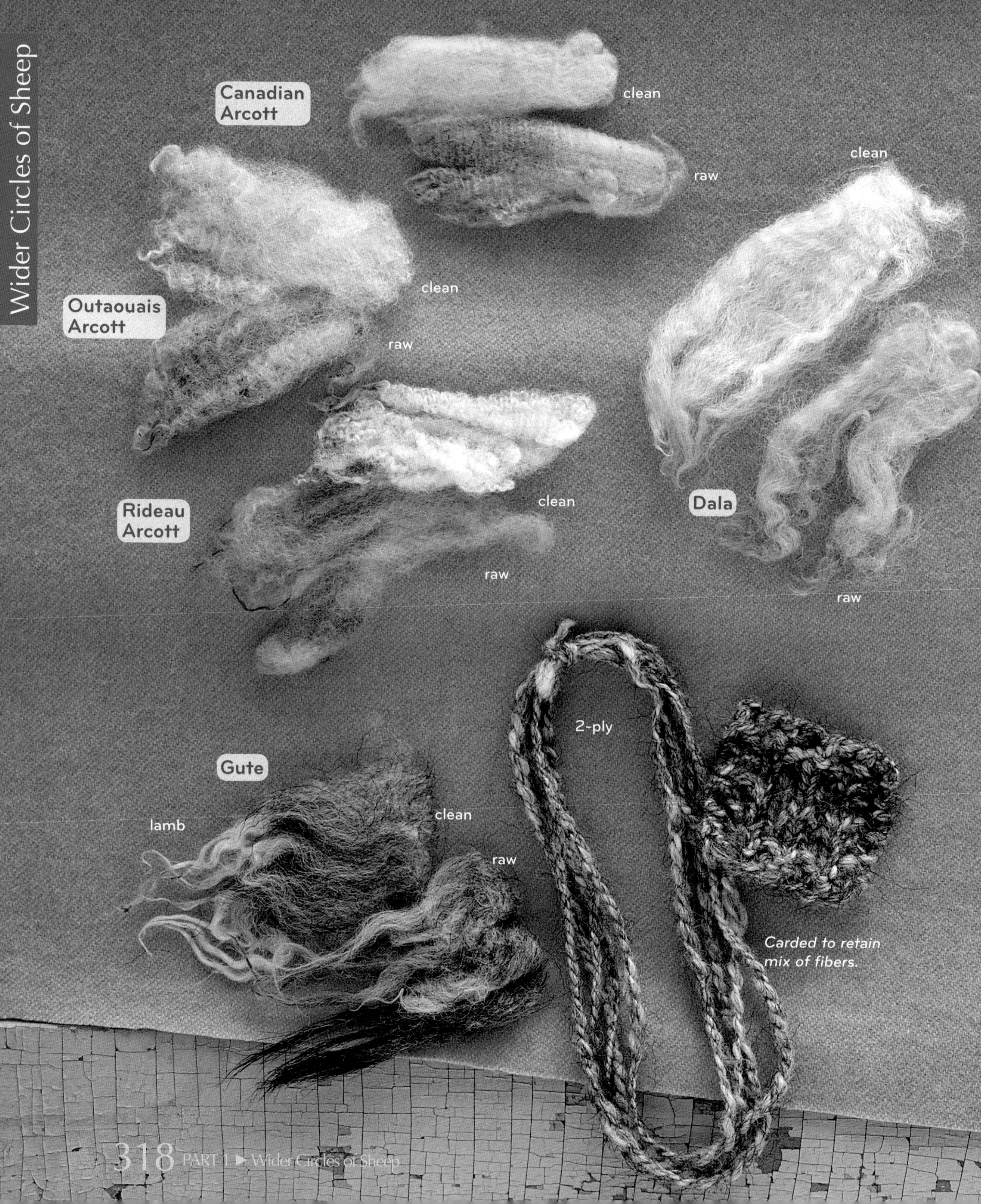
Canadian Arcott
clean
raw
Outaouais Arcott
clean
raw
Rideau Arcott
clean
raw
Dala
clean
raw
Gute
lamb
clean
raw
2-ply
Carded to retain mix of fibers.

singles

2-ply,
medium

2-ply,
light

singles, *from
picked light
gray locks*

Pellsau/C1 (50/50)

batt

C1 is a Norwegian commercial grade, a blend of wools like Dala, Rygya, and Steigar. Batts are often used for felting.

Pellsau

clean

2-ply, *light,
from light gray
locks*

2-ply, *medium,
from mixed
gray locks*

raw

The raw locks display luster not apparent in the batt, in which the orientation of the fibers gets jumbled up.

Racka

Racka

Conservation Breed

Hungary. Native to the plains of Hungary, the double-coated and distinctively curly-horned Racka breed was displaced by Merinos during the eighteenth and nineteenth centuries. By the middle of the twentieth century, it had become endangered, its decline accelerated by losses during World War II. In 1983, a breed society was established to support the survival of the Racka.

About 60 percent of Racka sheep are white (shading into and out of browns) and 40 percent are black. Lambs are born with very curly, lustrous wool, initially much like Karakul (see page 276). Adults grow long, curly fleeces of variable quality, with undercoats between 16 and 30 microns (an exceptionally broad range) and outercoats between 40 and 60 microns, about 8 to 12 inches (20.5–30.5 cm) long. An annual fleece, between 4½ and 11 pounds (2–5 kg) with a yield of 60 to 70 percent, consists of about two-thirds outercoat and one-third undercoat. Both types of fiber will felt.

Roslag

Conservation Breed

Sweden. The Roslag sheep (Roslagsfår) is a landrace that was once widely kept for its wool. In 1995, only about 40 of these sheep remained; due to conservation efforts since then, the population has been increasing. About 90 percent of the sheep are white, and about 10 percent are black with white markings. Their fleeces are double coated, of varying quality that may include lots of kemp, and can be up to 15 inches (38 cm) long.

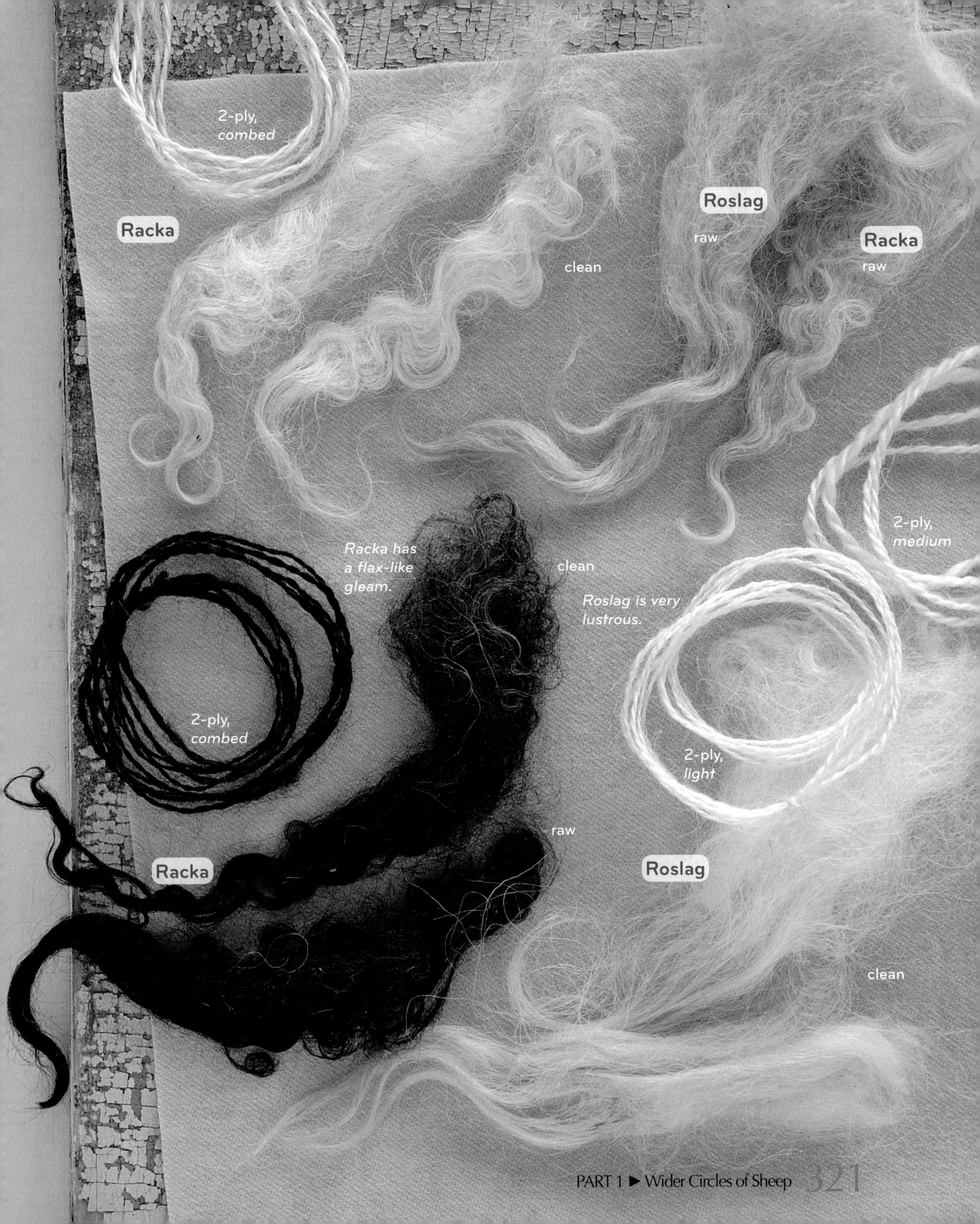
2-ply,
combed
Racka
Roslag
raw
Racka
raw
clean
2-ply,
medium
Racka has
a flax-like
gleam.
clean
Roslag is very
lustrous.
2-ply,
combed
2-ply,
light
raw
Racka
Roslag
clean

Both strongly double-coated (above) and relatively single-coated locks (below) were part of the same fleece.

Rya

Conservation Breed

Sweden. *Rya* originally referred to a type of long, strong wool; now it is an identified, double-coated breed of Swedish landrace sheep that includes some Norwegian Spelsau (at right) heritage. The wool can be 6 inches (15 cm) long on a three-month-old lamb and up to 12 inches (30.5 cm) on an adult. The hair coat should make up about half the fleece and should be lustrous and have a well-defined, broad crimp. Most Rya sheep are white, with some gray, black, and brown animals. Rya is an excellent rug wool.

Rygya

Norway. Established as a breed in 1924, the Rygya incorporated landrace bloodlines (Spelsau) with infusions of British breeds, such as Cheviot and Leicester (which also contributed to the development of the Swedish Dala sheep) along with Merino and Southdown. Additional breeding adjustments have included Texels, Finns, Dala sheep, and Steigar sheep (a meat breed that came into being at about the same time as, and in a similar way to, the Dala sheep). Rygyas produce both meat and wool. Most are white; a few are black. Average fiber diameters are in the vicinity of 37 microns (spinning count 40s), and fleece weights are around 8 pounds (3.6 kg). The lustrous wool is less bulky than that of Dala sheep.

Spelsau (Old and New)

Conservation Breed (Old Spelsau)

Norway. For several of the types of sheep in this section, there are two separate but closely related breeds: an older one, and a more recent, "improved" version with more modern bloodlines incorporated. That's the case with the Spelsaus. Both old and new types are dual coated, with a fine undercoat and a coarser, sturdier outercoat, both of which can be felted.

Sheep closely resembling the old Spelsau (also called Spaelsau, Villsau, or Old Norwegian) have been around for at least three thousand years, putting the origins for this type right back there in the prehistoric mists and mysteries with the Soay of St. Kilda (which may have been in their locale for four thousand years). In any case, they've certainly been around since Viking times, so well over a thousand years. They almost died out in the twentieth century.

The old Spelsau shed their wool annually in the summer, yielding an average of 3 pounds (1.4 kg) of fiber; they are often shorn at about the molting time. Fleece colors include white, black, grays, and browns. The dual coats are about the same length. They can be spun together into a sturdy yarn or separated, with the smooth, lustrous outercoat spun into durable yarn for rugs and weather-resistant garments and the undercoat made into more personal textiles.

These animals, hardly the models for the defenseless stereotype often applied to sheep, can survive in areas also inhabited by predators and can also be used as conservation grazers, eating undesirable vegetation and leaving native plants to flourish. Their wool, too, has

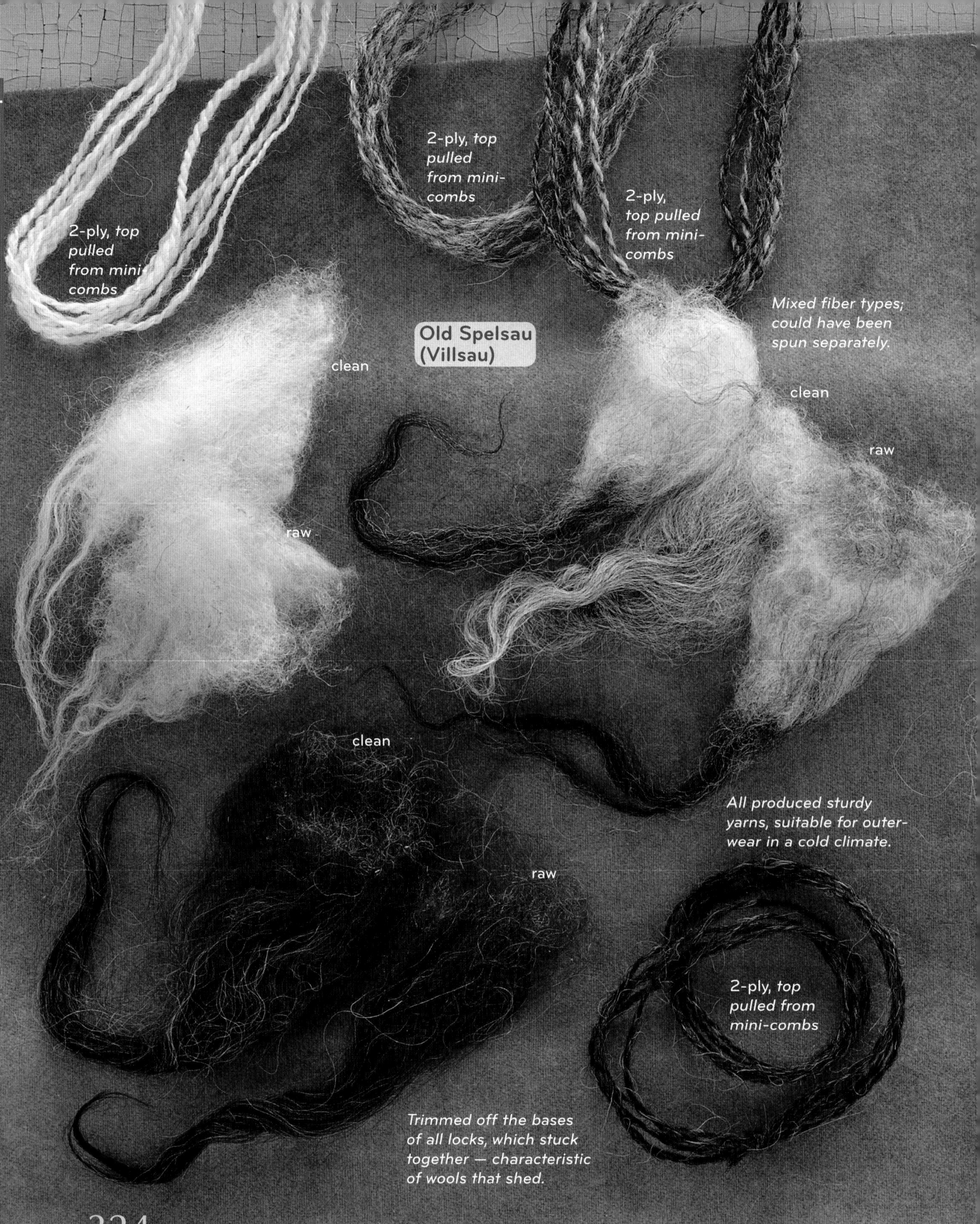

2-ply, top pulled from mini-combs

2-ply, top pulled from mini-combs

2-ply, top pulled from mini-combs

Old Spelsau (Villsau)

Mixed fiber types; could have been spun separately.

clean

clean

raw

raw

clean

All produced sturdy yarns, suitable for outer-wear in a cold climate.

raw

2-ply, top pulled from mini-combs

Trimmed off the bases of all locks, which stuck together — characteristic of wools that shed.

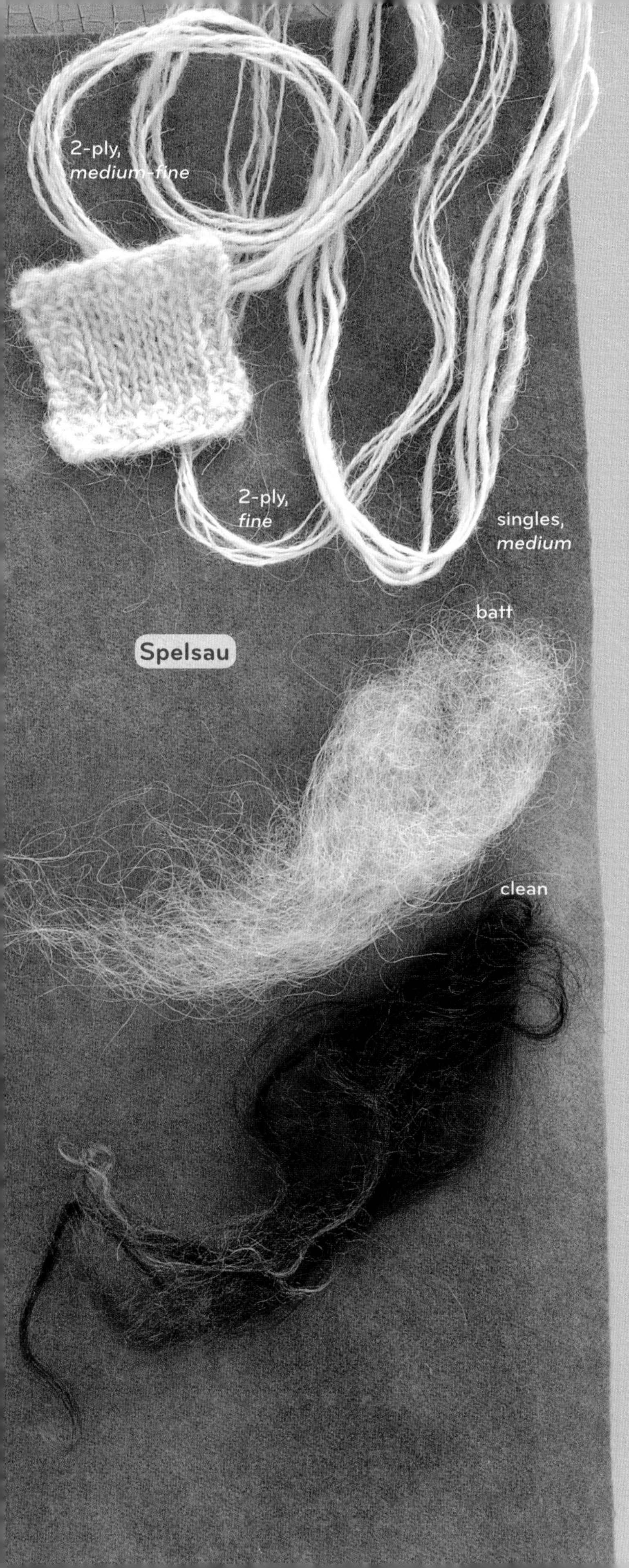

a staunch history, having been used to make the sails for Viking ships. A reconstruction of a ship of this type, called the *Ottar* — built and owned by the Viking Ship Museum in Roskilde, Denmark — boasts a massive Spelsau sail, completed with rigging made from hemp and horsehair. Spelsau wool has been used since the Renaissance for weaving tapestries and still is a yarn of choice for fine-art weavings.

The newer Spelsau also include genes from Icelandic, Faroese, and Finn sheep, infused through modern breeding initiatives. The introduction of other bloodlines has resulted in larger sheep, accompanied by a decline in wool quality, with increased micron counts and wider separation of fiber types in both coats. Most new Spelsau are white, although their ancestors' black, gray, and brown colors also continue to appear. The undercoat can be quite fine, and the outercoat a good deal coarser and twice as long, or even more. Fleece weights run between 4½ and 6½ pounds (2-3 kg).

Steigar

Norway. Like the Dala sheep and the Rygya, the Steigar originated in the late nineteenth and early twentieth centuries from crossing Spelsau with British breeds, including Cheviot, Leicester, and the now-extinct Sutherland. Texel and Finn blood were introduced in the late twentieth century. The Steigar is a commercial breed, producing both meat and white wool, around 37 microns in fiber diameter (spinning count 40s) and with a fleece weight of about 7¼ pounds (3.3 kg). Like the Dala and the Rygya, purebred Steigar sheep are being absorbed into the general population of Norwegian white sheep.

Svärdsjö

Conservation Breed

Sweden. The Svärdsjö is one strain of several called "forest sheep," each of which is geographically and genetically unique — and very rare. The others are Åsen, Värmland, Helsinge, and Gestrike. The Svärdsjö are unique among these populations for being polled. Wool qualities vary, and colors include whites, grays, blacks, and browns.

Swedish Finewool

Conservation Breed

Sweden. Swedish Finewool sheep trace their roots to Finland, but the breed has been associated with Sweden for a long time. The fleece consists of a single type of fine white, black, or brown fiber, crimpy and lustrous, with good elasticity. Conservation breeding efforts since the 1930s have emphasized the wool quality, aiming for fiber diameters in the range of 15 to 25 microns (spinning counts 58s–finer than 80s). The samples we sent for lab testing produced results between 20 and 28 microns (spinning counts 56s–70s), with fiber lengths between 1¾ and 2½ inches (4.5–6.5 cm).

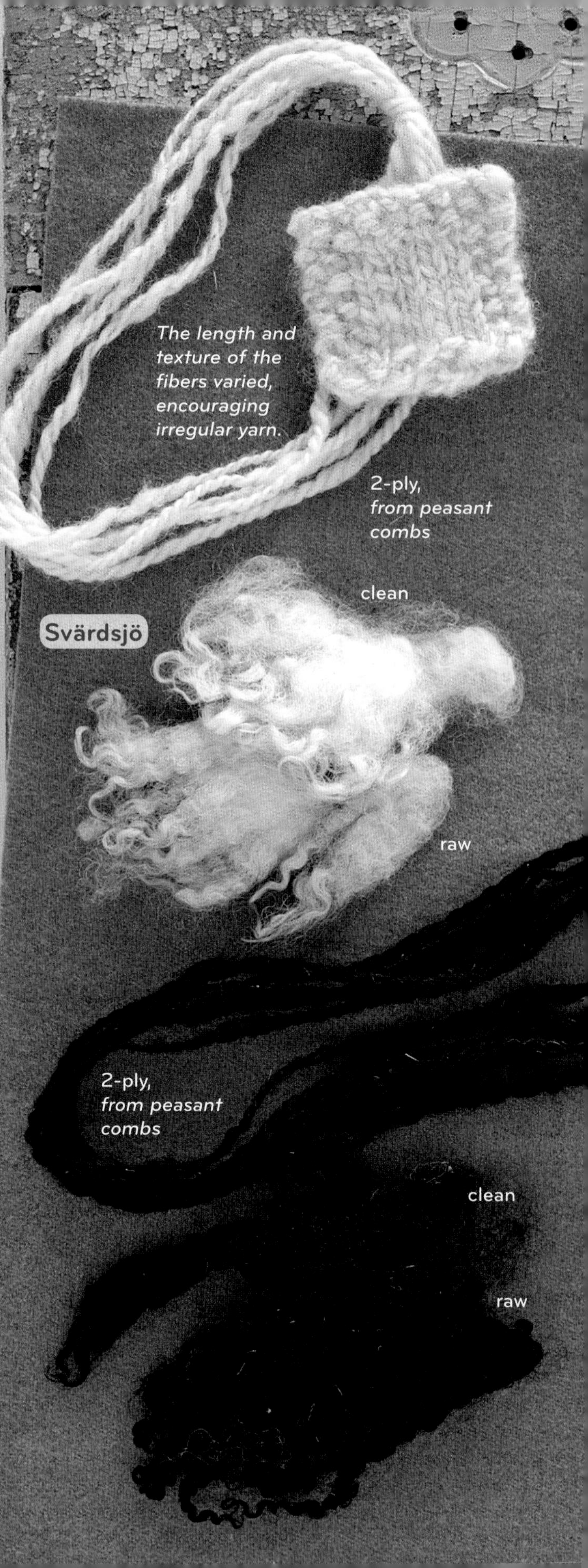

The length and texture of the fibers varied, encouraging irregular yarn.

Swedish Finewool

2-ply, spun directly from mini-combs

2-ply, spun directly from mini-combs

2-ply, spun directly from mini-combs

Has a purling crimp like Bluefaced Leicester's. Soft luster was evident to greater or lesser degrees on the different colors.

clean

raw

clean

raw

clean

raw

Mashams, Mules, and Other Crosses

Just as in the dog world, where there are purebred dogs and mixed-breed mutts, the sheep world too has breeds and crossbreeds. Some crossbreeds are like mutts, mixtures of many different bloodlines with little intention involved in breeding decisions. Yet other crosses are more refined, made by a shepherd with very definite breeding outcomes in mind.

Why would a shepherd develop crossbreeds when there are already so many really cool breeds? Well, just as mutts are often hardier and healthier than purebred dogs, the crosses may be hardier and better performing than their purebred parents, a tendency known as *hybrid vigor.* Thus crosses are made to improve certain production traits, such as speed of growth or number of lambs born per ewe.

From a fiber standpoint, many crosses produce luscious fleeces. Whether crossbred fleeces are more luscious than those of the parent breeds can only be decided by the person working with the fiber. We can't document the untold number of crosses you might find when buying directly from breeders, but there are two named crosses you may hear about at some juncture: Mashams and mules.

Masham. This cross, produced by breeding a Teeswater ram (or sometimes the similar Wensleydale) to either a Dalesbred or a Swaledale ewe (or now and then a Rough Fell ewe), has been employed regularly for at least a hundred years in the hill country of northern England. Ewe lambs from these crossings are raised as breeding ewes in commercial flocks because they have great hardiness, longevity, heavy milking qualities, strong mothering instincts, and high prolificacy. These Masham ewes also grow nice fleeces with long staples

Masham ewe and lambs

Welsh Mule

and good luster. The fiber characteristics vary, but the overall feel is along the lines of the English Longwool breeds.

You won't find Mashams on the American side of the Atlantic, because the Dalesbred, Swaledale, and Rough Fell breeds aren't found here at all, and as for the Teeswaters and Wensleydales that provide the ram side of this equation — well, that's a complex issue without a simple answer, and we talked about it on page 118. If you ever travel in Britain and scope out the sheep scene, however, you'll definitely hear about Mashams, and some Masham fleece does make its way to fiber shows in North America.

Mules. The term *mule* is used for a crossbred ewe that's part of a systematic three-way crossing practice developed in Britain. The goal of this cross was to yield the highest-quality lambs for meat production off of pasture — in other words, the goal is true grass-fed meat that is not fattened on corn in a feedlot. But the mules born of this plan also grow excellent fiber.

The crossing sequence works like this: A Bluefaced Leicester ram is bred to ewes that are suited to the locality where the flock is being raised; commonly these ewes come from the Blackfaced Mountain family, the Cheviot family, the Welsh Mountain family, or the Clun Forest breed. The ewe lambs out of these

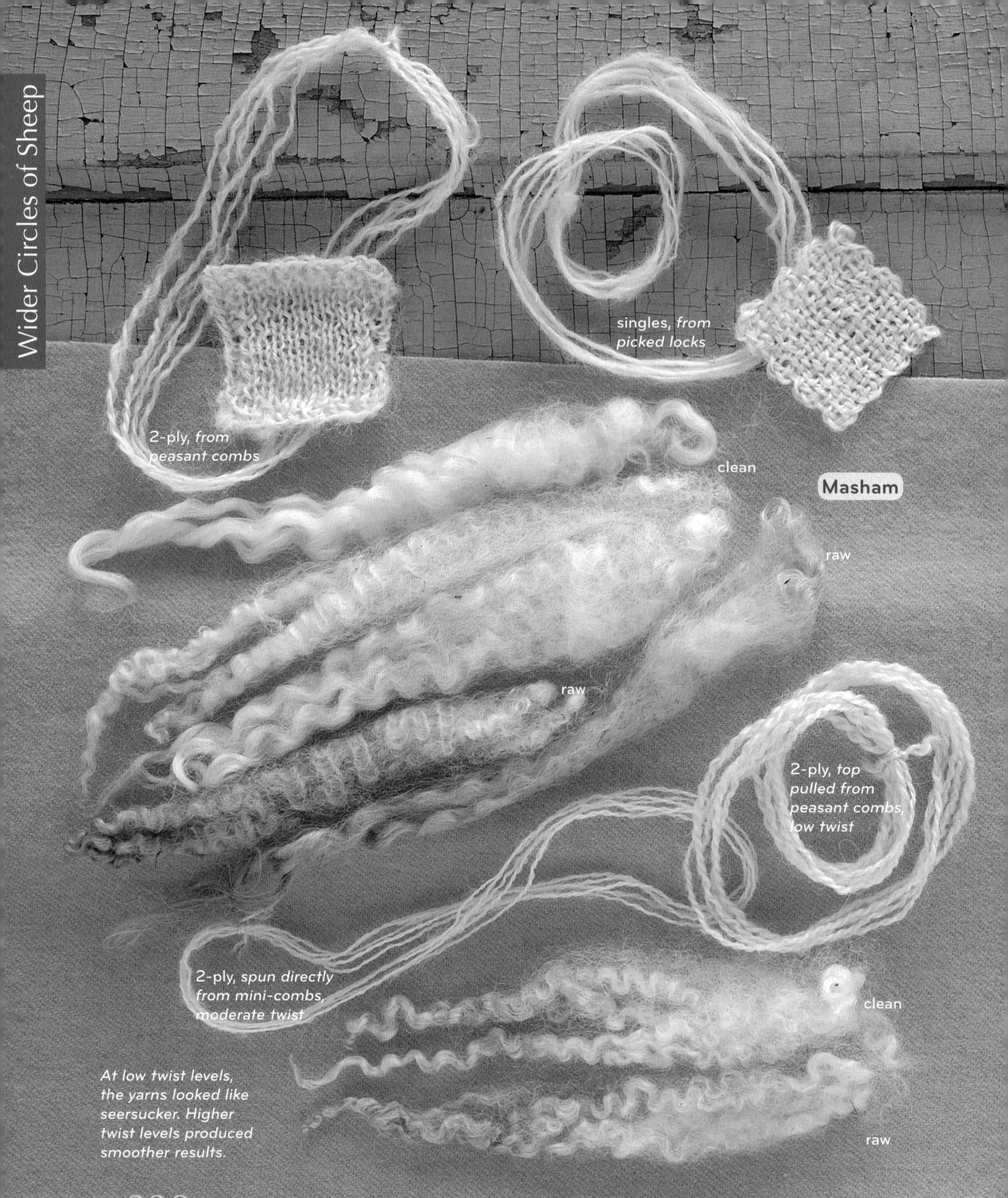

Masham

2-ply, from peasant combs

singles, from picked locks

clean

raw

raw

2-ply, top pulled from peasant combs, low twist

2-ply, spun directly from mini-combs, moderate twist

clean

raw

At low twist levels, the yarns looked like seersucker. Higher twist levels produced smoother results.

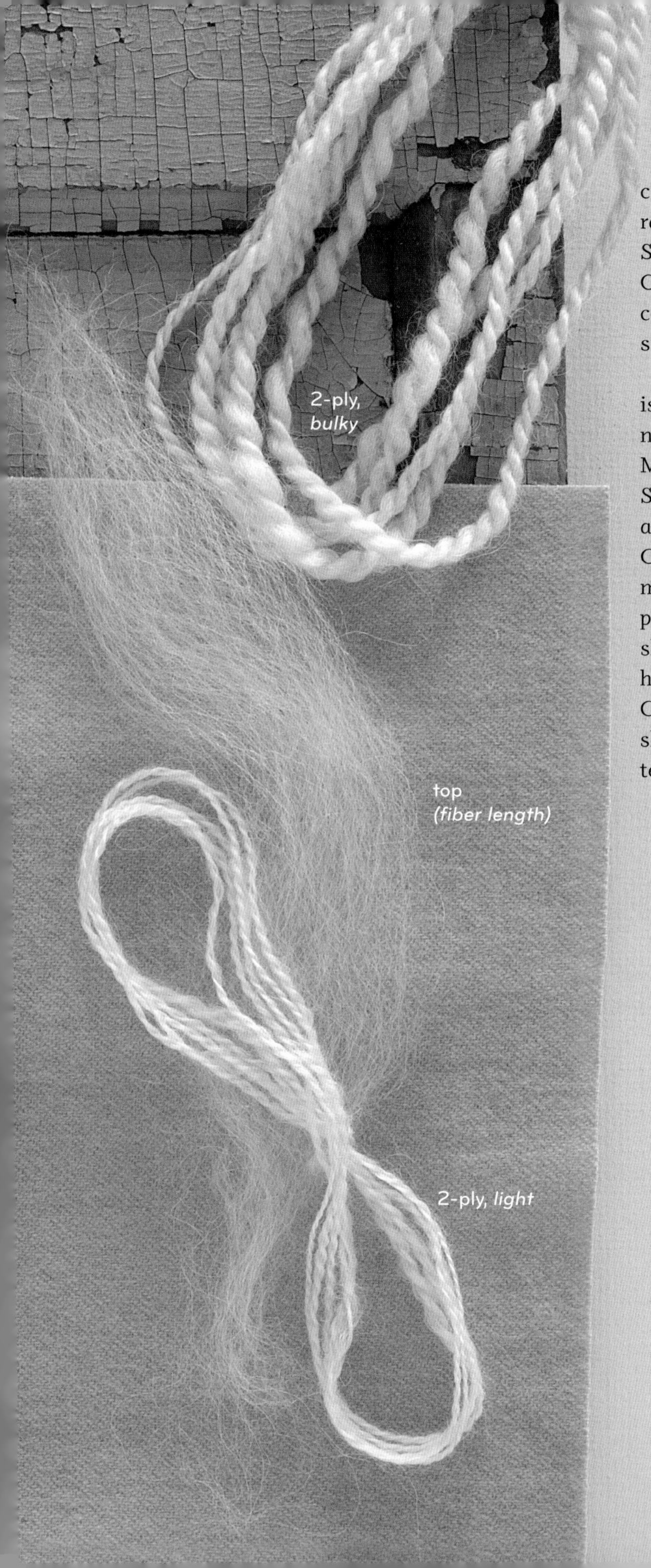

2-ply, *bulky*

top *(fiber length)*

2-ply, *light*

crosses are the mules; they're later crossed to rams of the larger meat breeds (such as Texel or Suffolks) to produce lambs for the meat market. Obviously mule sheep, unlike the mules that come from crossing horses and donkeys, are not sterile!

The traditional approach for naming mules is to combine some form of the dam's breed name with the word *mule*. For example, a Scotch Mule is a commercial ewe whose dam was a Scottish Blackface; a Welsh Mule's dam was a Welsh Mountain; and a Cheviot Mule had a Cheviot dam. There is now a growing movement of shepherds using the mule system to produce grass-fed lamb among North American shepherds, so you may come across some homegrown mule fiber in the United States or Canada. The Bluefaced Leicester ram's influence shows up in the length, texture, and crimp patterns of mule fleeces.

PART 2 Other Species: *The Rest of the Menagerie*

All animals are equal. But some animals are more equal than others.

— George Orwell, *Animal Farm*

In the previous chapters, we highlighted the many varieties of sheep. But of course, sheep aren't the only critters that meet our need (and craving) for natural fibers. Though not as numerous as the sheep, their distant cousins, the goats, provide some of the most well-known fibers; can you say *cashmere* and *mohair* and the new, surprising *pygora*? And members of the camel family, especially the alpacas and llamas, are true sweethearts of the fiber world.

Then there are the fibers of yaks and musk oxen and bison, oh my! These wild and semi-wild beasts provide lush luxury fibers and more. Some horses, cows, and donkeys also get into the act. Some dogs produce undercoats that rival the rarest fluff. And yes, sometimes you can even get yarn from a cat (make sure its fur is as long as possible — compared to other spinnable fibers, it will still be short — and be prepared to deal with serious slipperiness). As for bunnies, rabbits provide us with scrumptiously soft angora, and if the only angora you've experienced is in mass-produced clothing, you haven't experienced angora yet! Indeed, there is a marvelous menagerie for fiber lovers to choose from and explore.

The fibers of the animal groups that follow, like the fibers grown by sheep, are built from proteins. This means that if you are into dyeing, these fibers can be colored with the same synthetic or natural dyes you would use on wool. Although we'll talk about numerical count systems with regard to mohair goats, those systems have been historically applied to wools. Our notes on the other types of fibers covered in this section do not include either Bradford counts or USDA grades (see pages 11-13).

Angora Goats
Spanish Goat, a cashmere producer
Kiko Goat, another producer of cashmere

Goats

Goats have always played a critical role in animal agriculture. There are between 500 and 700 million domestic goats around the globe, raised as multipurpose animals that provide not only fiber but also meat, milk, skins, and transportation. Goats are used as pack animals through much of their primary range. They are known as the "poor man's cows," because they can live on hardscrabble lands that couldn't support real cows. Although goats are widely raised in other parts of the world, they are the least common animals in North American agriculture. That said, you shouldn't find it too tough to locate someone raising and marketing fiber from goats. In fact, there's such a wide variety of goat fiber that it includes a range of fineness, softness, length, and sturdiness that's nearly as extensive as the territory covered by sheep. With a few exceptions, goat-breeding for fiber has been less orderly than breeding for fiber qualities in sheep, so spinners can expect a lot of variability in the raw fiber.

There are eight wild goat species around the world, but just one domestic species (*Capra hircus*) that produces all our abundant variety of fiber. That species developed from the bezoar ibex (*Capra aegagrus*), a wild goat that still exists in its original form as well. The latest archaeological evidence suggests that goats were first domesticated about 10,000 years ago. From their first domestication site, currently thought to be in the Zagros Mountains of western Iran, goats quickly spread around the Middle East and into Africa and Asia.

The bezoars are considered vulnerable — at risk of extinction — by the World Conservation Union and are on the Red List of Threatened Species prepared by the International Union for Conservation of Nature and Natural Resources (IUCN). They are still found in the wild on several Greek islands, including Crete, as well as in Turkey, Pakistan, and Turkmenistan.

Bezoars are fairly large goats: Big males weigh as much as 300 pounds (136 kg) and stand as tall as 4 feet (1.2 m) at the withers. Bezoars come in various colors and patterns, but brown, black, red, gray, and white are the common base colors. Both males and females have beards, as well as tall, scimitar-curved horns that curl slightly to the back. Like their

An original. The wild bezoar goat is still found in a small portion of its historic range. It is considered vulnerable to extinction.

Mountain Goats

We live in Colorado, a state that's home to a small population of wild mountain goats (pictured above). These are a species native to the Americas, found at the highest and craggiest elevations of the Rocky Mountains. This type of goat did not influence the gene pool of domestic goats at all (it's not even in the same genus, although it is in the same biological family).

Yet if you are lucky enough to come across some of the fluff (shown below) while hiking in the high country, you'll discover it is a spinnable fiber. Deb has found mountain goat hair snagged on brush and finger-spun it on the spot. Collecting enough to make more than a bit of accent yarn would take a lot of effort. Carol's friend Swithin Dick gathered a handful and let us share it with you.

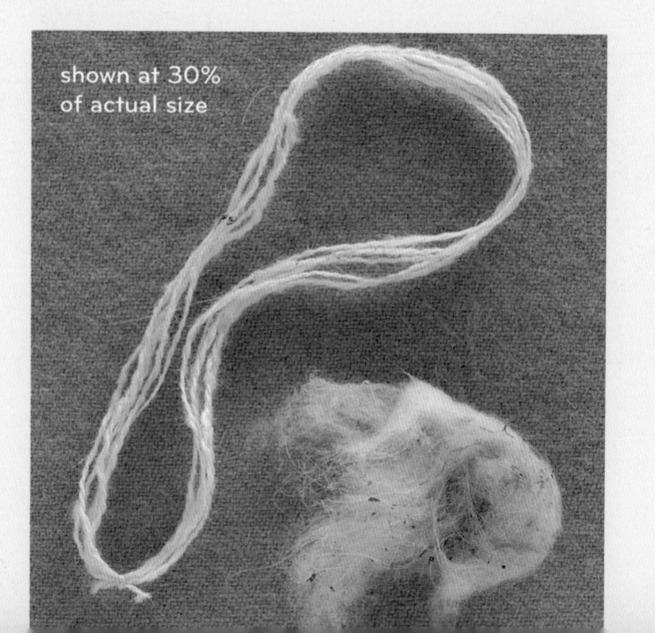

domestic descendants, the bezoars are gregarious, with females and young animals living in herds of 50 or more, and males living in smaller herds that disperse during the breeding season.

Interestingly, next to cats, goats are the quickest critters to go feral if given a chance, and there are bezoar-type wild goats in some parts of Europe where bezoars didn't exist traditionally. These are thought to be formerly domesticated goats that have gone through a feral transition and reverted to a wild type.

For fiber aficionados, there are two fundamental types of goats: cashmere producing and mohair producing. Mohair is produced exclusively by the Angora breed, but cashmere can be produced by all the other breeds of goats. You might be asking, "Don't angora fibers come from Angora goats?" The answer is no, although in casual speech *angora* can refer to these goats' hair. Angora fiber comes from Angora rabbits (which we talk about on page 404), while mohair comes from Angora goats.

You may also encounter other goat fibers such as cashgora and pygora and pycazz, but all these goat options come back to the two original sources, mohair and cashmere. These diverse and elegant goat-fiber alternatives derive from crossings (and sometimes recrossings) that involve mohair, cashmere, or both types of goat.

A note about goat fiber in general, and billy goat fleeces in particular: If you plan to work with the raw material instead of prepared fiber or spun yarn, know that the fiber smells like goat — in the case of billy wool, a *lot* like goat — but the odor washes out with soap and water, or products specifically designed for washing raw fiber prior to spinning. If you acquire a billy fleece, you will probably want to wash it outside or in a basement or garage. Be thorough in your cleansing and rinsing, and it will be fine. Fleeces shorn during breeding season may be more redolent than those shorn at other times.

Mohair

BOTH NAMES THAT these goats are known by — mohair and Angora — reflect their origins in the area that is now Turkey. *Mukhayyar*, the etymological parent of *mohair*, is an Arabic word for "choice" or "select." It also means "cloth made of goat's hair," although to be choice, it would need to be the hair of a special goat! Which, of course, it was, since it referred to goats that came into existence in the region called Anatolia, in a city that was then called Angora and is now the Turkish capital of Ankara. Ankara is also the original home of Angora cats and Angora rabbits (see page 404 for more on these "fiberrific" rabbits). The presence of Angora goats in Turkey has been documented since around 1500 BCE.

In 1849, James Davis, of South Carolina, first brought Angora goats (as well as the first Brahman cattle) into the United States. Davis was a renowned cotton farmer, and he had gone to Turkey as a representative of the United States government to work on improving cotton production there. The sultan of Turkey was so grateful to Davis that he gave him the goats as a present. Several additional shipments of goats made their way to North America between 1849 and 1881, but when a new sultan came to power he outlawed the export of Angora goats — under penalty of death.

Mohair goats are cute, plucky little creatures with fairly mellow personalities. Thanks to their curly locks, they look as if they just got out of the beauty parlor with a fresh perm. Those sweet locks are produced by no other goat breed.

Because mohair goats are not very hardy, most of the larger flocks in North America are found in the southern parts of the United States (especially in Texas). When mohair goats are raised in northern climes, where it's cold and wet, they require extremely diligent care to survive. One theory about why they are more fragile than most other goats suggests that the constant and fast growth of their coats is to blame. Adult Angoras easily grow at least three-quarters of an inch (2 cm) of fiber per month throughout the year, and top-producing animals generate as much as 25 percent of their body weight in fiber each year. Fortunately, most mohair goats are shorn twice a year, so they're actually carrying, at most, only half the weight of a year's fleece at any given time. Nonetheless, the metabolic load of generating that much fluff requires a tremendous amount of energy.

Very young goats normally grow the finest, softest fiber with the highest commercial value. An animal's first shearing happens when the fleece reaches a length of about 4 inches (10 cm), which is usually when it's about six months old.

The classing systems for mohair are not consistent, although the basic terms are: Kid mohair (very fine), yearling (midrange), and adult (sturdier); see Fiber Classes for Mohair, page 339, for more details. Commercial producers focus on white fiber, but there are a number of smaller-scale breeders producing naturally colored Angoras. These come in a range of cream and fawn colors, reddish browns, rich grays, and blacks.

Fiber Close Up

If you thought you'd dropped down Alice's rabbit hole into Wonderland with the sheep breeds, you can do it again here with mohair. The fiber of this unique goat breed encompasses a universe of possibilities. It's like having three different breeds in one, separated primarily by age rather than parentage. Within an overall amazingly consistent global population of mohair goats, family lines and unique breeding populations nonetheless have their own fiber personalities. Speaking generally, bucks tend to have coarser (as well as heavier) fleeces than does. On an individual animal, the prime handspinning fibers come from the sides.

Mohair fiber is full of both useful properties and apparent contradictions, many of which trace their existence to the pattern of the scales on the fibers. Thinner, smoother, and larger than those of wool fibers, these scales give mohair stellar luster, a smooth and even slippery feel, and a strong resistance to felting (although if you want to mat it together firmly, it will oblige).

Although mohair is chemically similar to wool, it does not have the same type of springy elasticity that wools display (to a greater or lesser degree, of course, among the wools) because it has no crimp, although it does have waves. Nonetheless, the individual fibers are elastic, with the ability to stretch out by almost an extra third and then return to their original length. This gives mohair fabrics great draping qualities (that can turn to drooping if these tendencies aren't factored into the design process).

Ways of Thinking about Mohair Classes and Grades

Microns: 16 17 18 19 20 21 22 23 24 25 26 27 28 29 30 31 32 33 34 35 36 37 38 39 40 41 42 43 44 45 46 47 48

Mohair Council of America

Class	Microns
superfine kid	24–26μ
fine kid	27–28μ
good kid	29–30μ
superfine yearling	31–32μ
good yearling	33–34μ
superfine adult	35–36μ
good adult	37–39μ

Australian Mohair Marketing

Class	Microns
superfine kid	< 23μ
kid	23–27μ
strong kid	27–29.5μ
young goat	29.5–32μ
adult	32–36μ

MOHAIR (grease)	USDA grades: average micron count	USDA grades: standard deviation
Finer than 40s	< 23.01	7.2
40s	23.01–25.00	7.6
36s	25.01–27.00	8.0
32s	27.01–29.00	8.4
30s	29.01–31.00	8.8
28s	31.01–33.00	9.2
26s	33.01–35.00	9.6
24s	35.01–37.00	10.0
22s	37.01–39.00	10.5
20s	39.01–41.00	11.0
18s	41.01–43.00	11.5
Coarser than 18s	43.01+	

For comparison:

Worsted (wool) spinning counts: Finer than 80s, 80s, 70s, 64s, 62s, 60s, 58s, 56s, 54s, 50s, 48s, 46s, 44s, 40s, 36s, Coarser than 36s

Fiber Classes for Mohair

Classing, or grading, of mohair happens differently depending on where it's being done and which system is being employed. Terms used traditionally in Texas — from finest to coarsest — include *kid*, *yearling*, *fine adult*, and *adult*, with variants reflecting whether the fiber is from a fall or spring clip. The Mohair Council of America uses the terms *superfine kid*, *fine kid*, *good kid*, *superfine yearling*, *good yearling*, *superfine adult*, and *adult* to span a range from 24 to 39 microns. Australian Mohair Marketing's grades go from finer than 23 microns to 36 microns with *superfine kid*, *kid*, *strong kid*, *young goat*, and *adult*. These systems can't be neatly lined up next to each other. The Australians' superfine kid doesn't exist on the Mohair Council of America's scale, and U.S. superfine kid is the Aussies' kid. The grades are similarly out of whack all along.

To make things more confusing, the somewhat outdated numerical grading system, with its origins in the Bradford counts and familiar from the world of wool, has also been applied to mohair, although it doesn't work the same way. We are not including numerical count assignments for mohair (or for any of the fibers in this section) because they have not been commonly used or understood in relation to these fibers by people outside of industry. However, you may come across references to mohair grades that look like Bradford counts, and you may need to know about the extreme differences between wool counts and mohair counts.

Here's the overview of mohair grading: As you'll remember from the introduction to wool (see page 11), the numerical count system was initially based on the hypothetical number of 560-yard (512 m) hanks of yarn that could be spun from one pound (0.5 kg) of clean fiber. Mohair, a slippery fiber without wool's idiosyncratic and cohesive surface-scale structure or crimp patterns, can't be spun as fine as wool. Consequently, mohair's 18s to 40s grades cover about the same fiber diameter range as wool's 36s to 62s grades.

Here's the working information you need: Speaking very approximately, micron counts for kid mohair range from the low 20s to about 30 microns; yearling or young goat is from 30 or 31 to between 32 and 34 microns; and adult mohair starts somewhere between 32 and 35 and goes on up toward 39 microns or so, for the most part.

Although these sound like age-related labels, the fiber is evaluated only on its qualities: An adult goat that grows very fine fiber will have its fleece classed as kid, while a yearling that produces coarse mohair will have its fleece bundled with the adults. Yarns for hand knitting, crochet, or weaving, unless otherwise labeled (for example, as kid mohair), tend to be made from adult fiber, in the range of 37 to 39 microns.

The bottom line: We recommend that you use a simple system that's easy to remember.

- Kid mohair is fine (as fine as Merino).
- Yearling mohair is medium (like the many midrange wools).
- Adult is the heaviest (corresponding more closely to the English Longwools).
- They are all sleek and shiny and drape well.

Mohair fabrics resist creasing (and wrinkles), and they tend not to attract or hold dirt. Like wool, mohair can absorb a great deal of moisture before it feels wet, yet the slick surfaces of the fibers also make it naturally water repellent.

Exceptionally durable, adult mohair can be the perfect material for constructing upholstery, carpets, and other materials that need to be elegant, dirt resistant, and hard wearing — like the fabrics in the first-class section of the *Queen Elizabeth 2* cruise ship. Mohair fabric doesn't pill as shorter fibers might; if it's not well spun, however, it might shed. (The same is true, to a greater degree, of angora from rabbits.) Adult mohair can also be used for more personal projects, of course, but its extraordinary strength makes it appropriate for applications you might not immediately consider.

Mohair Types

Up until the 1960s, mohair goats were grouped depending on the type of locks in which their hair grew: the B Type had flat, wavy, bulky locks, and the C Type had tight curls or ringlets. C Type goats grew finer, lighter fleeces. A third type, without a name, was "woolly" and undesirable. According to Kathryn Ross Chastant, who grew up in a longtime mohair-ranching family in Texas and wrote extensively about the fiber for *Spin-Off* magazine, lack of an economic incentive for the finer fleeces led to a blurring of the lines between the two types, with a loss of the fine end of the spectrum in exchange for heavier clips. However, hand-textile artisans in particular will still see, and be sensitive to, these historic differences in lock formation.

Despite its durability when in use, mohair is more vulnerable than wool to damage during processing. The fiber can be permanently stained by the originating goat's urine and by some types of soils and vegetable matter. When kept wet for too long at too high a temperature, in a solution that is too acid or, more dangerously, too alkaline, mohair will lose its luster, it may yellow, and it may also turn brittle later. Keep this in mind when you're scouring or dyeing this fiber, but don't panic. General, conservative washing procedures work fine; we follow our normal methods for cleansing protein fibers. A safe temperature is 140°F (60°C), although, with care, that can be exceeded for short periods without ill effects; the grease on mohair has a melting point of 102°F (39°C).

Mohair tends to be naturally dense, but it can be spun to be either sleek or fluffy. It's a natural for creating novelty yarns, like bouclés, wraps, loops, slubs, and so on. You can brush a mohair surface to bring up an enthusiastic halo. Although commercial yarns are often brushed before their sale, unbrushed yarns can be easier to work with, and it's simpler for the individual artisan to brush finished fabric than yarn. Brush gently, as you would your own (or a child's) hair; a hairbrush works fine, preferably one with natural bristles. In the old days, people used the prickly dried heads of a teasel plant, like fuller's teasel (*Dipsacus sativus*), for the job, often setting the teasels in a wooden frame to keep the spikes from prickling the brusher's fingers. You can also fluff up the fabric in the dryer on low (or no) heat. This doesn't produce results as dramatic as raising the nap by brushing, but it can be a useful trick.

Breeding for Color

Thousands of years of careful, selective breeding have produced the uniquely clear, brilliant white of the classic mohair goat. So what's with the colors, a recent and strong development in the mohair realm? The Colored Angora Goat Record was established in 1992, less than 20 years ago, and was replaced by the Colored Angora Goat Registry in 1999. This is relatively new stuff.

Genes for color are hidden deeply within mohair goats' genetic DNA, beneath the dominant combinations that overwhelmingly produce white in the main population. We know this to be a fact, because an occasional color-revealing purebred Angora goat is born (and, even more rarely, it is not culled). However, the cultivation of color over the last couple of decades has come primarily through the introduction of non-Angora blood — which unfortunately brings with it the potential for kemp and the deterioration of fiber quality. The breeders' magic includes the goal of producing fiber as nice as classic white mohair that also displays a spectrum of colors from tan to reddish, and from gray to black. We say "magic" because the color genetics of mohair goats are not understood nearly as well as those of other types of goats. Breeding programs involve careful study combined with thought, guesswork, experimentation, and hope.

Because of this experimentation, the early lines of colored fleeces contained kemp and medullated fibers, both of which are stiffer and coarser than the rest of the fleece. Registration of colored Angoras requires that there be no kemp, a sign that breeders are becoming increasingly successful at manifesting their vision for their animals. (In practice, there may be a very tiny amount of kemp in a mohair fleece of any type.)

Remember not to be completely swept off your feet by a specific color of mohair; always match the fiber quality to your project. In a *Spin-Off* magazine article, spinner and historian Carol Huebscher Rhoades aptly reminds us that "the perfect color in the wrong grade will leave you dissatisfied." That's only going to be a problem if you're an impulse buyer, unwilling to select carefully. We found a blissful array of colored mohair for our samples, in an exciting variety of colors and a full selection of qualities, from kid to adult. The continued success of breeding programs for colored mohair, like that of almost all natural-colored fibers, rests on the shoulders, and in the hands, of individual craftspeople. Fortunately, there's a lot of creative inspiration to be found in the diverse fleeces of mohair goats.

Mohair blanket from Tall Grass Farm

Mohair Facts

► **FLEECE WEIGHT**
Normally shorn twice a year, with each shearing yielding:
Kids 1½–3½ pounds (0.7–1.6 kg)
Young does 3½–7 pounds (1.6–3.2 kg)
Young bucks 12½–15 pounds (5.7–6.8 kg)
Adult does 4–8½ pounds (1.8–3.9 kg),
Adult bucks 10–25 pounds (4.5–11.3 kg)
Yield of clean, usable fiber is usually 80–90 percent, rarely as low as 60 percent

► **STAPLE LENGTH**
For each six-month shearing: from 3 inches (7.5 cm) for the youngest goats, up to 4–6 inches (10–15 cm) for adults

► **FIBER DIAMETERS**
Generally from 24 microns for kids, up to 39 for adults, with some kids as fine as 20 microns and some adults as coarse as 46 microns. (See Fiber Classes for Mohair, page 339.)

► **LOCK CHARACTERISTICS**
Locks are distinct, with organized waves and pointed, curly tips. Kids are born with a mother coat that's about half kemp, but it sheds out quickly and is replaced by the finest mohair in time for the first shearing. Softness is a characteristic of kid mohair; luster and body, along with fiber diameter, increase with age. An adult fleece should have minimal kemp, ideally no more than 1 percent.

► **NATURAL COLORS**
Customarily and commercially, a clear and brilliant white. Relatively recent breeding for natural colors has begun to produce excellent results with two primary color groups, red and black, with variations in each, and diverse color patterns in the animals as well. Colors usually lighten with the animal's age, although there are exceptions to this.

Hand-dyed mohair locks

Kid mohair skeins from Louet

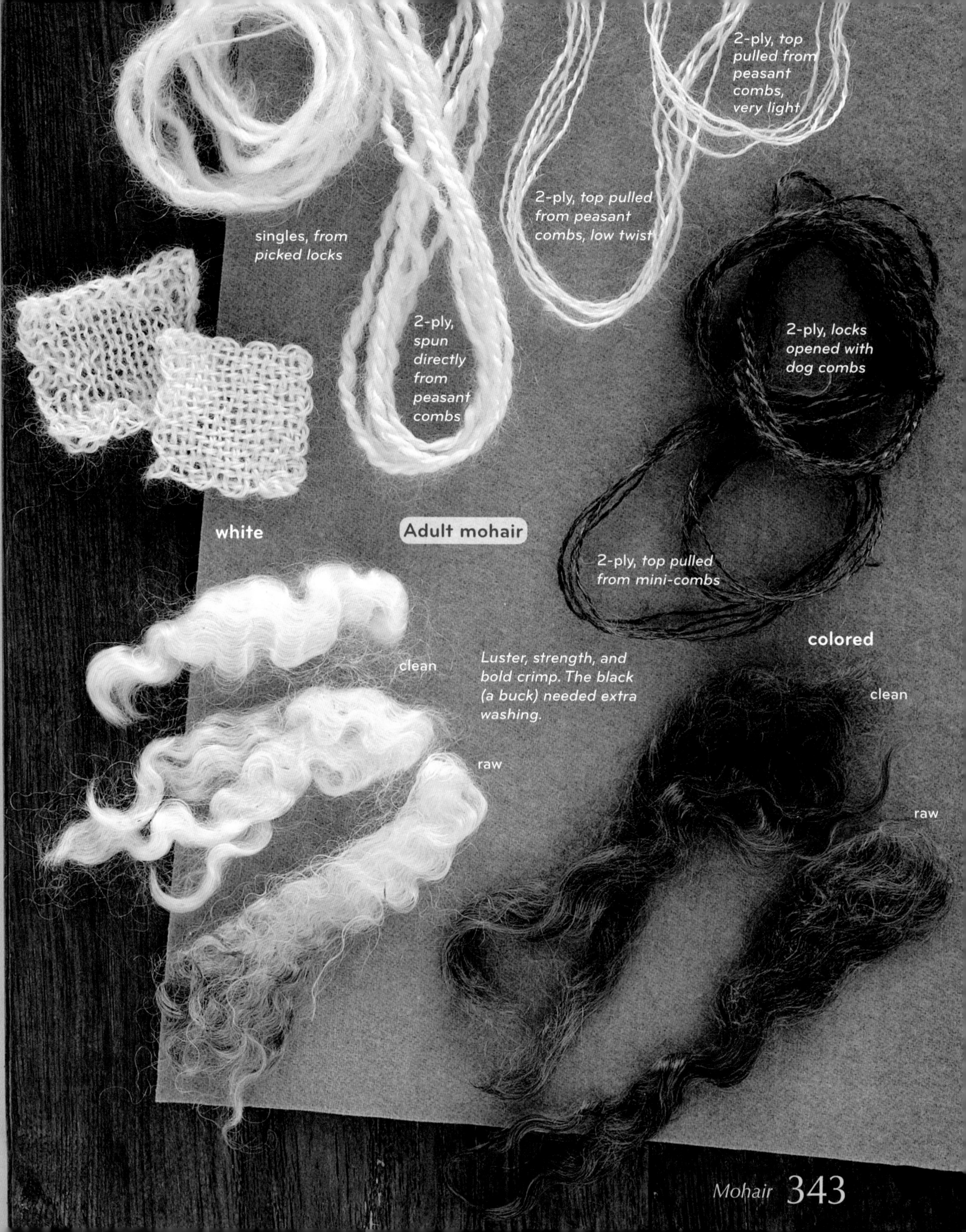

2-ply, *top pulled from peasant combs, very light*

singles, *from picked locks*

2-ply, *top pulled from peasant combs, low twist*

2-ply, *spun directly from peasant combs*

2-ply, *locks opened with dog combs*

white

Adult mohair

2-ply, *top pulled from mini-combs*

colored

clean

Luster, strength, and bold crimp. The black (a buck) needed extra washing.

clean

raw

raw

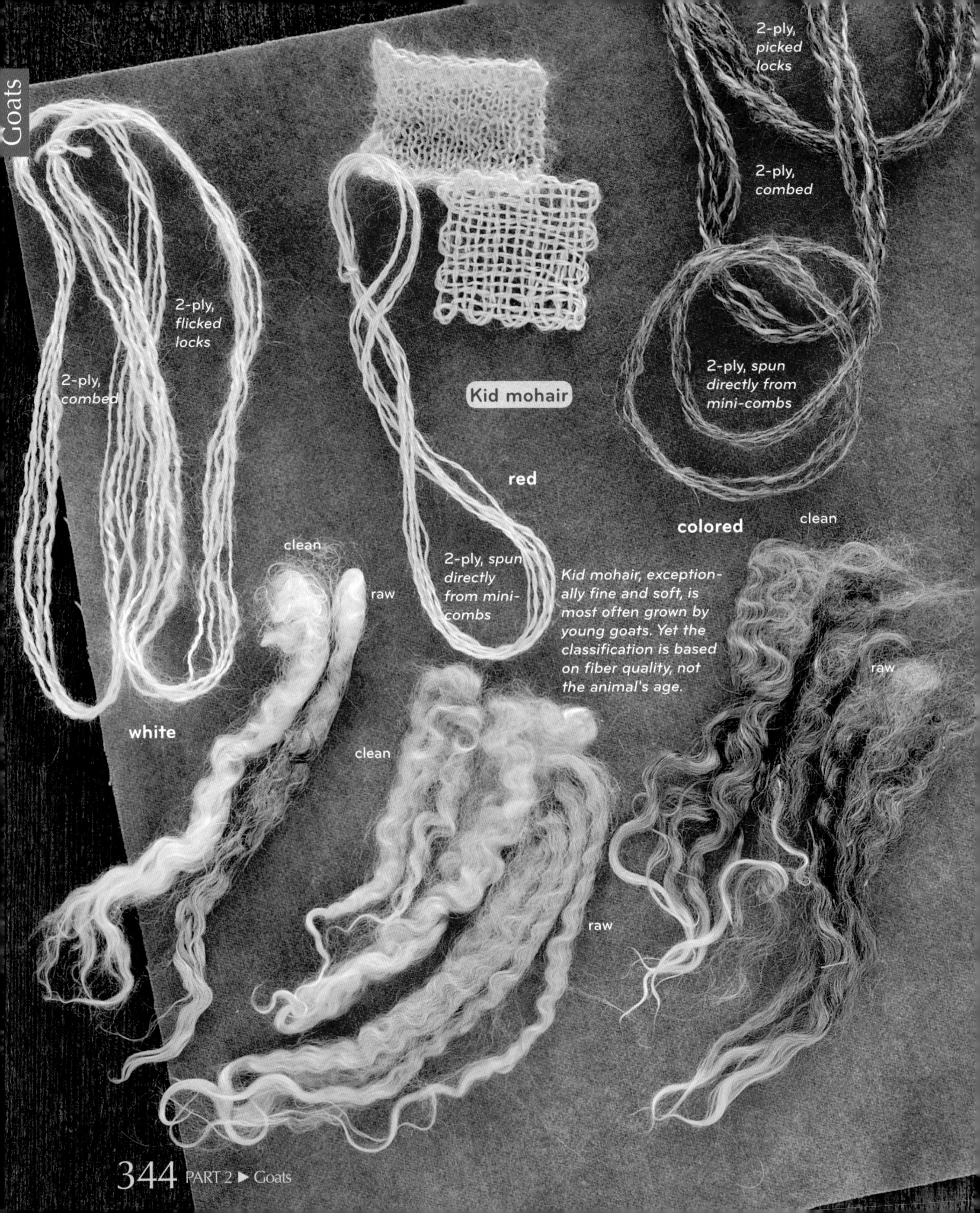

Kid mohair, exceptionally fine and soft, is most often grown by young goats. Yet the classification is based on fiber quality, not the animal's age.

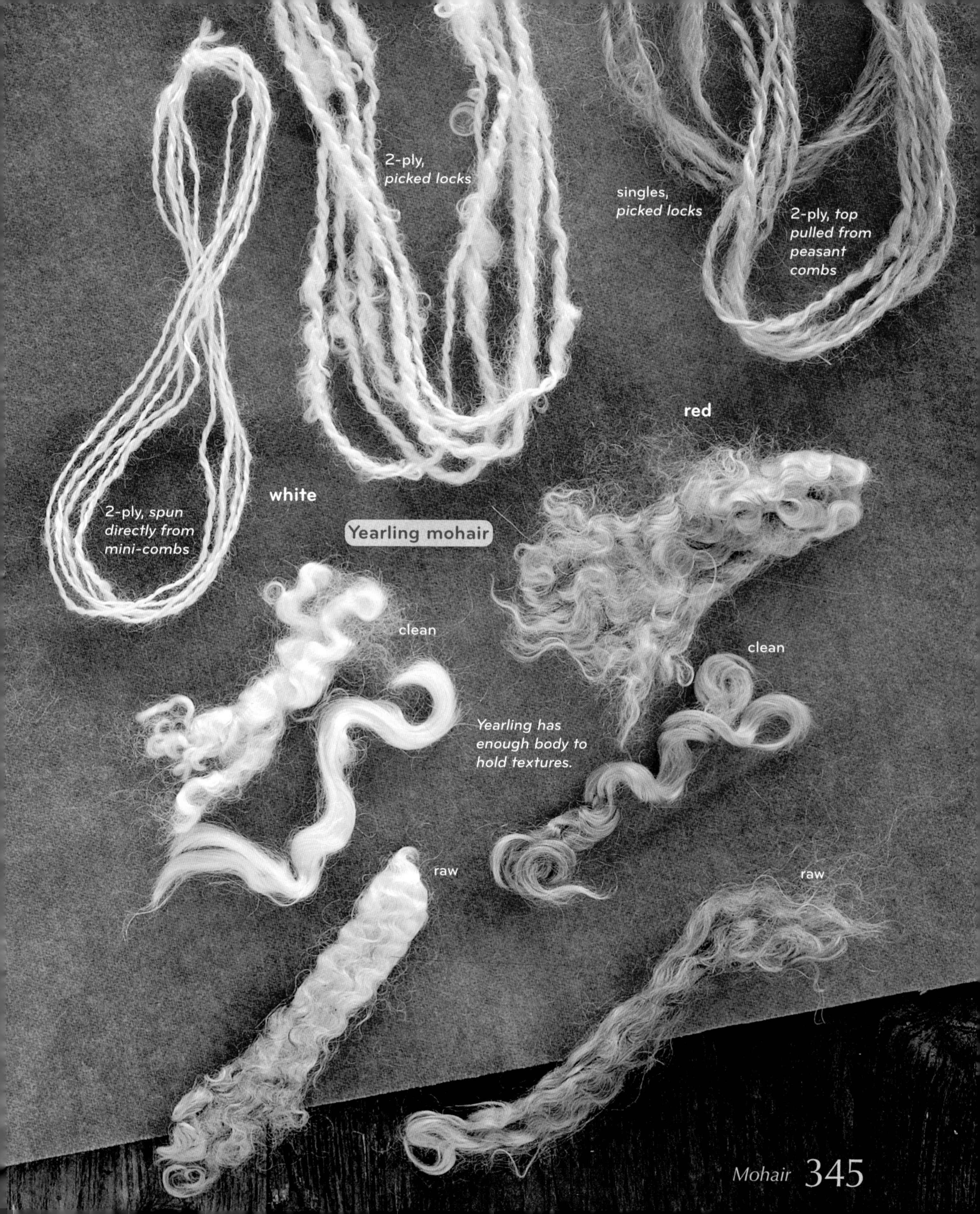
2-ply,
picked locks
singles,
picked locks
2-ply, top
pulled from
peasant
combs
red
white
2-ply, spun
directly from
mini-combs
Yearling mohair
clean
clean
Yearling has
enough body to
hold textures.
raw
raw

Using Mohair Fiber

Dyeing. Mohair takes dye readily and with sparkling results, and the colors have a liveliness to them that debatably cannot be exceeded by any other natural fiber (aficionados of silk and a few of the sheep breeds will be on the other side of the heated discussion, but the mohair folks will have a good case).

Fiber preparation and spinning tips

- Look for a fleece that is neither too dry nor too oily. The former may be matted and dull; the latter may be gummy and hard to clean.
- Depending on fiber length, mohair can be prepared by hand or machine carding (use carding cloth that matches the fiber's coarseness or fineness), combed, or spun directly from loosened locks.
- If you're new to mohair, it may feel exceptionally slippery in drafting. Loosen the draw-in on your wheel. Be sure to put in enough twist to hold your yarn together, but not so much that it becomes wiry. Adequate twist will also keep the slick fibers from shedding out of the yarn later.
- In very low-moisture areas, you may want to use a spritz of water, with or without a dash of something to control static electricity, such as olive oil (although in general mohair is static resistant).
- If you want to blend mohair with wools, which can contribute wool's elasticity and bounce, be sure to match the wool quality to the mohair quality in both fiber diameter and length.

Knitting, crocheting, and weaving. Mohair takes to any technique you want to use with it: knitting, crochet, weaving, laces, whatever. The finest fibers can be used for next-to-the-skin or baby garments. The midrange ones make durable and elegant clothing (which some folks will tolerate next to their skin and some won't), as well as pillows, blankets, and other household comforts. The coarsest fibers want to be rugs, bags, upholstery, and other sturdy textiles. If you want your mohair to develop a fluffy surface, spin for lightness and leave enough room between your stitches or in your weaving sett for the fibers to expand. If you want a sleek fabric, use a smooth-spun yarn with more twist and work it more closely.

Best known for. Luster, durability, length, a range of fiber-diameter qualities, and versatility (when the fiber is matched to the purpose). For use in doll hair and for making many of the highest-quality teddy bears. Clear white and now great colors also.

Cashmere

Toggenburg goat, a cashmere producer

CASHMERE IS THE MOST readily available luxury fiber, with such cachet that small percentages are often added to yarns or fabrics, for marketing and pricing reasons as much as for tactile ones. Followers of dramas in the knitting-yarn world may remember a series of lawsuits a few years ago concerning the cashmere content of a widely distributed yarn. (DNA analysis can now easily, if not cheaply, determine exactly what's what, even for fibers that have been bleached or dyed.) Given cashmere's appeal and the complicated international dealings involved in its production and distribution, this potential for misapplication of the fiber label is hardly surprising — and it's a reason to know about the stuff with your own fingers, and to go for higher percentages (of higher quality) in your textiles rather than less. In other words, if you're going to indulge, be informed and then don't be halfhearted about it! Go all the way to paradise — or else intentionally visit the pleasant, if less indulgent, nearby neighborhoods (like cashgora, which has its own section, starting on page 352).

The word *cashmere* comes from Kashmir, which originally referred to a high valley between two mountain ranges of the Himalayas on the Indian subcontinent, but today refers to a larger — and politically disputed — territory sandwiched between India, Pakistan, and China. Especially high-quality cashmere, and lots of it, is grown in this roof-of-the-world region where the globe's highest mountain peaks, such as Mount Everest (29,029 ft/8,848 m) and K2 (28,251 ft/8,611 m), reach for the heavens. Cashmere refers not to a breed of goats but to the downy underfiber that all goats, except the single-coated Angoras, produce as insulation against the cold of winter and then shed in the summer. Certain breeds and certain strains within breeds produce more cashmere than other breeds and strains, though goats (again, except Angoras) grow more down if they live in frigid climates than if they live in balmy ones.

Northern China produces about half to two-thirds of the world's cashmere, and Mongolia, Tibet, and Afghanistan have also been the sources of significant quantities, with additional bulk amounts coming from central Asia, Iran, Australia, and New Zealand. In North America, the main breeds being raised for cashmere production are Pygmy goats, small animals that can be readily kept as pets or in mini-flocks; Spanish goats, which are generally raised as meat animals and trace their ancestry to early imports by Spanish Conquistadors; and Myotonic, or Tennessee Fainting, goats (for more on these goats, see the box, I Feel Faint, on page 348). Through strategic selection of breeding stock from among feral goats, producers down under (in Australia and New Zealand) have spent the past several decades developing greatly improved strains of goats that are known for high-quality and abundant cashmere production. Some North American cashmere producers have been importing goats and semen from these animals to improve the cashmere production in their own flocks.

I Feel Faint

Myotonic goats share an unusual inherited condition called congenital myotonia. These animals — also known as Tennessee fainting goats, wooden-leg goats, nervous goats, stiff-leg goats, scare goats, or epileptic goats — stiffen up and fall over when they are startled or frightened. They remain conscious and recover quickly (usually in about 10 seconds). Myotonia, which in the case of these goats is caused by a recessive gene combination, is sometimes seen in other animals and in humans as well. A similar, and more familiar, defense mechanism happens when opossums "play possum," going into a short, involuntary coma to convince predators that they're already dead and therefore won't be good to eat.

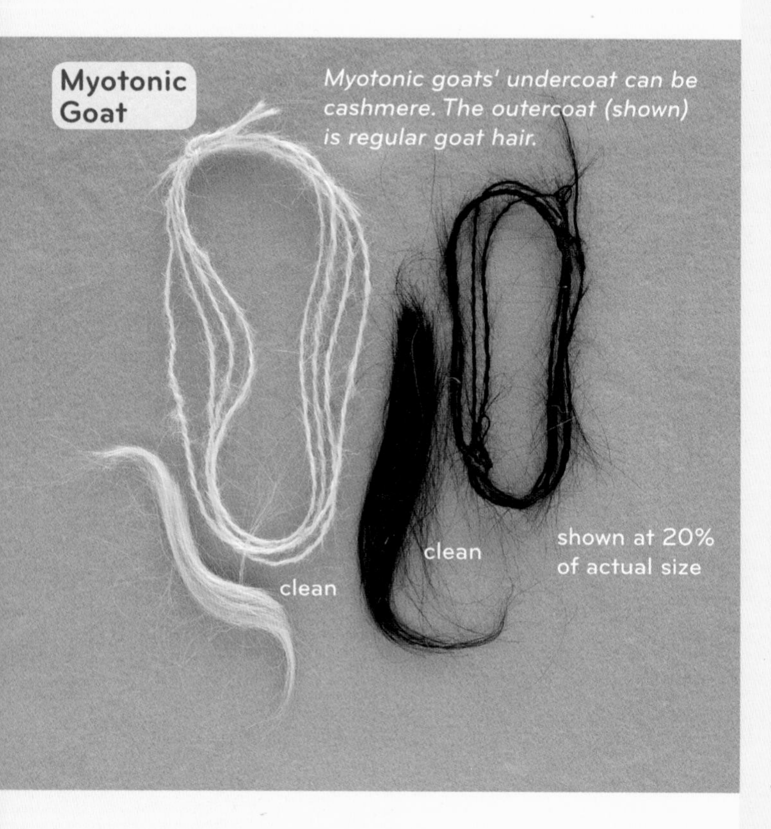

Myotonic goats' undercoat can be cashmere. The outercoat (shown) is regular goat hair.

When breeders here or abroad try to improve cashmere production, they select their breeding animals with the goals of maximizing four traits of this down: its quantity; its fineness; its appropriate character; and a clear distinction between the down and the guard hairs, so that dehairing becomes not only possible, but comparatively easy. Quantity is self-explanatory. We'll talk about the other qualities in turn.

Fineness provides one of the defining characteristics of down that's allowed to carry the name cashmere. Does cashmere have to be finer than 18 microns? 18.5? or 19? We're talking intervals of a half-millionth of a meter, but in the cashmere world increments that small do matter.

One set of firmly grounded qualifying measurements comes from three entities that do agree with each other on this subject: the International Wool Textile Organization (IWTO), the Cashmere and Camel Hair Manufacturers, and the United States government (in the U.S. Wool Products Labeling Act of 1939, as amended). All three, in slightly different words, require that the fibers be produced by goats; that they have an average fiber diameter of 19 microns or less with no more than 3 percent (by weight) of fibers measuring larger than 30 microns; and that they have a coefficient of variation not greater than 24 percent (a long and technical way of saying that the fibers have a specified amount of consistency). But there are also trading definitions in use that place the cutoff point at either 18 or 18.5 microns. The organizations responsible for maintaining the good name of Scottish cashmere, known for its exceptionally high quality, require that fibers measure under 16 or 16.5 microns.

In the industrial textile world, where most of the global trade in cashmere takes place, fibers between 14 and 16 microns are called hosiery grade and are used for knitwear, while those between 17 and 19 microns are used for weaving. Chinese, Mongolian, and U.S. cashmere can fall into either category, while Iranian and Afghani cashmeres are usually in the latter group, or possibly a bit coarser.

Appropriate character or style is another requirement for a fiber to qualify as cashmere. This involves the fiber's look and feel. For example, the desired crimp is irregular and three-dimensional and comes in the form of many tiny waves. This means yarns spin up to be light, airy, and warm.

Down vs. guard hairs. A clear distinction between the down and the guard hairs can make or break a fleece's entry into the realm of cashmere. Hairs are straight and stiffer than down. In order to get the true cashmere experience, you've got to remove the hair. The ideal fleece composition involves very fine down and remarkably coarse guard hairs, so the fiber types are easy to separate. Intermediate fibers — guard hairs that are relatively fine, possibly even with micron counts equivalent to some of the down — can sabotage an application for admission to the exclusive cashmere club.

Historically, cashmere fiber was harvested by plucking or combing out the shedding down in spring and summer. Because hair and down shed at different times, the hand-plucked down was fairly clean of hair, but today many cashmere producers shear their animals and then have the mass of fiber dehaired. Dehairing sheared fiber is never easy: it requires either extensive hand labor or specialized industrial equipment. Anything that gets in the way becomes a nonqualifier. (See our discussion of cashgora, on page 352, for information on what happens with fibers on the periphery of the cashmere classification; they have some attributes that may make them more attractive to use than pure cashmere in many cases, even if they're not *la crème de la crème*!)

The end result of fibers that meet the qualifications to be granted the cashmere crown is that they go forth into the world, bringing a floating lightness and treasured softness to appreciative and discerning humans. The winning fibers may display a touch of luster, although luster tends to increase with movement toward larger fiber diameters (and larger light-reflecting scales on the fiber surface), so the finest and softest cashmeres are the least likely to shine in the literal sense.

Speaking of finest, *pashmina* is a term for unusually fine cashmere, originally (and still) grown in the stark climate of the high Himalayas by Changthangi and Cheghu goats; some is apparently being grown now in Mongolia as well. Yields are light in more ways than one: an animal only produces 2 to 7 ounces (57-199 g) annually with average fiber diameters in the vicinity of 11.3 to 12.8 microns, although with relatively generous fiber lengths for fibers that fine — between 2 and 2¾ inches (5-7 cm).

Cashmere down does not have a reputation for felting, although if it's been blended (whether obviously or not) with wool or another fiber that does like to felt, a cashmere-containing fabric will go along for the ride.

Cashmere yarn from Sarah's Yarns

Cashmere Facts

▶ FLEECE WEIGHT

Given the number of types of goats that can grow cashmere, generalizations on total fiber weight and cashmere yield can only be rough at best. Because we want to provide some ballpark numbers here, let's say that a raw shorn fleece weight of 1–2½ pounds (0.5–1.1 kg) will yield 3–18 ounces (85–510 g) of down. This works out to between 20 and 60 percent down; a lot depends on the goat's fiber profile and the quality of the processing.

▶ STAPLE LENGTH

U.S., Australian, and New Zealand standards specify 1½–2½ inches (3.8–6.5 cm). International standards call for at least 1¼ inches (3.2 cm), and Chinese cashmere ranges from 1¼ to 1¾ inches (3.2–4.5 cm), while Mongolian cashmere is somewhat longer. Some commercial spinning conditions require 2–4 inches (5–10 cm), by which point the territory of cashgora is being approached, if not inhabited.

▶ FIBER DIAMETERS

Assuming the other factors qualify the fiber as cashmere (length, crimp pattern, and distinction between down and guard hairs), most of this premium down has average fiber diameters in the range of 14–19 microns. A few, notably those called pashmina, will be finer. Fleeces and fiber lots with coarser averages are classed as strong cashmere, cashgora, or caprine fine fiber (CFF), semiequivalent terms used in different locations. Strong cashmere has average fiber diameters of 19–22 microns along with clearly differentiated guard hairs. The coarse guard hairs (the ones that can be removed comparatively easily from all grades of cashmere) can be in the range of 50–100 microns.

▶ LOCK CHARACTERISTICS

Fine down, combed out or shorn, with easily identified and removed coarse guard hairs.

▶ NATURAL COLORS

White (the most commercially valuable, because it can be dyed to any color); light to medium to dark browns; and light to medium grays.

Using Cashmere Fiber

Dyeing. White dyes to any color, and the natural browns and grays add interesting undertones to any applied color.

Fiber preparation and spinning tips

- Dehairing cashmere is a task for the patient (or for long phone conversations, with a headset so both hands are free to work) or for machines, and some mechanical dehairing can damage the fiber. There is lovely prepared cashmere available for spinners.
- Cashmere can, depending on fiber length, be spun from the just-dehaired mass; carded; or combed. Because it's a soft, fine, and delicate fiber, its tendency to pill can be counteracted by gentle handling, along with selecting comparatively long fibers, spinning them into a fine yarn, and making sure there's enough twist to hold and protect the fibers within the yarn.
- Enough twist to get the yarn onto the spindle or bobbin is not necessarily enough twist to get the yarn off again. Also make sure you're using enough twist to secure the fibers in the strand but not so much that the yarn becomes wiry.
- Spin fine and then ply to build up heavier weights of yarn — keeping in mind the warmth factor. Heavy cashmere yarns will be beyond cozy.

Knitting, crocheting, and weaving. Delicacy and softness characterize cashmere; what you make from it will need to be lightweight and not subjected to lots of abrasion or wear, although it can handle significantly more than, say, English angora. Well-spun cashmere becomes a versatile yarn that can be used for understated fabrics (stockinette or plain weave or single crochet) that let the fiber speak, or can step back and let a lace pattern or twill show up, if through a haze of softness. Don't be afraid of cashmere for everyday wear, but if you make socks from it, plan to treat them gently and not jam them into your hiking boots.

Best known for. Softness, lightness, warmth, and luxurious delight, along with gently appealing natural colors. Cashmere deserves its standing as a classic fiber.

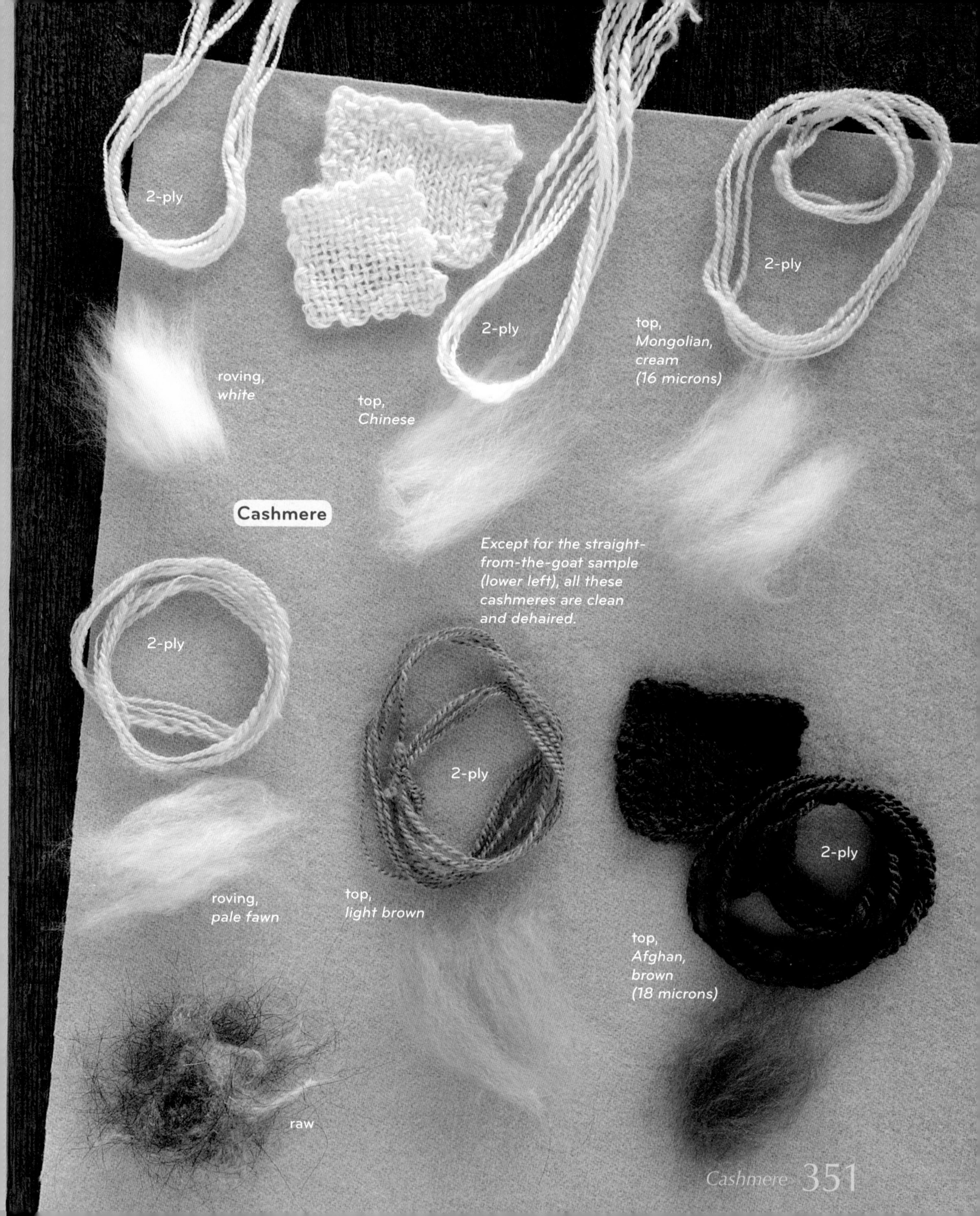

2-ply

2-ply

2-ply

roving,
white

top,
Chinese

top,
Mongolian,
cream
(16 microns)

Cashmere

Except for the straight-from-the-goat sample (lower left), all these cashmeres are clean and dehaired.

2-ply

2-ply

2-ply

roving,
pale fawn

top,
light brown

top,
Afghan,
brown
(18 microns)

raw

The Goat Crosses

Mohair offers strength, luster, smoothness, and drape. Down offers softness, lightness, fluffiness, and warmth. Wouldn't it be nice if goats grew a fiber that combined those qualities, and wouldn't it be nice if that fiber came in a rainbow of natural colors? Well, crosses between Angora goats and goats of other breeds (the cashmere producers) do yield really interesting fibers for handspinners, and sometimes for fiber folk who like to start with a ready-made skein of yarn. We liken this crossbreeding to chasing rainbows — and sometimes catching them.

Unfortunately, in the global commercial marketplace these crosses definitely reduce the value of the fiber, because it doesn't fit into the industrial processing scheme (mohair and cashmere on their own offer enough challenges to the mechanical systems). This means these crosses may never reach the critical mass of breeding animals to become truly distinct breeds that consistently reproduce the desired traits and the best fiber. And all that adds up to some perplexity among those who buy fiber and yarn, because new names keep cropping up attached to the latest fascinating discovery. Yet the batch of fiber X purchased last year may not resemble the batch of what is ostensibly the same type of fiber this year. Even two batches of fiber X sitting right next to each other may be radically different.

There are lots of reasons why this is the case, and while we can't give you a full scorecard for what to expect, we can introduce you to this part of the fiber world with enough information to keep you from getting dizzy.

The Big Picture

You'll likely encounter other crosses of goats for fiber production; we hope our discussion gives you a framework for understanding how they might come about and what the breeders' goals might be. If you find a strain or herd whose fiber you like, you can have a special and personal connection to the grower and the animals. Bred and processed well, it can be a fine material that is worth a premium price, and it can help you experience the allure that causes the breeders to chase their rainbows.

Cashgora

So is cashgora just a cashmere also-ran? No way! Though it doesn't ace the softness-and-lightness category, many of cashgora's qualities make it a better choice than cashmere: It's more durable (wears better), doesn't pill as much, and is easier to wash — and if you're a spinner, it may offer a more relaxed spinning experience.

Although we're using it to introduce the crossbred goats (as in the cross of a cashmere-producing goat with an Angora), the term *cashgora* may not always refer to fiber grown by crossbred goats. The word has been bandied about with serious ramifications since 1972, when it was used to describe goat fiber that didn't meet U.S. specifications for cashmere. Now there are two distinct and conflicting definitions for *cashgora*, with much confusion surrounding them.

One definition encompasses fibers that are similar to cashmere, but that don't qualify as cashmere for a handful of reasons. Usually they have larger average fiber diameters — likely between 19 and 23 microns — or they lack cashmere's "style," which means, to oversimplify

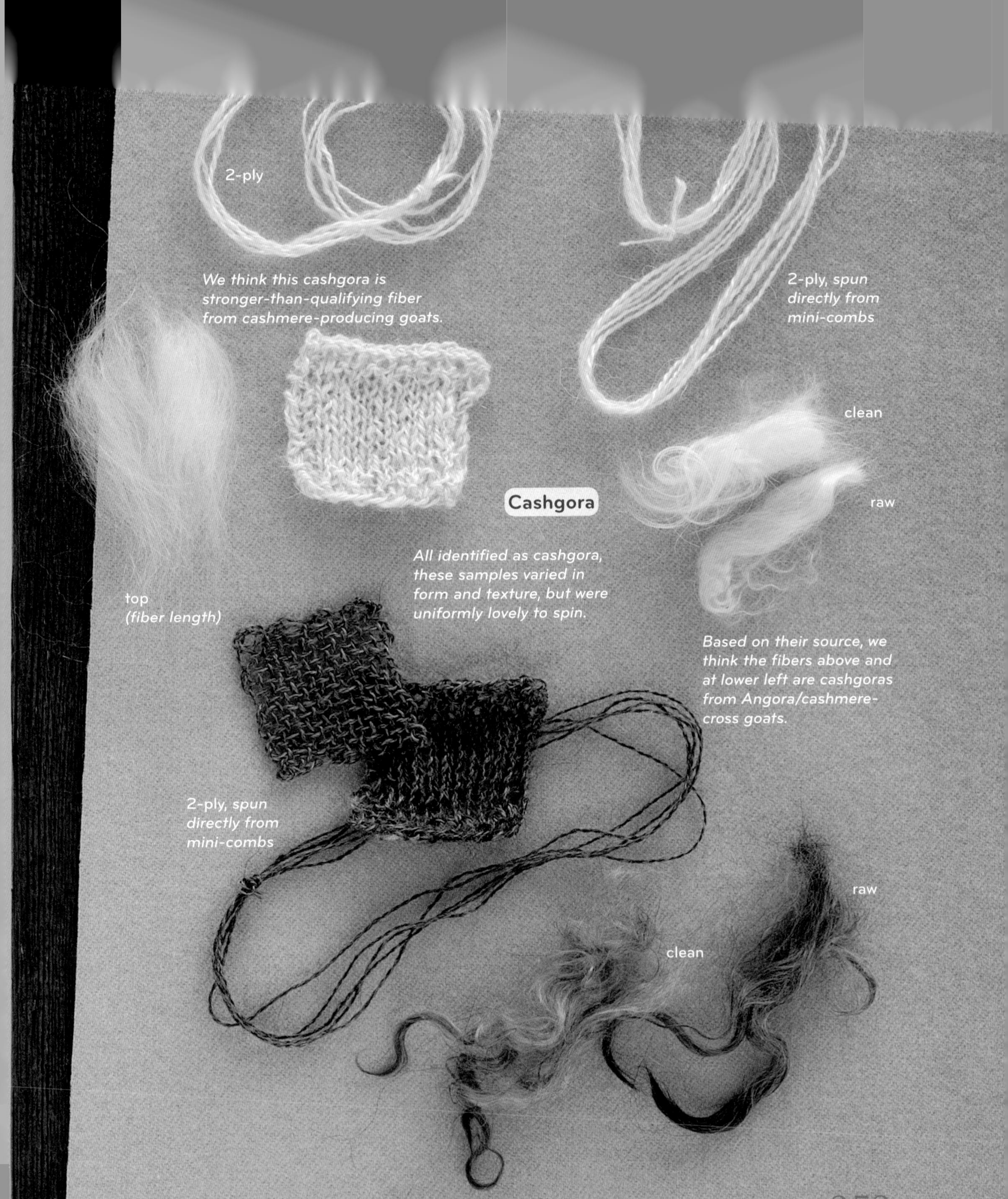

2-ply

We think this cashgora is stronger-than-qualifying fiber from cashmere-producing goats.

2-ply, spun directly from mini-combs

clean

Cashgora

raw

All identified as cashgora, these samples varied in form and texture, but were uniformly lovely to spin.

top (fiber length)

Based on their source, we think the fibers above and at lower left are cashgoras from Angora/cashmere-cross goats.

2-ply, spun directly from mini-combs

raw

clean

the idea of style and pinpoint the primary difference, that they are more loosely crimped. Within cashmere-producing herds, cashgora-style fibers (as opposed to cashmere-style fibers) are longer, straighter, more lustrous, and wavy (like mohair), not crimpy.

A fleece may also get dropped into the cashgora pile because it contains a third type of fiber, intermediate between the down and guard hair, making it difficult or impossible to differentiate between the two fiber types and thus successfully dehair the mass. The presence of these intermediate fibers, and their implications for dehairing, will disqualify it from being called cashmere, no matter how fine it is.

Sometimes there is differentiation between the fiber types, and the fine fibers are nicely crimpy, but the down simply isn't fine enough to qualify as cashmere. In some places, and in cases where the average fiber diameters are between 19 and 22 microns, this may end up being called strong cashmere (see our discussion of cashmere, page 348).

Transition from cashmere to cashgora may be the result of age. Like mohair goats, cashmere producers grow their finest fibers when they're kids and young goats. Goats — even young ones — that have not been on good-quality feed, or that were ill during the growing season, may produce fibers that don't qualify as cashmere.

The other prevalent definition of cashgora points to fiber grown by goats born from crosses between mohair- and cashmere-producing parents. Any cross of a cashmere goat with an Angora goat yields a cashgora cross, though some are now named crosses, such as those that follow.

Regardless of how a fiber has acquired the cashgora label, this very useful material occurs in the fiber-diameter range of many Merinos, tucked between cashmere and most of the mohairs except the finest kid mohair, with which it shares average fiber diameters but not overall feel.

Cashgora Facts

▶ FLEECE WEIGHT
Usable annual fiber yield per goat ranges from under 4 ounces (113 g) to 14 ounces (397 g)

▶ STAPLE LENGTH
Average 3–4 inches (7.5–10 cm)

▶ FIBER DIAMETERS
Average 18–23 microns, mostly 19–22 microns

▶ LOCK CHARACTERISTICS
Depending on where in the world, and under which system, it's being classed, cashgora either contains three types of fibers (down, guard hairs, and intermediate hairs) or consists of what might alternatively be called strong cashmere, with down (the cashgora portion) and coarse guard hairs but no intermediate hairs. Has some luster, not much crimp, and a smooth, silky feel.

▶ NATURAL COLORS
Mostly white, but also light grays and browns.

Using Cashgora Fiber

Dyeing. The fiber takes dyes nicely, with influence from the underlying natural color.

Fiber preparation and spinning tips. If you're working from industrially produced top, condition it (gently loosen the fiber mass so the fibers will slip past each other) before you spin. Freshly hand-combed top won't have this problem. You can card cashgora, using fine carding cloth. Fine and slippery, cashgora needs more twist than you might think (so the yarn has enough integrity to come on and off the bobbin or spindle but not so much that your yarn becomes harsh or wiry). Keep your hands relaxed.

Knitting, crocheting, and weaving. Regardless of why it was classified as cashgora, the fiber is likely to be smoother and shinier than cashmere, with less elasticity and bounce and more of a tendency to drape well. Because of all the possibilities, experiment to determine what the fiber in your hands will do best.

Best known for. An overlooked, valuable alternative to cashmere: It is likely to have more durability and luster, less tendency to pill, and a silkier feel.

Pygora

Pygoras have been in development for more than a quarter century, which makes them old-timers in this discussion of recently developed fiber-goat crosses. Pygora fiber has been available to spinners for a number of years and pygora yarns can now be obtained in both natural and dyed colors. The vocabulary and methods of raising pygora establish foundations for understanding other similar breeding efforts.

The pygora breeding efforts began in the late 1970s, when Katharine Jorgensen, a schoolteacher from Oregon who was also well versed in fibers, saw colored mohair goats in Arizona. She yearned for a mohair in a blue-gray color she'd seen before in Pygmy goats (at the time, there were fewer colored mohair animals than there are now). She crossed mohair and Pygmy goats and, after five years, got her blue-gray and a whole lot more.

Pygoras originated through crossing mohair goats registered with the American Angora Goat Breeders' Association (AAGBA) and

Pygora

Type A, *as shorn*

2-ply

2-ply, *from hand dehaired*

The Type A pygoras resemble mohair (A for "Angora").

Type A, *as shorn*

Type A, *hand dehaired*

Type A, *as shorn*

2-ply

Type A/B, *commercially dehaired*

2-ply

Pygmy goats registered with the National Pygmy Goat Association (NPGA). These are the only genetics within the pygoras. In addition to the grays, the Pygmies contributed other colors: a full range from black to white, dark to light gray, pale to dark brown, and some light reds. Black appears in the animals but is usually carried in the guard hair, so the spinning fiber is lighter in color. The breed association is PBA, Pygora Breeders Association. Keep those initials in mind. You'll need them in a few minutes to allay confusion.

Three distinctively different types of fleeces are grown by pygoras, labeled Type A, Type B, and Type C. It's easy to remember the fiber emphasis of each type: A is Angora-like (or most like mohair), B is a blend, and C is cashmere-like. Each goat produces between 6 ounces (170 g) and 2 pounds (907 g) of usable fiber per year. Pygora does felt.

Type A. Like mohair, this fleece hangs in long, lustrous ringlets. It's sleek and silky and resembles mohair that's in the kid and young-goat range (usually below 28 microns) and can be up to 6 inches (15 cm) long. Ideally it's single coated, as mohair goats are, but in practice it usually has almost invisible fine guard hairs. Type A fleece, again like mohair, must be shorn.

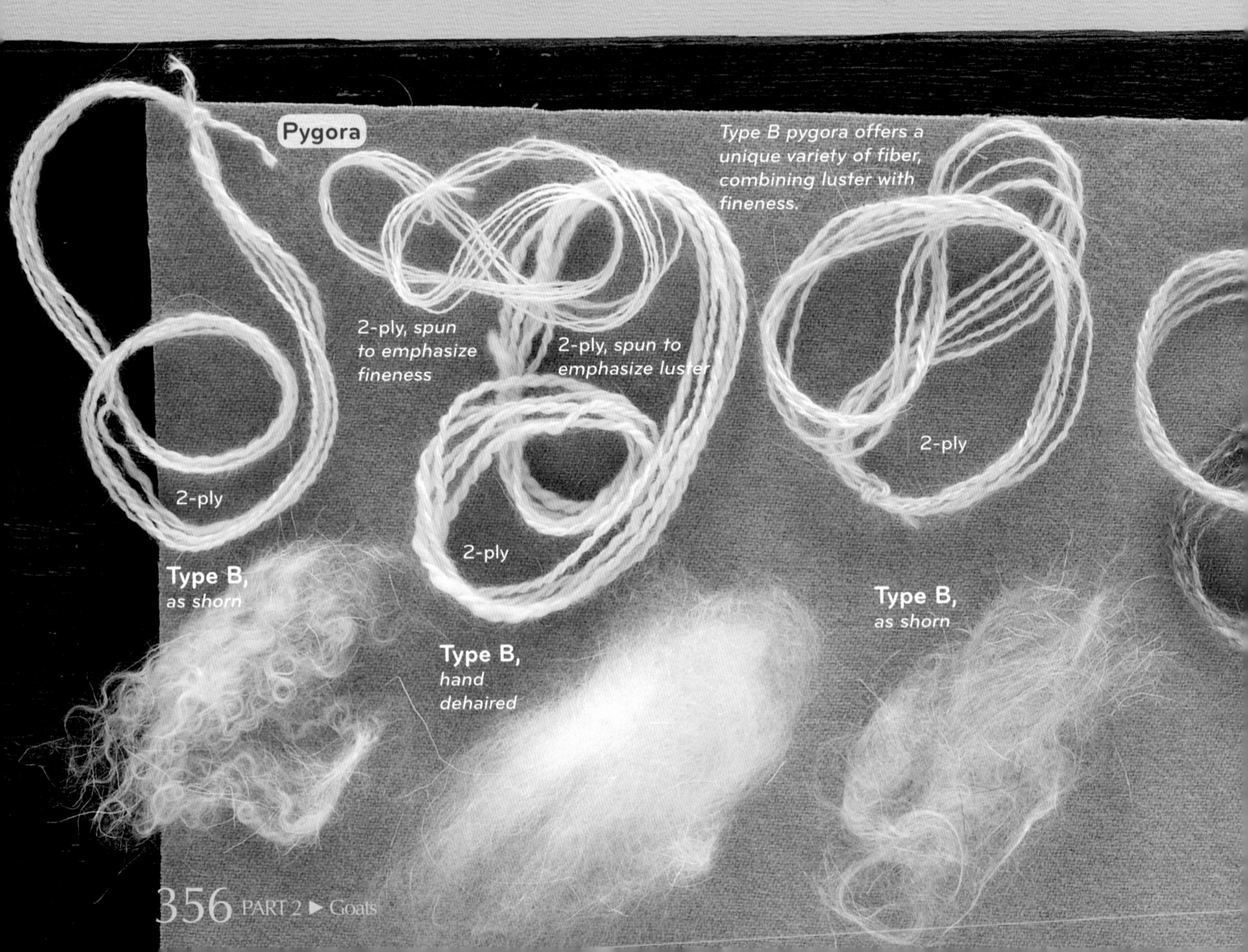

Type B. This fleece includes several types of fiber. A grower we spoke with said that this is her favorite "flavor" of pygora fiber, and in her opinion the undercoat combines the best of the mohair and cashmere worlds. There is at least one type of guard hair, and there may be two types, one coarser and one quite fine. The locks are curly, have some luster, and are between 3 and 6 inches (7.5–15 cm) long. The average fiber diameter is usually below 24 microns, so overall it's finer than the Type A fleece. Type B fleece can be shorn, combed out, or plucked when it releases seasonally.

Type C. This fleece can sometimes be accepted as true cashmere, with average fiber diameters tending to be below 18.5 microns (see our discussion of cashmere, page 348). Minimum acceptable length is 1 inch (2.5 cm), and fibers can be as long as 3 inches (7.5 cm). The luster vanishes at this stage, being replaced by crimp and a warm feel. Like other cashmere-type fleeces, there should be clear differentiation between the guard hair and the down to facilitate separation of the coats. Type C fleece can be shorn or combed out.

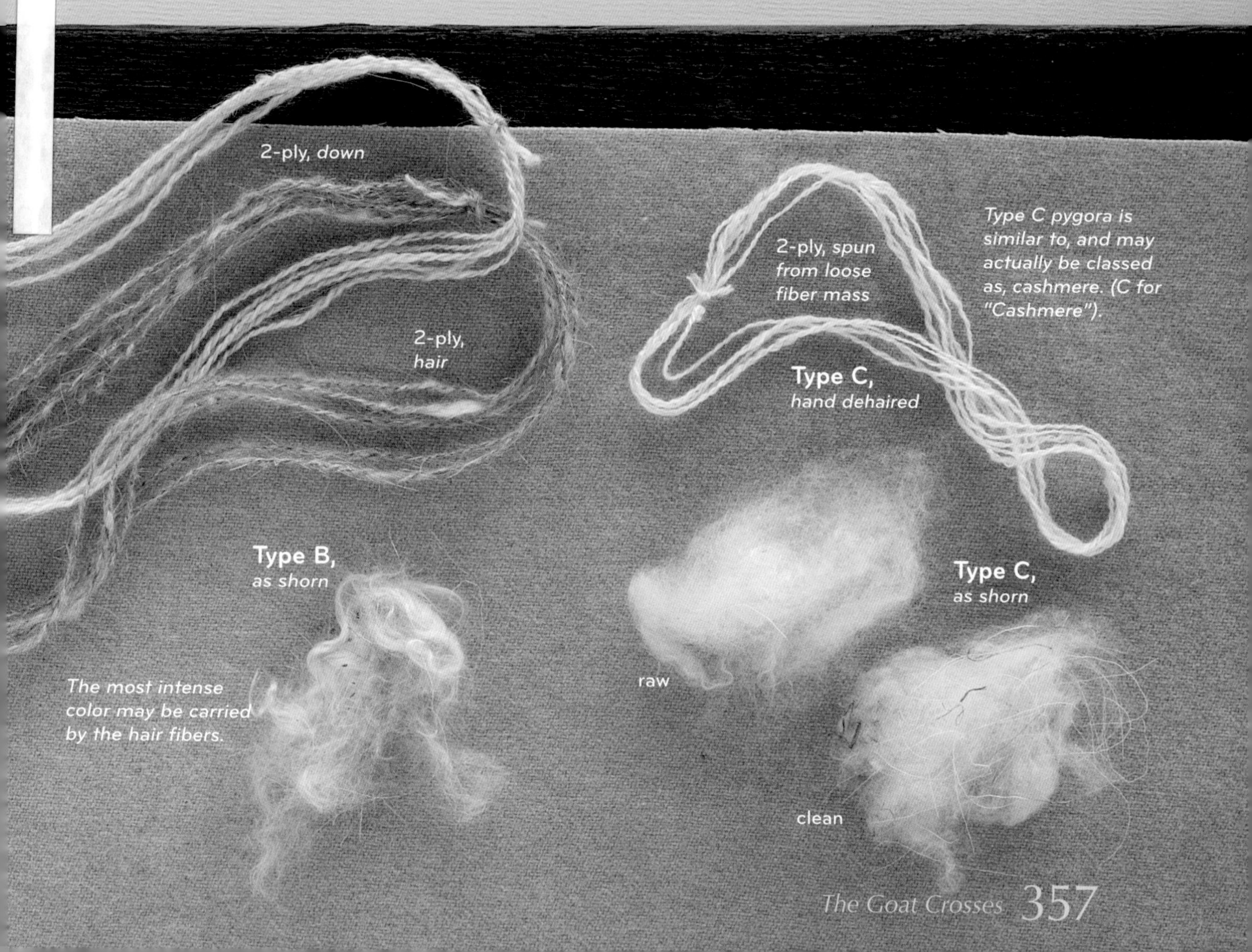

Type C pygora is similar to, and may actually be classed as, cashmere. (C for "Cashmere").

The most intense color may be carried by the hair fibers.

PCA

The fleece types defined for pygoras also occur in PCA goats. The name is just PCA: the *P* stands for both *Pygmy* and *pygora*, which are in this type's parentage; the *C* stands for *cashmere* and *color*, both of which are factors; and the *A* stands for *Angora*. The PCA Goat Registry coordinates this breeding effort.

How are PCA goats different from PBA (Pygora Breeders Association) goats? PBA pygoras, discussed in the last section, include only bloodlines from registered AAGBA (Angora) goats and NPGA (Pygmy) goats. PCA goats include those ancestries, plus crossing in from goats registered with the Colored Angora Goat Breeders Association (CAGBA), which came along after PBA was well underway.

PCA goats also have Types A, B, and C fleeces, defined in the same ways as the PBA pygora fleeces are.

If you're out in the world looking at fiber, a PCA Type B fleece should, for example, be similar to a PBA Type B fleece. However, as we've noted, the family trees that produced the two alternatives will have some differences.

Pycazz

Breeders looking for fiber as fine as cashmere combined with twice-a-year fiber crops (cashmere is gathered once a year, whereas mohair is shorn twice as often) began with Type B pygora and cashmere parents and named the offspring *Pycazz*. The ideal Pycazz grows a fleece with an average fiber diameter finer than 18.5 microns (cashmere territory; see page 348), but the acceptable range goes up to 24 microns.

Infusions of known cashmere-producing goats, Pygoras, and other types of fiber goats are acceptable in this endeavor, including both PBA and PCA animals of all fiber types, as long as the goals of fine fiber and extra harvest opportunities are kept in mind; at least one-quarter of the breeding has to reflect cashmere heritage. Direct cashmere/Angora crosses are not eligible for registration. Colors include white, blacks, and various shades of brown, with and without reddish tones.

Nigora

Under the auspices of the American Nigora Goat Breeders Association (ANGBA), this type comes about through the crossing of mohair goats with Nigerian Dwarf goats. What's missing, when compared with the types of goats we've already mentioned in this group, is the Pygmy component. The Nigerian Dwarfs contribute interesting fiber colors as well as distinctive color patterning on the animals. The goal is a small, smart, friendly goat with a wide array of fleece colors. The mohair goats can be from any of the standard registries, including the colored Angora registry; and an influx of other types, as defined by the Nigora registry, is also acceptable. Again, there are Type A, B, and C fleeces, as defined for Pygoras.

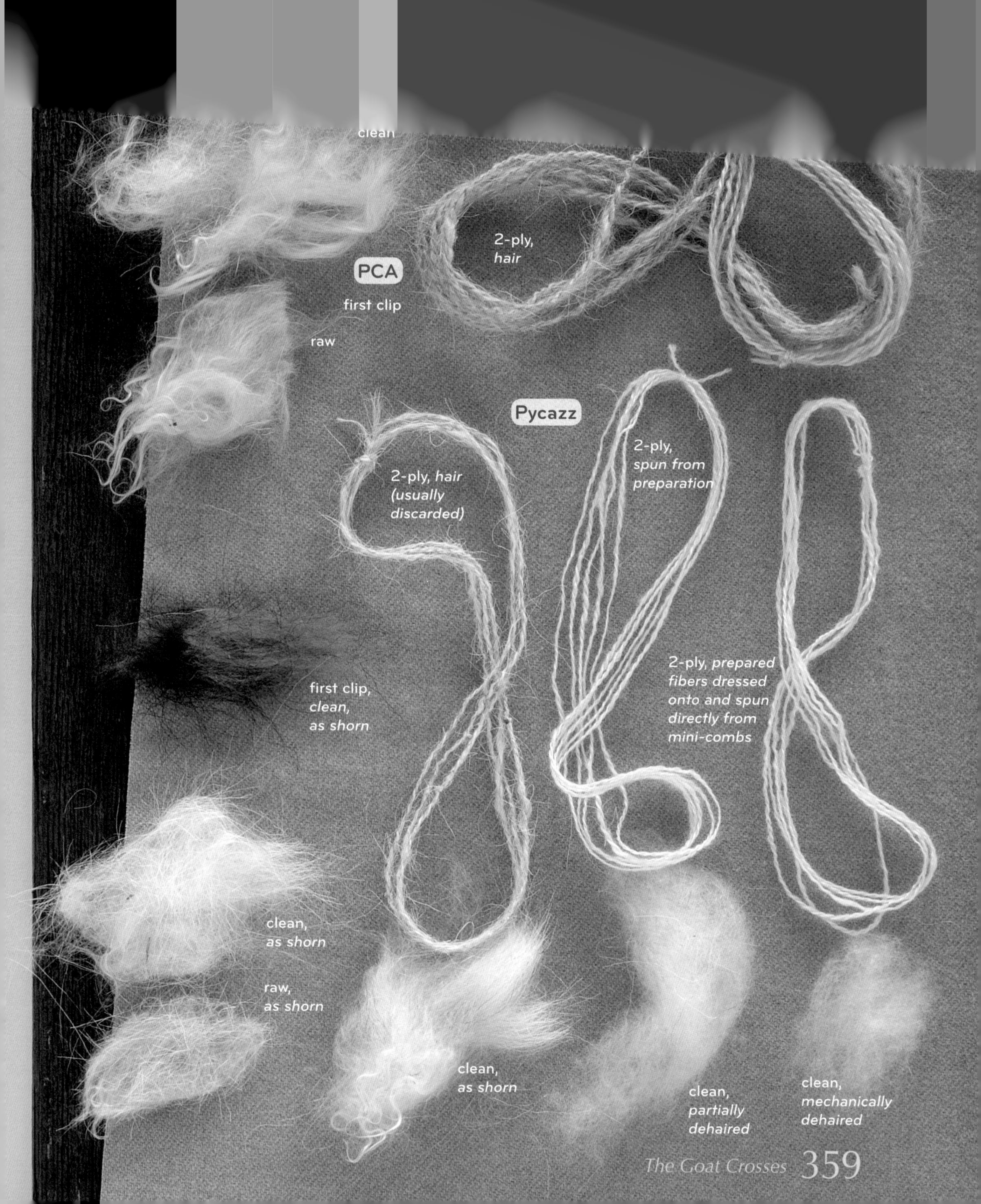
clean
2-ply,
hair
PCA
first clip
raw
Pycazz
2-ply, hair
(usually
discarded)
2-ply,
spun from
preparation
2-ply, prepared
fibers dressed
onto and spun
directly from
mini-combs
first clip,
clean,
as shorn
clean,
as shorn
raw,
as shorn
clean,
as shorn
clean,
partially
dehaired
clean,
mechanically
dehaired

Bactrian camels
Alpaca

Camelids

The camel family (known to scientists as the Camelidae) has a long history containing many intriguing events and wide geographic movements, resulting in two major branches that bring us fiber today. The family's oldest ancestors originated around 40 to 50 million years ago on the Great Plains of North America and were small animals, not much larger than jackrabbits. Over millions of years, the family tree branched and formed a number of subgroups, or tribes. One tribe included gazelle-like critters, while another included giraffe-like ones. The tribe known as Lamini ultimately gave rise to the New World camels, which migrated to South America about three million years ago. Another tribe, the Camelini, led to the Old World camels, which migrated to Asia across the Bering Land Bridge at about the same time.

The wild ancestors that gave rise to the camels we know today died out in North America 10,000 to 12,000 years ago, victims of climate change and hunting by the Lithic peoples, the first human inhabitants of the land masses now known as the Americas. Today we can see what became of the Lamini tribe by looking at their descendants: the alpacas, llamas, guanacos, and vicuñas of South America. We can also observe how the Camelini evolved through the Bactrian and dromedary camels of Asia and Africa. The good news for us is that all these camelids have usable fiber; the guanaco and vicuña have some of the softest and most luxurious fiber in the world!

As camelids evolved, they developed some unique traits. All members of this family have three stomachs, a situation that allows them to digest fibrous plants, like grasses and legumes. Known as pseudoruminants, the camelids are kissing cousins to the true ruminants, or four-stomached animals, such as sheep, goats, cows, bison, and yaks, as well as many wild herbivores ranging from antelope and deer to giraffes, musk oxen, and reindeer. In both pseudoruminants and true ruminants, one of the stomachs — the rumen — acts like a giant digester that converts grass to gas.

All ruminants chew cud, which means they burp up small wads of food that they re-chew. Once cud has been chewed, it can pass to the rumen, which is populated by millions of bacteria, fungi, and other microorganisms that help break down the fiber in the animals' diet into usable nutrients. We, with our single stomachs, would starve to death if we had to live on a diet composed mainly of grass, because we can't convert fiber to protein; with the help of those microorganisms, these guys do that magic trick.

Another trait that the camel family members share is funny feet. Each flat-shaped foot has a leathery pad on the bottom and two distinct toes with sizable toenails. This setup provides a lot of surface area, so camelids can travel over loose ground better than most critters. This relative light-footedness also explains llamas' popularity as pack animals, especially in vulnerable high-altitude environments. Their feet don't chew up the ground as they walk, like the hoofed feet of most other herbivores do.

Do Camels Really Spit?

All of the camelids can spit — and do, when annoyed, scared, or stressed — but happy and healthy camelids rarely indulge. The spit, which makes a statement that's hard to ignore, is actually a combination of saliva and cud (or regurgitated food).

Camelid Vocabulary

Calf. Baby camel.

Cama. Cross between female llama and male camel by artificial insemination, through a limited experiment undertaken in the United Arab Emirates from the late 1990s to early 2000s.

Ccara, Q'ara, or Qera. Classic style of llama with two subtypes: ccara (short wooled) and curaca (medium wooled).

Chulengo. Baby guanaco.

Cria. Baby alpaca, llama, or vicuña.

Huarizo or Wari. Hybrid from llama/alpaca cross (the term is derogatory, in reference to the animals, which are neither llamas nor alpacas).

Lanuda. Subtype of woolly llama (tampuli); heavy wooled with extra fringes of fiber on various body parts.

Tampuli. Woolly llama with two subtypes, tapada and lanuda.

Tapada. Heavy-wooled subtype of woolly llama (tampuli).

Tui. Yearling alpaca, or alpaca up to first shearing (like hogg or hogget in sheep).

Camelids also run in a way that strikes humans as comical, although it works well for them. They lift both feet on the same side simultaneously. Also, because of the way in which their hind legs join their bodies, when they lie down they fold their legs directly beneath themselves (a move called *kushing*) by bending both front and back knees.

We're not done with the camelids' unusual traits yet! They don't have horns or antlers, which are common in other ruminant species; their red blood cells are oval in shape, whereas most other mammals have round red blood cells; and all members of the extended clan have long necks. The Old World camelids have humps (one for the dromedary, and two for the Bactrian camel); the New World camelids do not.

Camelids are genetically close enough that they can interbreed, creating hybrid offspring (similar to the mules that come from interbreeding horses and donkeys). New World and Old World camels interbreed naturally within their own tribes; however, the only interbreeding across the gap between New World and Old World camels has occurred through artificial insemination.

Vicuñas, guanacos, and Bactrian camels are found in the wild, where they live in groups, each made up of a male with a harem. Groups of young and aged males that lose their harems will gather together. Bactrian camels once roamed throughout a large swath of Asia, but today less than a thousand animals are thought to remain in the wild in China and Mongolia. They are considered critically endangered, though there is also a domesticated strain of Bactrian camels that is not endangered.

Vicuñas were recognized as critically endangered in the 1960s, when the population had shrunk to fewer than seven thousand animals. To protect the remaining population, trade in the fiber became illegal. Conservation

Guanaco

efforts have succeeded in bringing the species back from the brink of extinction, and it has rebounded to over a quarter-million animals in Argentina, Bolivia, Chile, and Peru. Thanks to the improvement in population numbers and conservation management efforts, international trade in vicuña fiber is again legal, and small quantities are finding their way into the delighted hands of textile artisans. It is still illegal to transport the animals, a fact that has led to the development of a specialized North American variant of the alpaca known as the paco-vicuña (see page 369).

Guanacos are also threatened by hunting and by habitat loss, though their numbers have never plummeted as badly as those of the vicuñas.

Camelid Fiber Overview

The fiber produced by camelids is technically hair because of its structure, although the softer fibers are often referred to as wool. Some of the animals are single coated, some are double coated, and some have coats that contain gradations of fiber that are not clearly differentiated. To a greater (camel), lesser (alpaca, guanaco, and vicuña), or intermediate (llama) extent, camelid fleeces may contain medullated fibers. Some of these, known as guard hairs, are very coarse and stiff. They are a plus when the fibers are made into camel halters or rugs and a big negative in blankets or sweaters. Guard hairs can be hand-picked from a fleece, though sorting can be quite a time-consuming job, depending on the fleece.

Within the New World camelids, some breeding programs have successfully produced animals who grow long, shiny, smooth fibers that are not medullated, and that occur in the same micron count ranges as the wool-like fibers. The word applied to alpacas and llamas that produce this type of hair is *suri*. Baby camelids have the softest and most desirable fiber, and you'll often see it marketed as such. Expect to pay considerably more for "baby" fiber or yarn.

New World Camelids

Most of the South American camelid fiber we can get our hands on as yarn or spinning fiber is from alpacas and llamas. Vicuña and guanaco fibers are both very fine, and very scarce. Chasing down exactly what these fibers are like is complicated. Although all of the New World camelids originated with South American stock, they're now found throughout the world, with different husbandry and marketing practices from place to place; this leads to highly variable fiber and yarn. There are strong regional biases in what types of fibers are produced and how they are processed.

South American Camelid Family Tree

This is a rough sketch of the current understanding of family relationships among the South American camelids, based on the suggestions of recent DNA analysis. A lot of hybridization has occurred, especially (but not exclusively) among populations identified as llama. This matters dramatically when we talk about llama fiber and plays into the discussion of alpaca as well.

- Vicuña and guanaco (both wild) are primarily independent and distinct populations.
- Most contemporary alpacas show some hybridization with llamas. This affects fiber production, although alpaca fiber is far more consistent and easier to categorize than llama fiber.
- Llamas produce the most complex and diverse array of fiber types, the result of a greater history with, and ongoing experience of, hybridization.

Note: Our South American camelid interpretations are based primarily on the work of Jane Wheeler, Eric Hoffman, and Mike Safley.

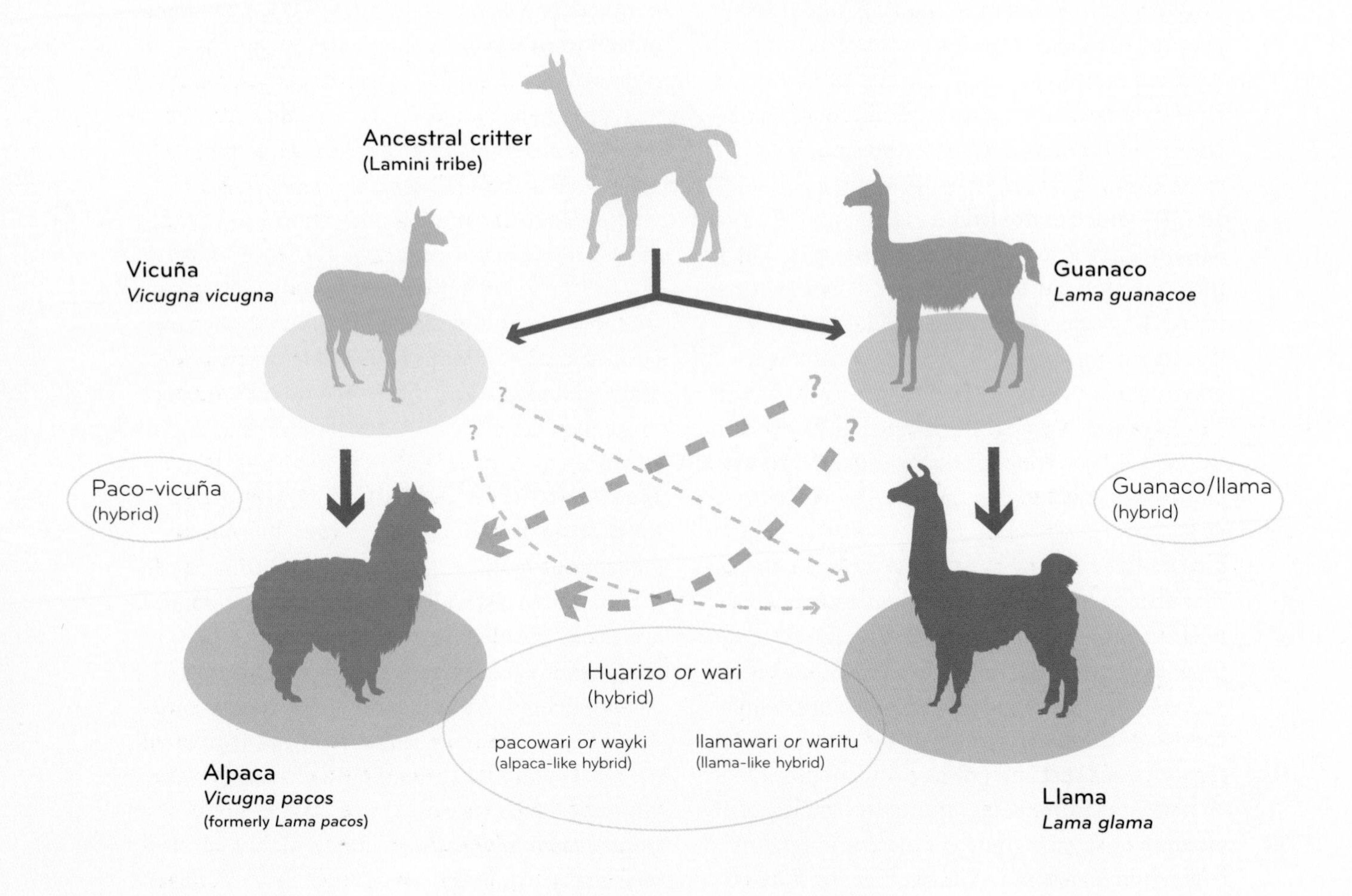

It's even more confounding that South American producers label fibers as alpaca or llama based on the fineness of the fiber, with no relationship to the animal that grew the fluff! Fiber grown in the 30-micron range by an alpaca gets labeled llama, and fiber grown in the 20-micron range by a llama suddenly gets put into the alpaca category. (See the chart on page 366.) Most llama and alpaca fiber and yarn in North America comes from the North American animal pool, though imported fibers and yarns are also available.

If the names applied to the fibers are sometimes interchanged, just what is the real difference between alpacas and llamas? They are distinct domesticated species. The best scientific information, using DNA, indicates that alpacas were first domesticated at least six to seven thousand years ago in the Andes mountains of Peru. Not long ago, everyone thought that the alpaca, like the llama, descended from the wild guanaco. DNA analysis now strongly suggests that the original parent of the alpaca was the vicuña.

The llama was domesticated from the guanaco shortly after the alpaca was domesticated, also in the Peruvian Andes. By the time the Spanish arrived in South America during the 1500s, giant herds of alpacas and llamas spread over much of South America. There were numerous recognizable breeds of each, some of which had much finer fiber and showed greater uniformity than is typically found in mature alpacas and llamas today.

But within one hundred years of the Spanish conquest of the Incan Empire, "native llama and alpaca herds virtually disappeared," according to Dr. Jane Wheeler, an archaeozoologist (an archaeologist who specializes in the early relationships between humans and animals). Wheeler, one of the world's leading experts on the history of llamas and alpacas, is using DNA testing and other techniques to try and bring back the fine-fleeced llamas and alpacas of the past by helping growers to select for those traits.

Peruvian weavers have constructed some of the world's most intricate and beautiful fabrics, working with alpaca and llama fiber and in recent times with some wool as well.

Much of what we know about these ancient alpacas and llamas comes from an amazing archaeological find in 1986, when 26 perfectly preserved alpaca and llama mummies dating back more than a thousand years were discovered in El Yaral, Peru. These preconquest discoveries show alpaca fiber in the 18-micron range and llama fiber in the 23-micron range. At that time, both types of animals were single coated, and the fiber was consistent in fineness and color. (Some coarse-fiber llamas, with single-coated fleeces in the 33-micron range, were also found.)

Modern animals have lost the consistency shown by their ancestors. Present-day alpaca fibers range from 9 to 88 microns, and llama fibers run from 8 to 144 microns. Fibers in both groups are more mixed than they used to be, and guard hair exists in both present-day groups, although it is much more common in llamas.

Alpacas are smaller than llamas, and whereas llamas are primarily used as beasts of burden in their native lands, alpacas are mainly raised for their fiber. Although neither is raised as a meat animal today, older animals may still be butchered for food by indigenous people in the Andes. Prior to the introduction of European meat animals, these were important food animals.

Fiber Classes for New World Camelids

Traditional South American alpaca herders called the first shearing "baby," and everything after that just "alpaca." The major industrial processors in Peru class camelid fiber in several categories. The women who do this work grade the fiber by sight and feel, although the companies that employ them assign micron-count descriptions to what the women's sensitive fingers know. Each company has slightly different classifications, and you won't know which system you are dealing with unless you know which company assigned it.

Here's a summary of how some of the systems overlap, along with guidelines for how the grades compare to preconquest camelid fibers and to other contemporary luxury fibers:

	SYSTEM 1	SYSTEM 2	REFINEMENT OF SURI CLASSIFICATION	COMPARISONS
12µ				
13µ				Vicuña (13µ)
14µ				Chinese cashmere (13–15µ)
15µ				
16µ				
17µ				
18µ				Preconquest alpaca (18µ) cashmere (18µ)
19µ	Royal (less than 19 or 19.5µ)*			
20µ			Suri, baby (20µ)	
21µ		Baby (20–22µ)		
22µ	Baby (22.5µ)			
23µ				Preconquest fine llama(23µ)
24µ				Guanaco (24.6µ)
25µ		Superfine (25.5µ)		
26µ	Superfine (26µ)		Suri, superfine (26–27µ)	
27µ		Suri (27µ)** Adult (27.5µ)		
28µ				
29µ				
30µ				
31µ	Adult: Huarizo (31.5µ)***	Huarizo (30–32µ)***		
32µ			Suri, coarse (32µ)	
33µ	Adult: Coarse (31–34µ)	Llama (34µ)***		Preconquest coarse llama (33µ)
34µ	Adult: Mixed pieces (33µ)	Coarse (34+µ)		
35µ	Very coarse (35+µ)			

International Alpaca Mark (not greater than 28µ) [applies to 12µ–28µ]

Classification of raw fiber in Peru is done subjectively, by the trained sight evaluation and touch of the classers; the micron-measurement breakdowns above reflect objective descriptions of their fingers' knowledge. The symbol µ represents microns.

* Different graders differ on the micron definition of royal alpaca, when it is recognized at all.

** The 27-micron suri grade has been broken down into three suri grades: baby, superfine, and coarse.

*** The huarizo and llama classes in systems 1 and 2 reflect fiber grade, not the animal that grew the fiber. The fiber may have been produced by an alpaca, a llama, or a hybrid.

Alpaca

ALPACAS HAVE BEEN bred for fiber quality for thousands of years. Following the Spanish invasion of South America, growers were rewarded more for quantity than for quality, which has consequently declined since the sixteenth century (until very recently). But even if the alpaca's fiber isn't what it was at the height of the Incan Empire, it still solidly holds its place as one of the most exquisite natural fibers — if you know how to manage it. Its dominant distinguishing characteristic is usually listed as drape, which is correct but unimaginative and not especially helpful. Eric Hoffman, author of *The Complete Alpaca Book,* observes with insight that "when it comes to alpaca, the subjective qualities are what separates it from sheep and almost all other natural-fiber animals." Alpaca fiber is supple, sensuous, and swingy, with a hand unlike anything else you will experience.

This observation applies to the fiber from both varieties of alpaca, the huacaya (woolly type, top photo) and the suri (with long, sleek locks, bottom photo). Between 90 and 93 percent of alpacas are huacaya, and only 7 to 10 percent are suri. Both types of fiber are soft, and in most cases they do not contain guard hair, although there may be thicker hairs on certain parts of the body. Alpaca will felt.

Peruvian mills sort alpaca for 22 natural shades and either 6 or 7 quality grades (depending on the system being used). About half the yield of both huacaya and suri alpaca falls into the range of 20 to 26 microns (baby or superfine). The colors pretty much reflect the spectrum for huacaya fiber, since only a very small percentage of the suri is colored.

Alpaca is heavy compared to wool and gets dramatically heavier as the micron count increases. If alpaca garments aren't carefully constructed, they will sag. Getting the right gauge (in knitting or crochet) or sett (in weaving) is critical to the success of a project: too loose and the yarns will move around and droop; too tight and the fabric will be stiff (keep this in mind if you want a stiff result); just right and you'll think you're in fiber heaven.

Huacaya

Suri

Spinning Alpaca Blends

Prepared alpaca top is almost always huacaya (because there's so much more huacaya than suri). Although a pure-alpaca yarn can be one of life's great delights, adding a percentage of compatible wool during preparation will make this fiber easier to spin and will increase the elasticity of the yarn and of the finished textile. Consider also blending with mohair, silk, or angora rabbit, none of which will increase elasticity, but they may add other qualities you'll like.

Alpaca Facts

▶ **FLEECE WEIGHT**
Usually 2–6 pounds (0.9–2.7 kg); some animals produce 12 pounds (5.4 kg) or more

▶ **STAPLE LENGTH**
Huacaya 2–6 inches (5–15 cm)
Suri up to 11 inches (28 cm) per year

▶ **FIBER DIAMETERS**
Huacaya and suri 15–35 microns, with average micron counts in the mid-20s. Most alpacas do not have noticeable guard hairs.

▶ **LOCK CHARACTERISTICS**
Huacaya alpaca's overall impression is fluffy, even though it spins up with alpaca's characteristic qualities of sleek density and shine. Suri hangs in long, exceptionally lustrous, curling locks. There are varying amounts of crimplike profile in individual huacaya fibers, and suri is bred to have no crimp.

▶ **NATURAL COLORS**
White; a range of browns from very light to dark, reddish browns (including cinnamon and a rosy color); black; a range of grays from silvery to charcoal; and spotted. The vast majority of suri alpacas in North America are white or light brown; suri animals do exist in the full array of colors, and medium or dark browns are the next most likely shades to appear.

Alpaca scarf from eBay

Spindle-spun alpaca from CTTC

The Center for Traditional Textiles of Cusco (CTTC) is working to help Andean peoples rediscover and maintain the highest standards of their unique fiber-related heritage. Learn more at *www.cttccusco.org*.

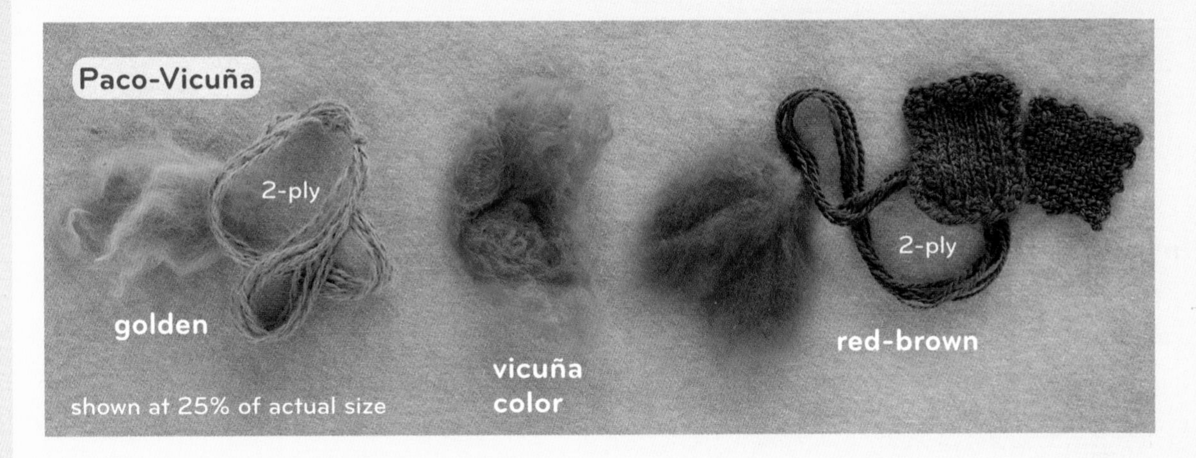

Paco-Vicuña: U.S. Style

Paco-vicuña means two different things, but contemporary fiber folk are likely to encounter one only. The first meaning, and the least likely to come up, refers to an alpaca/vicuña cross. Because there are no vicuñas in North America, this only occurs by accident in South America; allowing this to happen is not endorsed by the people working to keep vicuñas from extinction. The second meaning denotes a precisely defined type of alpaca that originated in the United States through a precisely constructed breeding program initiated by Phil, Chris, and Dave Switzer, of Switzer-Land Alpacas in Estes Park, Colorado.

The Switzers came up with the paco-vicuña idea after they'd gotten a lot of experience raising alpacas and taken a number of research trips to South America. They knew that alpacas are the domesticated descendants of wild vicuñas. They thought, What if there were a way to cultivate vicuña qualities (like exceptionally fine fiber in attractive reddish colors) in a strain of alpacas that would be, unlike vicuñas, relatively calm and easy to work with?

It's illegal to export vicuñas from South America, but through DNA testing the Switzers located animals that were definitely alpacas (so they could be moved to North America) yet had strong vicuña traits. They brought these animals north and started breeding toward their vision, with a clear concept that takes into account all aspects of the animals — fiber, physical structure, and behavior.

The breeding schematic for North American paco-vicuñas recognizes six types, ranging from Type 6 (closest to alpacas) to Type 1 (closest to vicuñas). The goal is a minimum fiber length of 3½ inches (9 cm). Shorn every year are Types 6 (20–26 microns) and 5 (18–23 microns); shorn every year or every other year are Types 4 and 3 (both 16–23 microns); shorn every two to three years are Types 2 (13–20 microns) and 1 (12–18 microns).

The finest paco-vicuña fibers are exceptionally fine — well within the average diameters of the world's very finest fibers. Paco-vicuñas have a wider variety of natural colors than their wild ancestors, including light and dark variations of the traditional vicuña shade of soft, reddish brown, whites, and blacks.

Paco-vicuñas are alpacas. But they're very special alpacas, closely related in many ways to vicuñas' ancestors, that produce fine, long, beautifully colored fiber. The more generous length makes paco-vicuña a bit easier to spin than other very fine fibers.

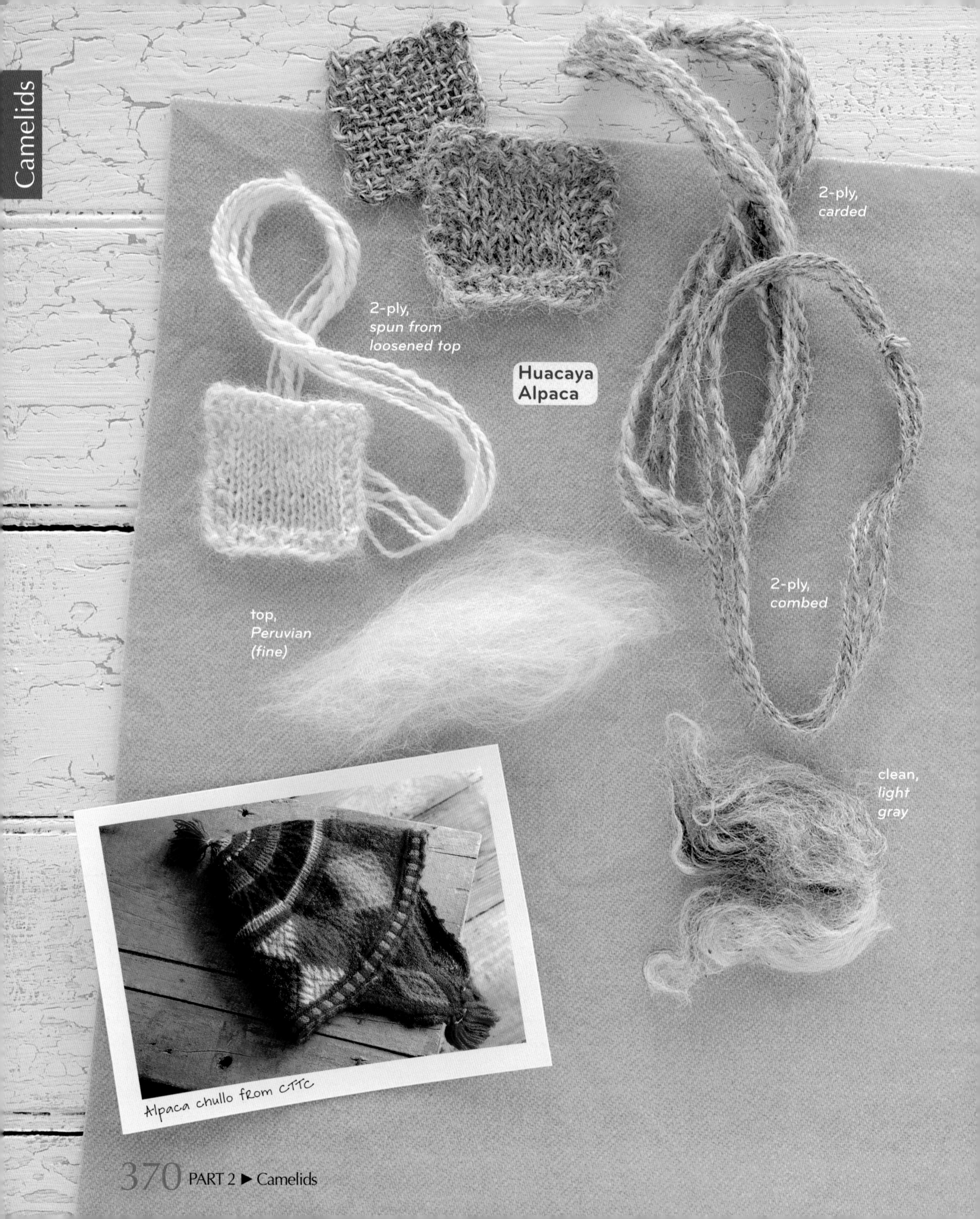
Huacaya Alpaca
2-ply, spun from loosened top
2-ply, carded
top, Peruvian (fine)
2-ply, combed
clean, light gray
Alpaca chullo from CTTC

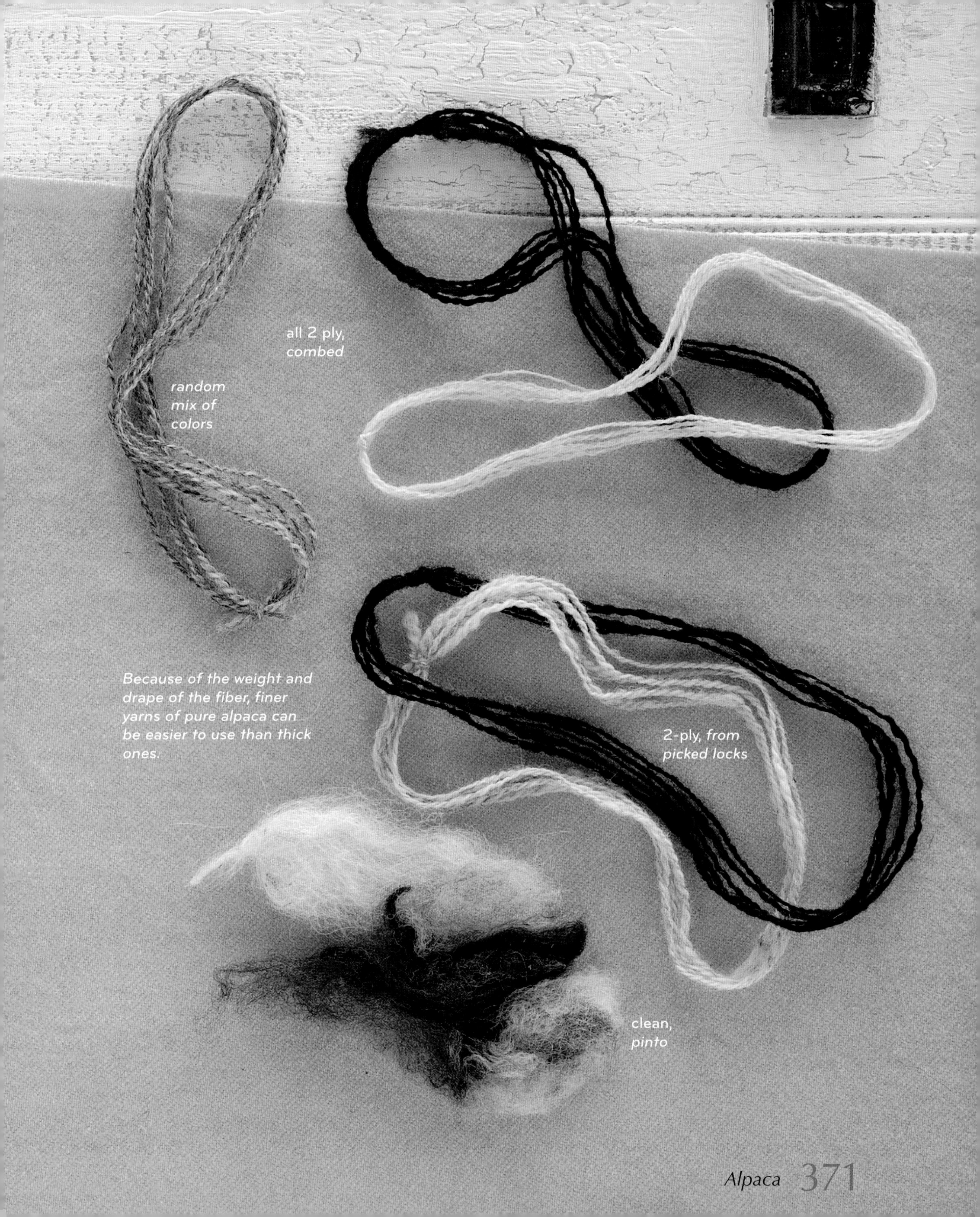

all 2 ply,
combed

random
mix of
colors

Because of the weight and drape of the fiber, finer yarns of pure alpaca can be easier to use than thick ones.

2-ply, *from*
picked locks

clean,
pinto

Suri Alpaca

The extreme length and fineness make this a challenging fiber to spin. Misting with water helped control flyaway tendencies (spritzing, so the fiber actually gets wet, causes it to clump).

2-ply, spun directly from mini-combs

all 2-ply, combed

pinto

white

clean

clean

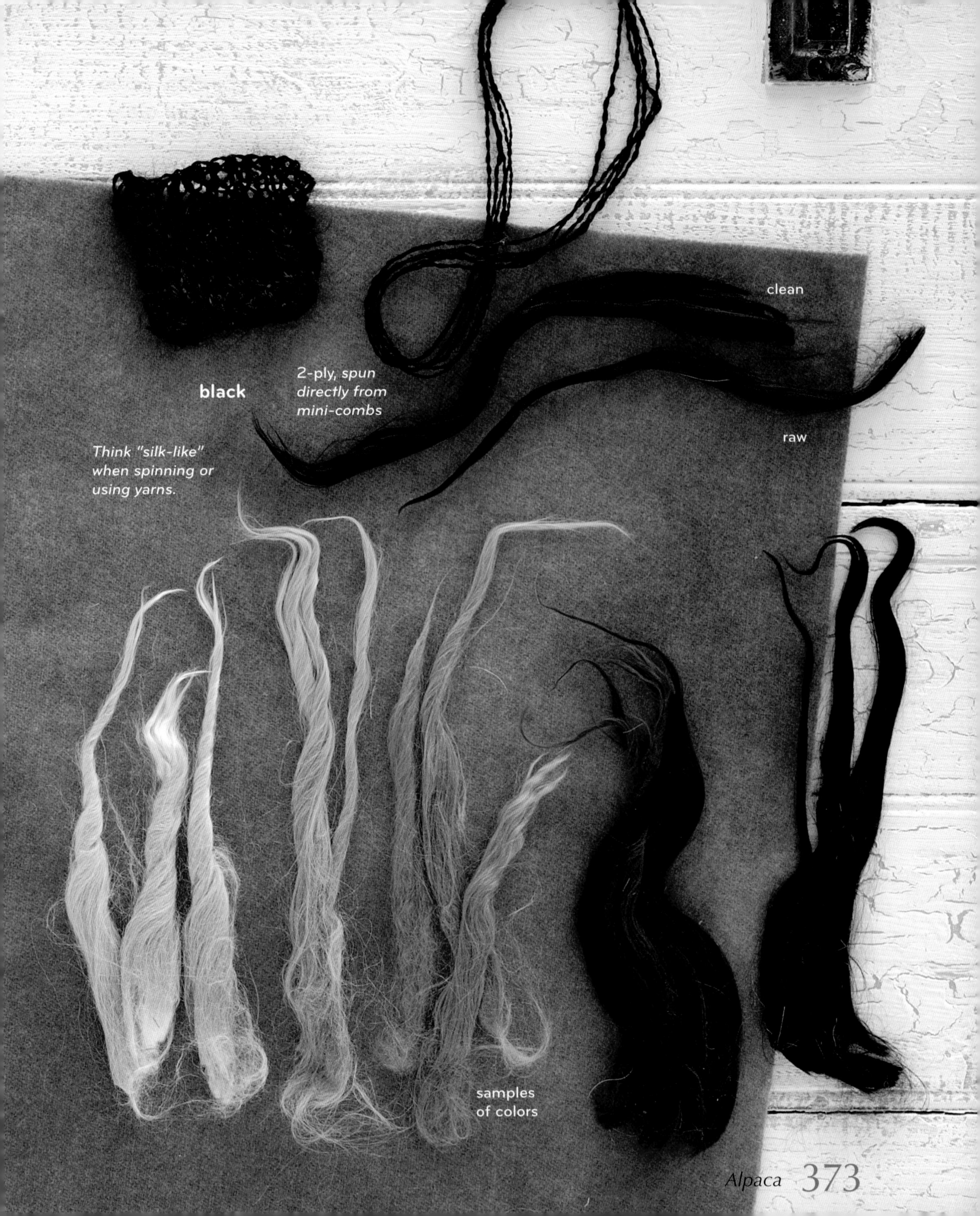
black
2-ply, spun directly from mini-combs
clean
raw
Think "silk-like" when spinning or using yarns.
samples of colors

Using Alpaca Fiber

Dyeing. Use dyes that will work on wool. Lustrous alpaca will gleam with natural or dyed colors. Of course, the underlying natural color will affect the dyed results.

Fiber preparation and spinning tips. Alpaca is slipperier than wool and will need a bit more twist than a comparable wool yarn, but be careful not to put in so much twist that you make the yarn stiff and wiry. Use carders with fairly fine teeth, possibly cotton carders, which have both fine and closely spaced teeth. You'll most likely card huacaya, although longer fleeces may be flicked or combed. We found that suri was much trickier to spin than huacaya, and our impressions are supported by reports from spinners who have extensive experience with South American camelids, including Kaye Collins and Chris Switzer, both of whom are experts in this realm. Suri fiber feels exceptionally fine and soft (even in comparison to huacaya that may technically have the same fiber diameters) and has no crimp. Its tendency to felt becomes evident in the locks, which need to be gently teased apart. In theory, you can choose a wide variety of preparation options for suri alpaca: Pick the locks to open and spin them; open up the fleece and then hand- or drum-card it, using fine carding cloth; or flick or comb the locks. We found that mini combs worked especially well at keeping the fibers in order and opening the locks. Yet even yarns that appeared to have been evenly spun looked distinctly uneven after washing — the opposite of the normal effect. Plan to spend some time getting to know suri alpaca.

Knitting, crocheting, and weaving. Make sure you have the right gauge or sett for your fabric, so that your finished project either holds its shape or has the type of movement that you want it to. Some huacaya fleeces may have some elasticity, but elasticity is not generally an alpaca high point; suri will have next to none. Expect huacaya yarns to bloom when you wash them. Reports on suri alpaca suggest that it has a mind of its own: It resists being twisted sufficiently in spinning, and fabrics made of it may show unevenness or curling. If you are lucky enough to get your hands on suri alpaca, experiment and see what happens for you; expect an interesting experience, and let the fiber teach you how to use it, planning to enjoy its idiosyncrasies as it does so.

Best known for. Delightful smooth, supple feel, and wide range of natural colors. Alpaca is a strong fiber, and durability is extremely high in the coarser varieties. It's not elastic (it doesn't bounce back into shape as many, but certainly not all, wools do).

Alpaca yarns — purple from handknitting.com and brown from Local Harvest/Red Rosa Farm

Llama

THERE ARE THREE TYPES of llamas in South America today — a nonwoolly type, known for its sparse fiber growth and hair-free face; a woolly type with a heavy fleece and fiber growing on its forehead and ears; and a third intermediate type. The North American llamas have been bred from a limited number of imported animals, first brought here by zoos beginning in the 1800s. No one's sure exactly how many llamas are in North America today, but a breed association shows 157,000 registered animals in 2006, and the population could easily be double that number when unregistered animals are included.

As a fiber provider, llama is a trip and a half. If you look at one fleece at a time, there's no big deal, but if you look at the species from a fiber-producing angle, it's like stumbling upon a smorgasbord of possibilities. Pretty much anything you want, the llama can provide. The question is which llama and when, and the trick is in getting a grip on reality while you're looking for the fiber you want, or working with the fiber you have.

That's one of the reasons the llama has been overshadowed in the contemporary textile world by the other New World camelids, the alpaca and even the rare and expensive vicuña and guanaco. The other reason is that llamas — wonderful creatures that they are — adore vegetable matter. Their fleeces are magnets for hay, straw, seeds, and twigs, especially little bits of those things that get stuck in the mass of fibers and are hard to get out. Some growers have figured out how to minimize vegetable matter and other trash in fleece, but doing this requires hard work. Llama shows up in a few commercial yarns. It would probably show up more often if llamas produced an overall more-consistent type of fiber.

Are these drawbacks worth dealing with? Yes! Llama can be terrific fun to work with, and it can teach you a lot about fiber in general.

Many llamas still shed their fiber naturally, whereas alpacas — and now, in some strains, certain llamas — must be shorn. Because of the ability to shed, some llama fiber can be gathered by brushing, which yields mostly downy

undercoat. Shearing, the alternative, takes off the full range of fibers, mixed as they grew together.

Llamas can be single coated, producing a relatively soft, fine, and consistent body of fiber, or double coated, with a fine undercoat and a hairy outercoat. Trying to make a clean distinction between single- and double-coated llamas can be misleading, though. The types of fiber within a fleece may gradually move from finer to coarser, with the full range of textures in between, so in some fleeces it won't be easy to just pull the outercoat away from the undercoat. Woolly-type llama felts; the coarser fibers can be included in the mix, but they won't felt on their own. You can change the quality of the felt by blending with some sheep's wool.

North American breeders now also offer suri llama, which produces high-luster, long, fine fiber that hangs on the animal in locks (like suri alpaca). Although these fibers have the shine and length of guard hairs, the animals are bred to grow fibers with minimal medullation and no crimp.

Llama yarns from handknitting.com

Llama

2-ply, top pulled from peasant combs

clipped

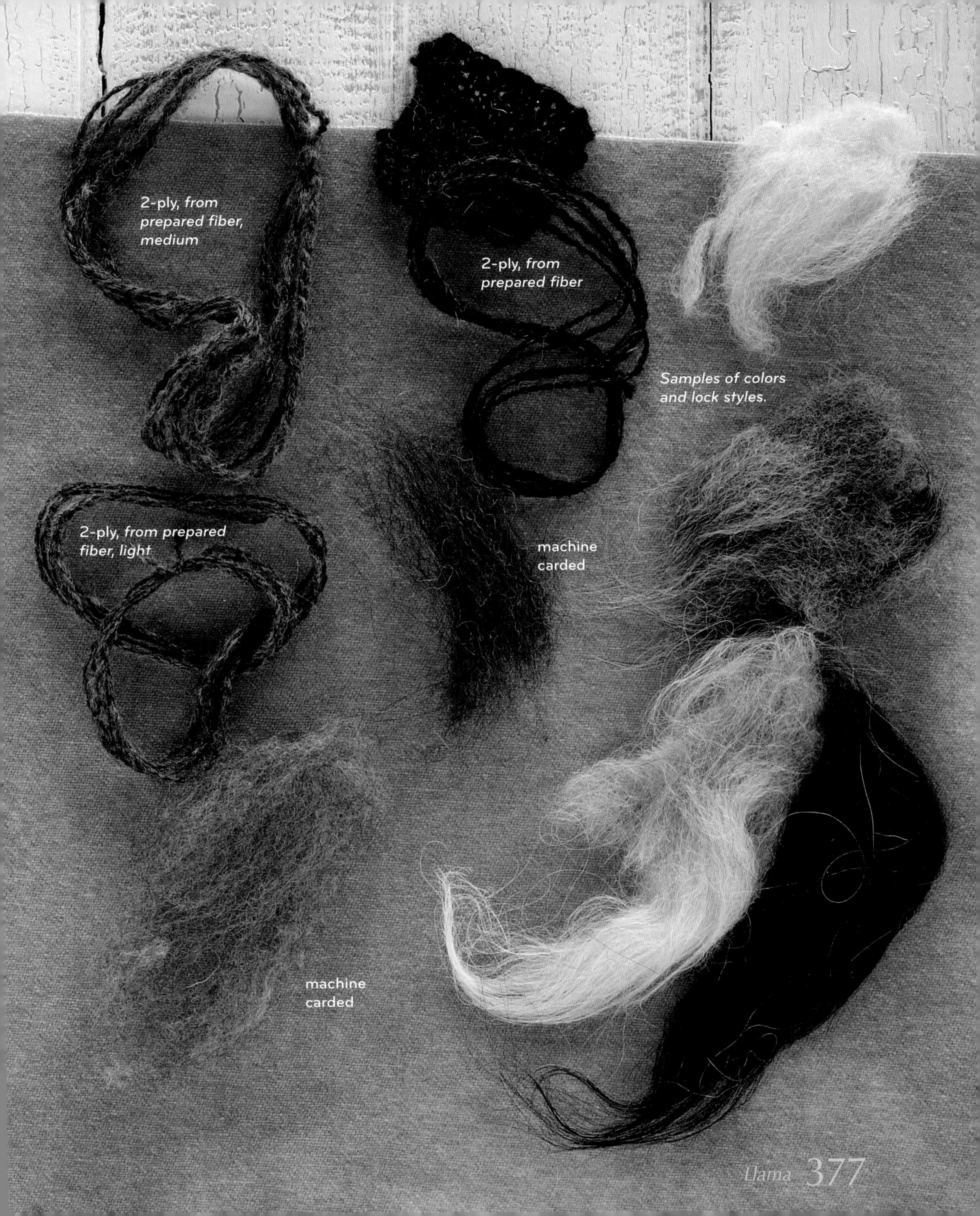

2-ply, *from prepared fiber, medium*

2-ply, *from prepared fiber*

Samples of colors and lock styles.

2-ply, *from prepared fiber, light*

machine carded

machine carded

Llama Facts

▶ **FLEECE WEIGHT**
Usually 2–5½ pounds (0.9–2.5 kg) from a woolly llama; 6–8 pounds (2.7–3.6 kg) from a suri llama; some llamas produce more than 15 pounds (6.8 kg)

▶ **STAPLE LENGTH**
3–8 inches (7.5 to 20.5 cm)

▶ **FIBER DIAMETERS**
16–45 microns, with occasional guard hairs that are even heavier

▶ **LOCK CHARACTERISTICS**
Crimp varies.

▶ **NATURAL COLORS**
White, cream, browns (light to dark), grays, and blacks.

Using Llama Fiber

Dyeing. Use dyes that will work on wool. Guard hairs won't show dyed colors like the finer fibers will.

Fiber preparation and spinning tips

- Depending on the quality of your fiber, you may want to use carders with fairly fine teeth, possibly cotton carders.
- Because of the mix of fiber lengths and qualities, combing may produce a lot of waste. If you want to keep the fibers aligned so you can spin for smoothness, consider flicking the locks.
- Blend fine llama with angora rabbit, fine wools, fine mohair, or silk, if you like. When the llama fleece contains coarse fibers, consider blending with a coarser mohair, silk, or wool, perhaps Romney, Lincoln, or another wool that matches your llama in fineness and length.
- Spinning treatment will depend on the length and overall quality of the fibers. The woollier varieties can be spun like wools with similar qualities (say, Corriedale or one of the Down breeds, depending on fineness), and the sleeker fibers can be spun more like mohair or alpaca.

Knitting, crocheting, and weaving. There's so much variety in llama fiber that you'll be winging it, depending on what's in your hands. Don't expect as much elasticity as wools, but the finer llamas have more elasticity than similar alpacas.

Best known for. Variability! Plus color and versatility.

Guanaco

Conservation Species

There are about 600,000 guanacos living in South America, mostly on the high plains of Chile and Argentina, with smaller numbers found in the mountains of Ecuador, Peru, and Bolivia. A few hundred are currently being raised in North America as well. Before the Spaniards arrived, scientists estimate there were somewhere between 10 and 30 million guanacos.

Guanacos spend most of their time in open grasslands and are very hard to capture thanks to their speed: They can run up to 35 miles (56 km) per hour for extended periods. Almost immediately after birth a young guanaco can keep up this pace with its herd (generally a dozen or so animals). Guanacos' only natural predators are pumas (South American mountain lions) and humans.

Although their speed makes them hard to catch, once caught, guanacos tame easily. Darwin wrote a great deal about guanacos in *The Voyage of the Beagle.* "That they are curious is certain," he said, "for if a person lies on the ground, and plays strange antics, such as throwing up his feet in the air, they will almost always approach by degrees to reconnoiter him. On the mountains of Tierra del Fuego, I have more than once seen a guanaco, on being approached, not only neigh and squeal, but prance and leap about in the most ridiculous manner, apparently in defiance as a challenge. These animals are very easily domesticated, and I have seen some thus kept in northern Patagonia near a house, though not under any restraint."

Although guanacos grow to the size of small to medium llamas, the fiber yield is much less, perhaps 8 ounces (228 g) in a good year. They produce an exquisite undercoat, one of the finest and most consistent fibers in the world, almost as fine as vicuña, though it is one of the most difficult to obtain. They have cinnamon-brown body wool (varies in shade depending on the population), creamy underbellies, and gray faces.

Guanaco Facts

▶ **FLEECE WEIGHT**
1–1½ pounds (0.5–0.7 kg)

▶ **STAPLE LENGTH**
Undercoat 1¾–2¼ inches (4.5–5.7 cm)

▶ **FIBER DIAMETERS**
Double coated, with 5–20 percent guard hair (which is significantly coarser than the undercoat and must be removed); undercoat ranges from 16.5 to 24 microns, and most is 14–19 microns

▶ **LOCK CHARACTERISTICS**
Crimp varies.

▶ **NATURAL COLORS**
Body wool is cinnamon brown (southern population in South America) to lighter brown with golden highlights (northern population in South America).

Using Guanaco Fiber

Dyeing. Use dyes that will work on wool, planning colors that will be influenced by the underlying natural shade.

Fiber preparation and spinning tips. You will probably encounter this fiber in dehaired form, as a slightly disorganized but thoroughly carded mass. Spin directly from this preparation or, if you must prepare it further, use carders with fine teeth. Handle gently to avoid tangling. Our preparation, while generally very even, did have a few almost imperceptible clumps of fiber that wouldn't draft well; we just pulled them out and kept on. Spin lightly, as you would cashmere, qiviut, or other very fine fibers. Make sure your hands stay dry or the fibers won't draft well. Because this is a fine-diameter, short fiber, spin thin singles and ply to create heavier yarns.

Knitting, crocheting, and weaving. Because of the expense, you'll most likely want to make small items from guanaco. It's clearly soft enough for use next to the skin. It has enough body to show openwork patterns well, with a slight softening haze. While our fiber was not remarkably crimpy, the yarns have more puffiness than drape.

Best known for. Extreme fineness and scarcity of fiber; distinctive natural warm brown color.

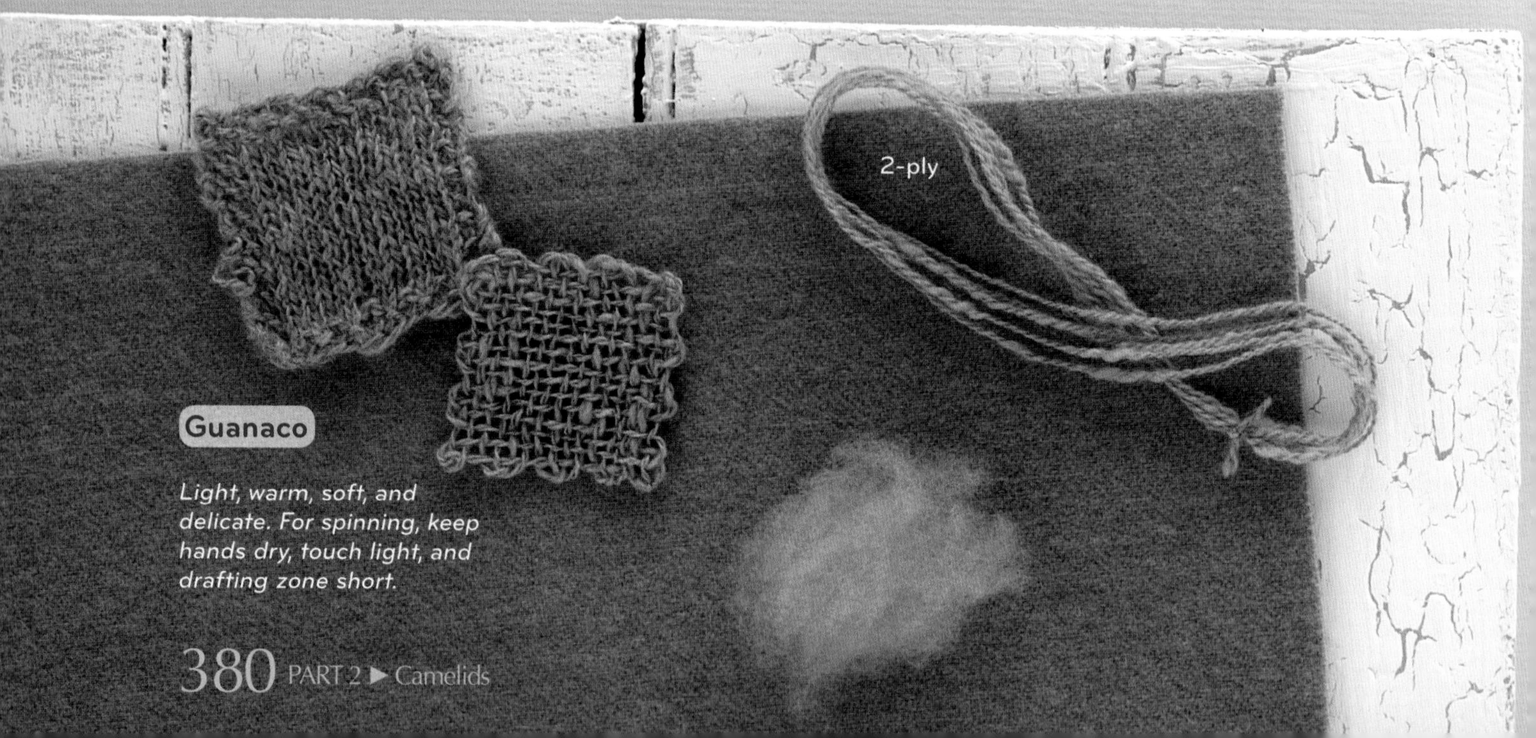

Guanaco

Light, warm, soft, and delicate. For spinning, keep hands dry, touch light, and drafting zone short.

Vicuña

Conservation Species

THE VICUÑA IS A WILD New World camelid that is thought to have developed almost two million years ago on the high plateau, or altiplano, of the Andes mountains, at an altitude of about 12,000 feet (3,660 m) above sea level. The vicuña, with its golden brown body and its off-white bib, lower sides, and underbelly, produces some of the finest and most desirable down fiber in the world, which ends up as a lovely reddish brown. (There is a color difference between the appearance of the animal and the shade of the harvested undercoat.) Vicuña down is also some of the most expensive fiber on earth, with prices for unspun fiber running in the hundreds of dollars per ounce!

Why so costly? Well, largely because the animals were driven to near-extinction, so export of their fiber was completely forbidden by international laws and treaties for decades. Even now that it's legal to move the fiber across international borders, the process requires paperwork and strict control.

When Europeans first arrived in South America, there were probably close to two million vicuñas. But these relatively shy and dainty animals (they are the smallest of the camelids, weighing less than 150 pounds, or 68 kg) were easy pickings for European hunters. By 1971, fewer than 10,000 vicuñas remained across Peru, Chile, Argentina, and Bolivia. However, thanks to international efforts to protect the species, today there are about a quarter million of these animals in the wild again, and indigenous people in the Andes are hand gathering vicuña wool for legal export.

Family groups form the basis of vicuña existence, and the typical family consists of one breeding male, with up to a dozen females and their young. Younger males are driven from the family unit at around eight to ten months of age, joining "boys' clubs" or "bachelor bands." Young females are also expelled from their family herds at about the same age, but they are generally accepted into the family herds of non-related breeding males.

During the time of the Incas, the Sapa Inca, or divine ruler, organized the capture of vicuñas in a collective hunt known as a *chaku.* Thousands of people would encircle an area that could stretch a hundred miles (161 km) across. They would move in until they could join hands, and then keep moving in, and in, and in, until the animals were tightly encircled by a human chain. Some animals were killed, but most were sheared and released.

Today there are some farmed flocks of vicuñas in Argentina, while Peru, Chile, and Bolivia are leaning toward a gathering system similar to the *chaku.*

Vicuña Facts

▶ **FLEECE WEIGHT**
½–1 pound (0.2–0.5 kg)

▶ **STAPLE LENGTH**
Usually shorn every other year, possibly even less frequently, for a length of 1–2 inches (2.5–5 cm)

▶ **FIBER DIAMETERS**
Double coated, 30 percent of which is guard hair; undercoat fiber diameter averages 8–13 microns

▶ **LOCK CHARACTERISTICS**
As a down fiber, can be obtained in a fluffy, clean, carded mass or roving.

▶ **NATURAL COLORS**
Body wool is dark cinnamon (northern population) to light beige (southern population).

Using Vicuña Fiber

Dyeing. The natural color is strong, distinctive, and lovely, and unless you know how to handle exceptionally fine fibers in a dyebath and have a clear idea of what dyed color you want and how to achieve those results, we suggest enjoying the original shade. If you must change that color, use dyes that will work on wool.

Fiber preparation and spinning tips. Vicuña fibers are exceptionally short and fine. Handle gently to avoid tangling. Spin the mass of fiber as you receive it, or use carders with fine teeth and maintain a delicate touch. Spin lightly, as you would cashmere, qiviut, or other very fine fibers. This was the only fiber for which we altered our sample-spinning equipment; consider using a charkha-style wheel, great wheel, or supported spindle. Spin very fine (because of the fiber length), and ply to achieve heavier yarns. Unless you have an abundant source, however, you will want to keep your finished yarns slender. To get the skills needed for spinning vicuña with a less expensive fiber, practice on some cotton. Vicuña is softer, lighter, and fluffier, but the techniques for managing it are similar.

Knitting, crocheting, and weaving. Vicuña textiles will be lightweight, warm, and deliciously soft. The surface of the fabric will be soft, like qiviut, rather than crisp, like silk or linen, which will affect your choice of pattern stitches. With well-constructed yarns, vicuña can be woven, although small looms of types that have no warping waste will be the logical tool choices because of cost. You'll probably opt to knit or crochet with it, staying in close touch with every inch of the yarn.

Best known for. Exquisitely fine, soft, rare — and expensive — fiber. Unusual, rich, reddish color.

Vicuña

Cloudlike, rare, and cinnamon-colored. Adopt a Zen approach when spinning or using.

Bactrian Camel

Conservation Species (in the Wild)

BACTRIAN CAMELS *(Camelus bactrianus)* are native to the steppes of northeast Asia (think Mongolia). They have two humps. At birth, a Bactrian calf weighs between 80 and 130 pounds (36–59 kg) but grows to about 1,600 pounds (725 kg), with adults standing up to 9 feet tall (2.8 m) at the tops of their humps. The Bactrians are long-lived animals, with life spans reaching between 40 and 50 years.

One interesting thing about Bactrian camels is that there are both domesticated and wild Bactrians, though the wild population is currently thought to be fewer than one thousand, making it one of the most critically endangered mammals on the planet. Wild Bactrian camels survive in a few pockets of remote terrain in the Gobi desert in Mongolia and China.

For centuries, people have assumed that the wild Bactrian was the predecessor of the domesticated Bactrians (which number close to 1.4 million). There are differences between the two types of animals, however, including the wild variety's ability to survive on saltwater, something its domesticated relatives can't do, and some differences in DNA. These factors have led scientists to think that the wild Bactrian might be a separate, yet similar-looking, species and that the domesticated Bactrian's forebears may be extinct. Wild Bactrian camels are smaller than their domesticated counterparts, and their humps are shaped differently.

Another point that interests scientists about the wild Bactrians is the fact that their range was in the midst of a Chinese nuclear testing area, yet the animals seem to have suffered no ill effects from the radiation. The wild Bactrians' main nemesis is hunters, who kill them for food or to remove them from grazing lands used for domestic camels. A continuing drought that has reduced oases in the desert has also sent the wild animals' numbers hurtling downward in recent years. The Chinese government established a protective reserve and is working with conservationists to keep the wild Bactrians from disappearing.

You might see a Bactrian camel for yourself. There are a small number of domestic Bactrians — perhaps five hundred — in North America today, mostly held by zoos or exotic animal breeders.

Although we think of "camel colored" as a specific shade, camels themselves come in a variety of browns, as well as gray and, rarely, white. Mongolian camel hair, which has been developed through focused breeding, tends to

Old World Camelids

When most of us hear the word *camel,* we think immediately of the Old World camelids — the Bactrian and dromedary camels that are found in Africa, the Middle East, and Asia. Both types are found in limited numbers in North America.

be the finest and is near cashmere in its fiber diameters. Baby-camel down can compete for softness with the most delicate cashmere.

Camels are double coated. The hair portion of the fiber harvest, which is most often removed from the raw fleece long before fiber folk in industrialized nations even hear about it, can be incredibly useful. Like yak hair, it can be made into warm and waterproof tents, yurts, and coats. It's also astonishingly strong, and good for weaving narrow, tough bands (like completely inelastic tie-down cords). Although camel down resists felting, the hair is incorporated into felts for the shelter-providing purposes we've mentioned, and it's also used to make bedding, since it's believed to ease the aches and pains of arthritis. (Angora rabbit fur is also known for this quality, although it's more practical to make items like wrist warmers, rather than mattresses, from angora.) Camel hair, if you can find enough, would make great leashes for dogs, halters for horses (or llamas, or a camel if you have one), or rugs.

Let's go back to the down, though, because that's what's most commonly available. Camels shed their undercoats in the spring, and the down can be gathered either by hand or by

The Story of the Weeping Camel

If you haven't seen the movie *The Story of the Weeping Camel,* it is well worth seeking out. The story revolves around a Mongolian family's efforts to save a rare white Bactrian camel calf abandoned by its dam. Part documentary and part dramatic interpretation, it gives an amazing glimpse of a world totally different from the one most readers of this book live in.

top

2-ply

Camel (baby)

2-ply

top

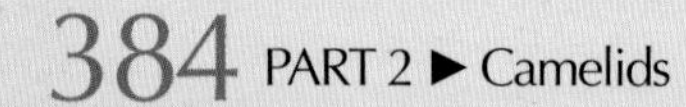

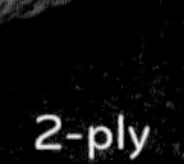

The down is what we most often find in yarns and spinning fibers; the hair can also be put to use, for ropes, belts, halters, or bags.

The baby camel at top left is as fine as cashmere.

combing during this molting season, or it can be shorn. If down gets collected through the natural season-related release, there's less hair to sort out. If the fiber is clipped off, the coat on the hump is left in place as a protection for the animal's climate-control system.

Bactrian Camel Facts

► **FLEECE WEIGHT**

Raw fleece 17–22 pounds (7.7–10 kg), sometimes as much as 28 pounds (12.7 kg) from a large male
Down 5–15 pounds (2.3–6.8 kg)

► **STAPLE LENGTH**

Hair 12–15 inches (30.5–38 cm)
Down 1–5 inches (2.5–12.5 cm), mostly 1–3 inches (2.5–7.5 cm)

► **FIBER DIAMETERS**

Double coated, with varying percentages of guard hair; undercoat fiber diameters average 19–24 microns for adults, and 16–18 microns for baby camel down

► **LOCK CHARACTERISTICS**

Shed out and gathered, combed, or shorn, the fibers present a jumbled mix of hair and down that needs to be separated.

► **NATURAL COLORS**

Light to dark browns are most common, although grays and whites do exist.

camel yarn from Greenberry House

Using Bactrian Camel Fiber

Dyeing. Use dyes that will work on wool, keeping in mind the effects of the underlying natural color.

Fiber preparation and spinning tips

- If you have down that has been separated but not prepared, you can hand-card or drum-card. You will likely find fine hand carders or a specialized drum carder helpful, or use mini-combs if the fibers are long enough.
- Prepared fiber, which is easy to locate as adult or baby camel, may flow smoothly, clump together, or fall apart in your hand as you're drafting. There's a lot of variety among camel downs. For some of the shortest ones we had, if we'd been spinning for something other than notes and pictures, we'd have re-carded the top on fine hand carders and rolled the batts around a knitting needle into *punis* (small, sausagelike forms normally used for cotton spinning) and used a long draw.
- As for hair, which you are less likely to be able to obtain unless you are beginning with the raw mass, spin it as it comes, or organize it with Viking combs or by flicking.

Knitting, crocheting, and weaving. Camel down makes a more compact, less lustrous, and more stable fabric than other fibers of similar length and fineness. That means it offers good stitch or weave definition, and an outstanding sense of body. Elasticity isn't its strong point. Baby camel down can be as fine as cashmere, but it won't have the same tendency to expand and contract (bounce), fluff up, or partially fill the holes in a knitted lace pattern. Those qualities could be just what you're looking for.

Best known for. Fine, sometimes very fine, but reasonably sturdy luxury fiber that is most often a soft brown with golden overtones.

Dromedary Camel

SAY CAMEL to most people, and the animal that jumps to their mind's eye is the Lawrence of Arabia type of camel, the dromedary — a single-humped beast of burden throughout northern Africa and the Middle East. Dromedaries can go for very long periods without a drink of water, though their hump, contrary to popular belief, doesn't store water. Instead, the hump stores fat, which is critical to survival in harsh desert climes.

Dromedary camels are better adapted to hotter climates, and thus produce less usable fiber, than Bactrian camels. Nonetheless, there are breeders of dromedaries in North America, and several collect and market fiber. As with all the other animals, there are different types of dromedaries. Some types have essentially no fiber. Yet an Arvana dromedary can produce between 4½ and 7 pounds (2–3.2 kg) of useful fluff, about 90 percent of which is down, with a clean yield of close to 80 percent. Arvanas live in Turkmenistan, Uzbekistan, Azerbaijan, and parts of Kazakhstan, as well as Afghanistan, Iran, and Turkey, where they provide the residents with a steady supply of transportation, milk, meat, and yarn.

Wild Camels of Australia

The wild ancestors of dromedary camels have long been extinct, so all dromedaries are domesticated, but in Australia, large herds of dromedary camels have gone feral. The animals, offspring of dromedaries brought to the continent in the late 1800s and early 1900s to access its desert areas, now exceed one million animals living across 40 percent of the land area.

Dromedary

Some breeds of dromedaries produce small amounts of usable fiber.

raw

English AngoRa Rabbit
MusK ox

Other Critters

Old MacDonald had a farm, and on that farm he had a — musk ox? Well, Old Mac may not have had a musk ox, but when you dig deeply into the world of animal fibers, it's amazing just how many farmed animals and pets you stumble across. In the case of musk oxen, they are farmed in Alaska by the Musk Ox Development Corporation, a nonprofit organization that then provides the fiber to an Alaskan women's cooperative called Oomingmak. The women use yarn made from this truly sumptuous fiber, known as qiviut (pronounced "kiv-ee-*uht*"), from the Eskimo word for down or underwool, to knit exquisite scarves and other accessories that they sell to produce family income.

No one would be surprised at Old MacDonald having cows (we scored some Scottish Highland locks, but not enough to spin). But you might be surprised to learn that fiber from some of the cows' distant bovine cousins, the bison (familiarly called buffalo in North America) and the yaks (long-haired clan members from the high mountains and steppes of Asia), are becoming readily available in both yarn and unspun-fiber forms from North American producers.

Breeds of cattle to look for, if you are a spinner, include not only the Scotch Highland, but also the Dexter and the Galloway breeds. Spinners can also seek out the fiber of horses (look for the American Curly) and some donkeys (especially the globally rare Poitou).

Rabbits make cute and cuddly pets, but a number of breeds provide beautiful fiber as well. There are various breeds of Angora rabbits found throughout the world, and two dwarf breeds (the Fuzzy Lop and the Jersey Wooly) can also produce usable fiber.

Then there are dogs, those family members often overlooked as producers of abundant fiber. Just because it routinely clogs the vacuum, we don't usually recognize that dog hair, along with angora rabbit, could be a mainstay of urban and rural fiber farms. Dog hair is often of a quality that rivals the expensive luxury materials that come preciously packaged.

Let's take a look at these alternative fiber suppliers, in (roughly) alphabetical order.

Samoyed

Bison

WHEN EUROPEAN Americans made their way to the Great Plains and the intermountain West during the nineteenth century, there were somewhere between 30 million and 80 million bison roaming the land. There were actually two distinct subspecies found in North America: the wood bison (*Bison bison athabascae*), which traditionally inhabited parts of Canada, and the plains bison (*Bison bison bison*), also sometimes referred to as the buffalo. Native Americans relied on bison for both utilitarian and cultural roles. Bison meat, bones, hides, and horns were used to support the tribes, and the animals symbolically played a part in the religious and spiritual lives of the people.

When Euro-Americans began their westward expansion in the mid-1800s, they nearly wiped the species out. As the nineteenth century gave way to the twentieth, the vast herds had been reduced to fewer than 1,500 animals. A herd of about 50 was living in the Yellowstone area, which had already been made into a national park; some were held in five private herds around the United States; and some were in small groups of free-roaming animals in the United States and Canada.

A handful of people were instrumental in saving the animals from extinction, including President Grover Cleveland, who signed into law the 1894 Yellowstone Wildlife Protection Act, which made it illegal to shoot wildlife within the park. William Hornaday, chief taxidermist of the Smithsonian Institution in the late 1800s and the first director of the National Zoo, wrote a book titled *The Extermination of the American Bison,* which helped sway public opinion in favor of conservation. He included bison as some of the first residents of the National Zoo. Today bison are back from the brink of extinction, and many people are raising them for their meat. A few are harvesting and selling their fiber as well.

Bison grow at least five types of fiber, shed in a big, mixed mass every spring. They have an outercoat of coarse, shiny black hair; coarse midrange hairs that bulk up the coat, improving its insulating qualities; two shorter types of guard hair; and fine, soft down. The latter is the fiber we're most inclined to find appealing for contemporary textile work, although the other coats, if you have access to them, can be used for ropes, as stuffing, or in artwork.

Bison only look like they are domesticated because currently almost all are raised on ranches. They're still semiwild animals who don't take kindly to shearing. Thus the bison fiber available to fiber artists is really a by-product of the buffalo meat industry. It is removed from the hides of slaughtered animals in an effort that aligns, coincidentally or not, with the traditional Native American philosophy of honoring the animals by not letting any useful part go to waste.

Fiber shed by bison in fields is lower quality, because it has been more weathered. Yet the most important reason not to be tempted to gather any shed fiber for yourself, should the occasion arise (as it may, if you are traveling around western North America), is that bison *are not tame,* and, in fact, are quite dangerous. Despite their massive bodies and somewhat closed-off-looking visages, they have excellent senses of sight, smell, and hearing — and they can run as fast as 37 miles (60 km) per hour. For comparison, a normal, healthy, panicked human, adrenaline pumping, can run less than half that fast, about 15 miles (24 km) per hour. (Top-level human athletes can sprint at about 27 miles, or 43 km, per hour; they'd lose, too.) In addition, a bison dramatically outweighs an adult human, has horns and heavy hoofs,

Bison

These samples represent just the down. The top fabric in each pair of larger swatches has been fulled: fiber from some sources fulls more readily than other types.

2-ply

clean

and can easily jump over the average fence — although it would be more likely to go through it. Enough said.

The down fibers do have scales and crimp, as wool does. Some knowledgeable folks report that buffalo down felts very easily for them, while others say it won't felt at all. We tested both fiber and a small woven sample from two different sources, and all felted quite quickly. This may not mean that all bison felts, but some of it certainly does.

Wool is well known for its ability to absorb water without feeling wet (which is handy when you're out in the rain), and tests indicate that at least some bison fiber has an even greater ability to do this than wool does. The guard hairs are hollow, which increases their insulation value and stiffness. Dehaired buffalo down often contains at least a small number of guard hairs.

Bison Facts

▶ **FLEECE WEIGHT**
A hide may yield as much as 3 to 4 pounds (1.4–1.8 kg) of raw fiber, although each hide may also end up producing only 1 pound (0.5 kg) of finished yarn.

▶ **STAPLE LENGTH**
About 1 inch (2.5 cm)

▶ **FIBER DIAMETERS**
Researchers at North Dakota State University have reported data for fibers from two bison. The guard hairs averaged 59 microns, with a full range from 21 to 110 microns. The down averaged roughly 18–22 microns, running from 12 to 29 microns overall.

▶ **LOCK CHARACTERISTICS**
At least five fiber types: very coarse, long outercoat; midlength, coarse, insulating coat; two undercoats of fine guard hairs; and one undercoat of down.

▶ **NATURAL COLORS**
Natural grayish or reddish brown.

Using Bison Fiber

Dyeing. The natural brown does not lend itself to overlaying by most other colors, and the fiber may change its feel when dyed, losing some softness — if you are interested in dyeing it, experiment with dye types and processes.

Fiber preparation and spinning tips

- Bison's short staple length guides all its preparation and spinning.
- If you have raw fiber, expect it to contain massive amounts of dirt that will turn to mud before it's washed out (this is similar to yak). The raw fiber will also contain all of the coats, simultaneously shed or gathered and in need of being separated.
- Once you've separated out the down, you'll note it has a tendency to form neps, resulting in a textured yarn because the neps make it difficult to draft very smoothly.
- Spin the down as it comes or prepare it with cotton carders; spin commercially prepared roving without any additional handling. Bison needs to be spun fine and with enough twist to hold the yarn together, even after plying, if you choose to ply.

Knitting, crocheting, and weaving. The crimp gives bison yarns some elasticity and loft, and the yarns will bloom when they are washed after the fabric is completed. Suitable for use with any construction method: knitting, crochet, weaving, or your choice.

Best known for. Short, quite fine, brown fiber that offers unusual durability for its level of softness.

Dog, Wolf, and Cat

WHEN MAN'S BEST FRIEND has any kind of hair, it sheds; we dog owners know that every time we run our vacuums. Is dog hair good to spin? That depends on the dog! Although sheep hold the record for variety of fibers, there's also a lot of diversity among pups. Some dog hair can be an exquisite, readily available, and (most often) free luxury fiber. Some is slick and short and will smell like dog again every time it gets wet. Some dogs, like many other down-producing species, are double coated, and the down needs to be separated from the hair. Brush your dog and this happens automatically. Poodles and other breeds often recommended to folks with allergies don't *shed* enough hair to be useful as at-home fiber sources, although their clipped hair can be spun. The few truly hairless breeds (like the Chinese Crested, the American Hairless Terrier, and a small handful of others) really don't have anything to offer. But for the most part, the canine realm abounds in fiber.

Commercial production hasn't caught onto this resource yet, so chances are good that if you're considering dog hair, sometimes called *chiengora,* you're either a spinner or you know one. The more the fluff feels like down, the more likely you will love it for your finished product, wondering why people spend big bucks on more famous fine stuff. The more your dog-produced fiber feels like hair, the harder it will be to spin and the greater its tendency to remind you of your muddy best friend when you go out in the rain.

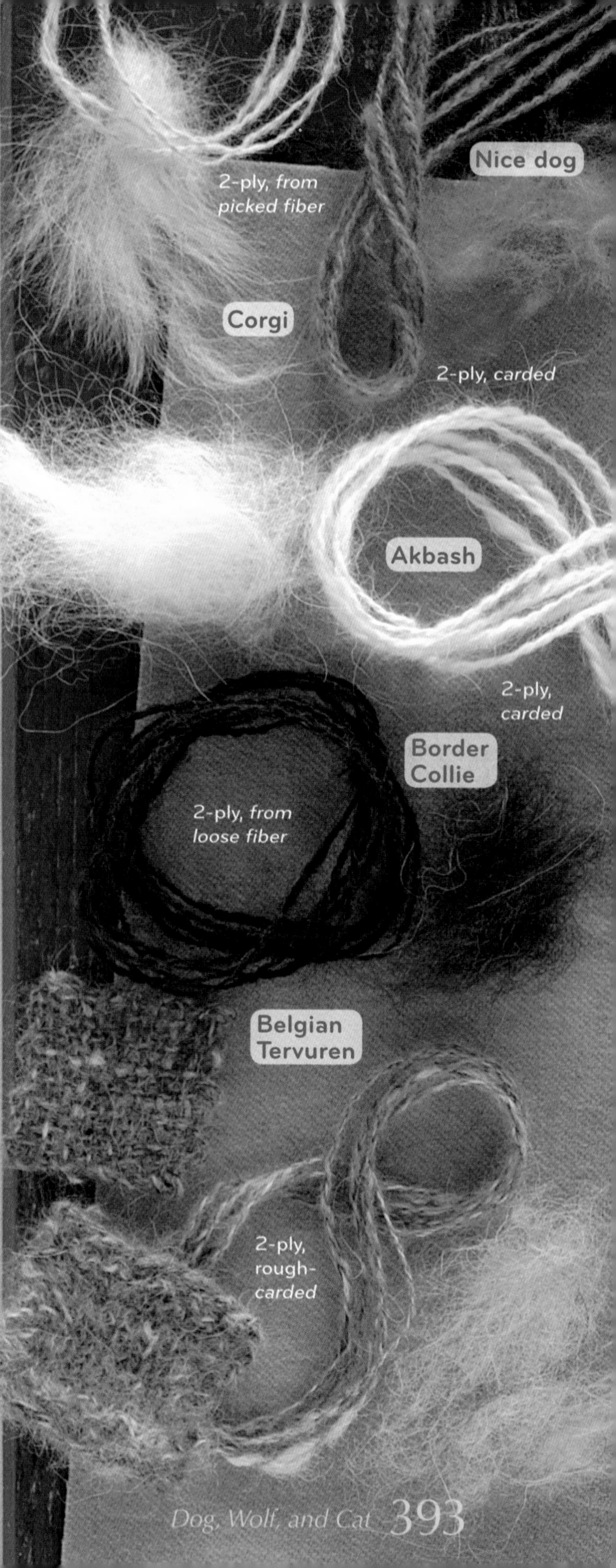

Most of our samples of dog down come from herding and livestock-guardian dogs, because that's what we have ready access to. Some other breeds, like Samoyeds, might actually be worth acquiring for their high-quality fiber, although for perfect bliss you want to be sure to match any dog's personality with your family's. Samoyeds can be exuberant, boisterous, and stubborn, so if you want a "flock" of Sammies you need to be, too — or know how to accommodate those behaviors in the dog. We enjoy Stanley Coren's book *Why We Love the Dogs We Do* as a resource for happily-ever-afters with canines. Perhaps a Great Pyrenees or an Afghan hound would suit you better than a Samoyed or a rescue Border collie. Golden retrievers produce delightful down; if you know how nice the fiber is, you can celebrate when they blow their coats (or experience massive, seasonal shedding) instead of becoming impatient with the overabundance.

Dog down is short. Some other types of dog hair may be fine and quite long, while still others run the gamut of length while embodying more coarseness. In any case, dogs' fiber is usually slippery. Composed of protein, as are all the other fibers in this book, dog hair takes well to the same sorts of dyes as sheep's wool, and just like the other hair/down combinations, the down will take dye better than the hair, and the natural color influences the final shade. In spinning yarn and constructing fabric, think as you would about angora rabbit, camel down, or yak. Use enough twist to make a secure yarn and minimize shedding; don't expect much natural elasticity, but do anticipate a halo.

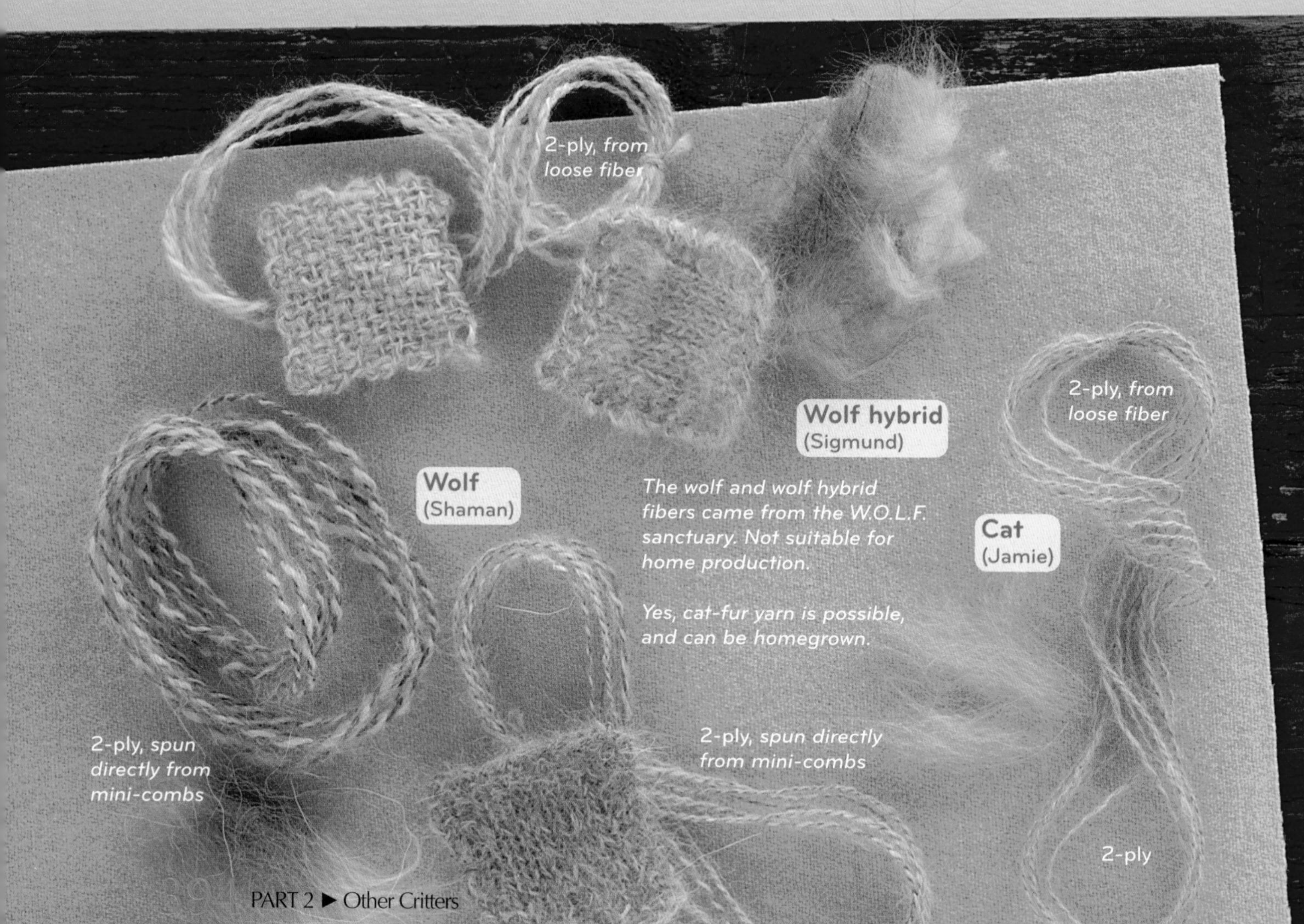

Fur and Pelt Animals

FROM TIME TO TIME, you will find, in yarn or spinnable form, fur from a variety of pelt animals such as beaver, mink, muskrat, chinchilla (a rodent native to the Andes Mountains), and possum (from the common Australian brushtail possum, not the North American opossum). In recent years a number of yarns have shown up in the marketplace that are a blend of fur with either cashmere or Merino. Here's the skinny based on our research into fur for fiber lovers.

Beaver and muskrat fur have traditionally been used to make felt hats. The fibers are described as soft and silky, ranging in color from light to medium brown, with Canadian beavers' fur also coming in a brown with bluish overtones. Beaver fur has a fiber diameter around 16 microns. Although we haven't seen these fibers in spinning preparations, as we write this there is at least one manufacturer including them in a yarn blend.

Chinchilla shows up from time to time in spinning fiber. We've seen it (once, long ago) as pure chinchilla, although, because the fiber is relatively short and slick, it's usually blended with fine wool, silk, or both. If you spin one of these preparations, be sure to make a thin enough yarn to securely catch the chinchilla fibers in the twist so your finished product will hold together and wear decently. (Just a note: We didn't find any available yarn with real chinchilla in it. The widely distributed commercial yarn from Berroco named Chinchilla is 100 percent rayon.)

Mink is being harvested in Inner Mongolia (by brushing or shearing, not killing, the animals) and then combined with cashmere to produce novel luxury yarns. These yarns, generally 70 percent mink, are said to include only the animal's undercoat, whereas fur stoles are made from entire pelts, guard hairs and all. With no elasticity, mink-focus yarns make fabrics that drape well.

To us, the dominant story for this section involves the New Zealand possum, whose fur now appears frequently in yarns and also, less frequently, as spinning fiber (usually mixed with Merino). We think this fiber will be readily available for a number of years, unlike the other fur and pelt fibers, which come and go and are not likely to become common parts of the textile scene.

The common brushtail possums were native to Australia, where their population is controlled by a plethora of predators, among them owls, wild canines (dingoes and foxes), cats, and pythons. In Australia, because of predator pressure, the common brushtail possum is a protected species. On the other hand, that same common brushtail possum was brought to New Zealand in 1837 to start a fur industry and the resulting environmental situation could not be more different. As often happens when

Mink

we humans introduce a new species to a different ecosystem, the possum went out of control due to a total lack of natural predators and has become a noxious pest. Possums have been destroying New Zealand's native vegetation and wildlife and threatening its agriculture with disease. According to data from fall 2009, about 30 million common brushtail possums live in New Zealand. That's actually a major improvement from the 1980s. And just for comparison, New Zealand is home to about 4.3 million humans; so for each human, there are about 7 possums. There used to be between 16 and 21. Imagine.

The possums sleep all day and then wake up to spend each night eating (collectively) thousands of tons of plants and trees, mostly the young native ones that never get to grow up and become forests. That's not all. These possums also opportunistically kill native wildlife species, especially birds, and take over their homes. Yet more: They have become carriers of disease affecting New Zealand's dairy herds (bovine tuberculosis). Because of the destruction caused by the possums, the government of New Zealand formed the nonprofit National Possum Control Agencies in the early 1990s to coordinate the efforts of the government, other nonprofits (such as the World Wildlife Fund), and private citizens, with the goal of humanely getting a handle on the wildly burgeoning possum takeover.

Neither of us would generally use fibers from animals that are killed for their pelts, but with the overpopulation of these possums in an environment where they weren't meant to be, we see the use of the fiber harvested from the New Zealand possums that are being killed for ecological control as making the best of an unfortunate situation. Marketing the fiber provides economic support for the efforts to conserve the rest of New Zealand's natural environment, and in addition, most of the yarn manufacturers pledge a percentage of their income to conservation agencies.

So beyond the environmental issues, what's this New Zealand possum fur like? Turns out these critters have soft, workably long, appealing fiber. It comes in several colors. The most obvious are the grays, which range from light (near white) to silvery, and what's called black, which is really brownish with red overtones. In the Australian state of Tasmania, which is a separate island, there's also a lighter gold that is conspicuous to the predators there (meaning that few Tasmanian gold possums survive long), so gold likely shows up in the New Zealand possums, too, from time to time. The undercoat is a lighter color than the guard hairs. Possum fur is not currently being sorted by color or quality, although some animals have longer and softer fur than others, and sorting may come with additional marketing opportunities. Right now, the people who are working on critter control are just trying to use as much fur as they can and develop a system that will get the population back to reasonable levels (complete elimination of the invaders is, at this point, unrealistic to envision).

Possum fur is about 1 to 1½ inches long (2.5–3.8 cm). Because it's a hollow fiber, its insulation qualities are extremely high (it's very warm). Although it's soft, it tends to form a halo and not to pill very much. The tips of the fibers taper to almost nothing (1 to 2 microns), so the prickle factor should be nonexistent, although some individuals notice even this (another example of how any generalizations about what does and doesn't feel comfortable cannot be made for any fiber). Despite its luxury-like status, this fiber should perform well in a blend used for socks.

You'll almost certainly find possum blended. Possum is very smooth, and hollow fibers like this also tend to be a bit stiff and difficult to spin on their own — the reason we have historically seen reindeer hair used in many ways, but not as spun yarn. Most often possum is combined with Merino wool, because of the compatible fineness and short staple length. Yarns tend to include between 20 and 40 percent possum. It would be nice to be able to use up the fiber more quickly by increasing the percentages, but it has no crimp (therefore no bounce) and it doesn't take dye well, so the blends allow colors into the lineup as dyestuffs can infuse the surrounding fibers. Blending makes a huge positive contribution to the versatility of the finished yarn. Plucked possum is a higher quality, in both length and strength, than mechanically harvested fiber. Possum doesn't felt, so if what it's blended with doesn't either (naturally or because it's been treated not to, like superwash wools), then you'll end up with washable fabrics. In fact, the yarn in a possum fabric will bloom in its first washing and develop its surface halo with time and use.

Possum-Merino-blend yarns from handknitting.com

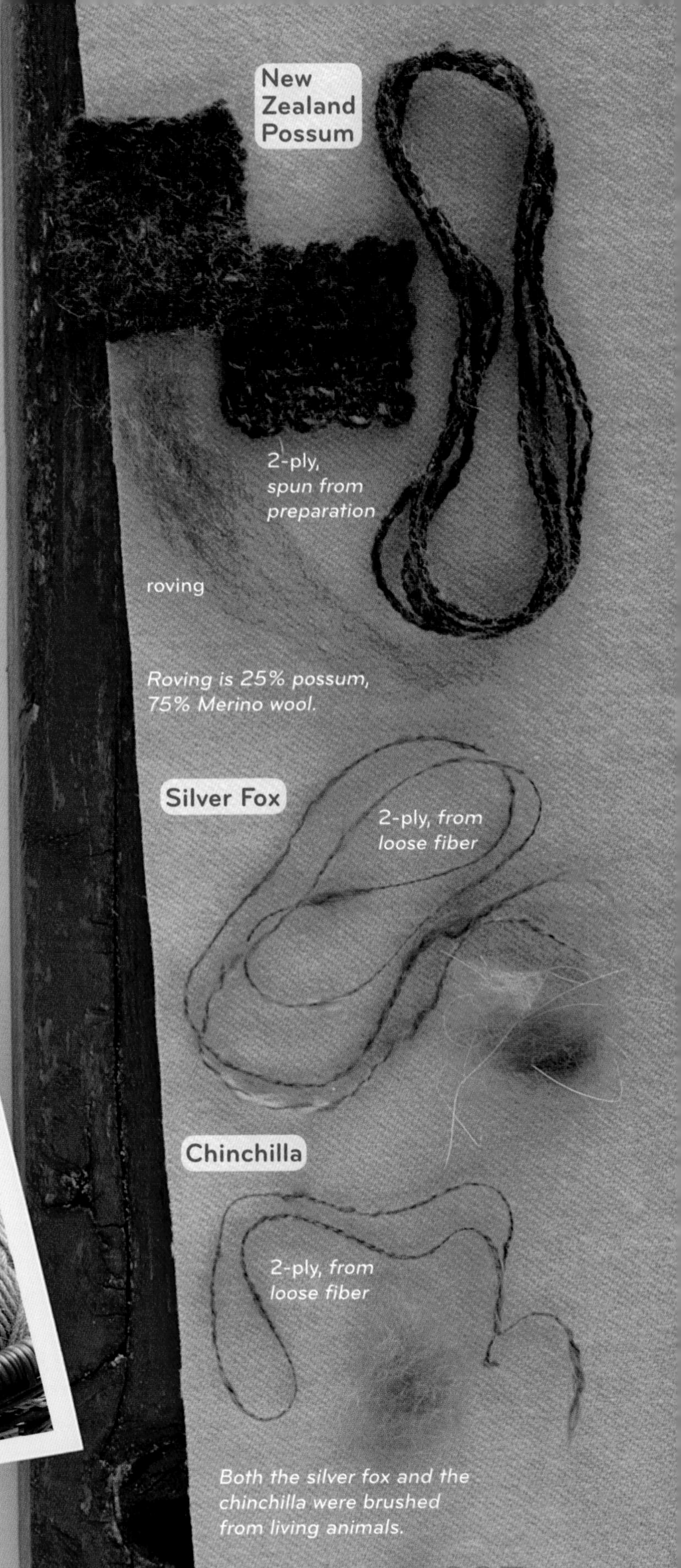

New Zealand Possum

2-ply, spun from preparation

roving

Roving is 25% possum, 75% Merino wool.

Silver Fox

2-ply, from loose fiber

Chinchilla

2-ply, from loose fiber

Both the silver fox and the chinchilla were brushed from living animals.

Horse

Horsehair may be the most unusual entry in this volume. Like several other animals we've considered, horses offer several types of fiber, but they don't grow mixed: They come from drastically different parts of the animal. There are brushings of body hair, short and possibly spinnable but definitely good for stuffing mattresses or upholstered items; more obviously there are the mane and tail hairs. Before nylon was invented, horsehair served many of the functions now taken over by that synthetic usurper. Horsehair was made into all-weather hats, woven into sieves, bound into brushes, and used as a stiffening braid in dressmaking. It even provided a support for some types of fine lace making, like Alençon.

Of the longer horse-provided strands, tail hair is strongest, and some craft techniques, like the braiding of bracelets, require that extra oomph. Mane hair is both softer and weaker, which means it's good in applications requiring more pliability and less robustness.

Glistening horsehair comes in a multitude of natural colors — black, deep or medium brown, white, reddish shades — and it can also be bleached or dyed. It doesn't yield well to spinning, being comparatively straight and stiff and slick, and it's hard to bend and knot. Despite

Cow and Reindeer

Fibers from cattle and reindeer generally don't spin well by themselves but can be (and traditionally have been) blended into wools to extend the softer fibers and add durability (and texture!). The cow hair we obtained was mostly about ½ inch (1.25 cm) long and stiff. The reindeer hair was even shorter and stiffer. Rigidity makes the fiber very difficult, or impossible, to spin. (Cottons of similar length can be spun, as can other animal fibers, like vicuña.) A lack of pliability in horsehair, which in the case of mane and tail hairs has much greater length on the plus side of the equation, results in the use of braiding, knotting, and special weaving techniques to produce textiles from that fiber.

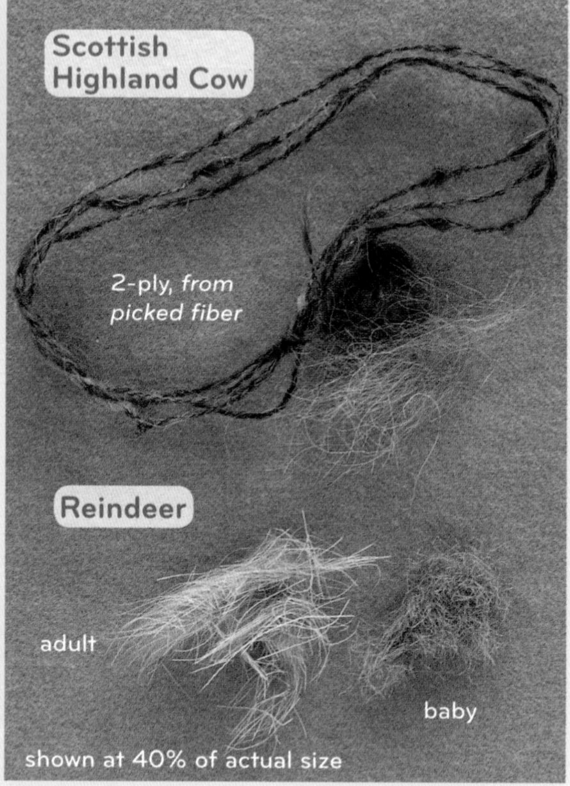

these limitations, creative and persistent crafters have devised wonderful ways to use it. Intricately patterned horses' bridles and halters have been, and still are, made from their own species' fiber; some sell for many thousands of dollars. They're braided or hitched, depending on the stiffness of the fiber. Of course, carefully selected horsehair stretched onto bows coaxes music from the string sections of orchestras.

Haircloth has been woven on narrow looms, warped in linen or cotton no wider than the hairs were long, each hair set in as weft in a direction that alternated from that of the previous hair, so the thin tips and thicker butts evened out in the long run. The standard width for such cloth was 27 inches (69 cm), for which one could get a reliable supply of wefts, even though some tail hair can be 40 inches (101.5 cm) long. In the eighteenth century, Chippendale and Hepplewhite chairs boasted hard-wearing and expensive horsehair upholstery.

In parts of Central Asia — including Uzbekistan, Turkmenistan, and Tajikistan — woven or looped-construction horsehair provided women with traditional privacy veils, called *chachvans,* as part of garments called *paranjas.* The veil allowed its wearer to see and to breathe freely while maintaining the culturally requisite degree of social decorum. When the veil was to be fashioned by looping, the horsehair was soaked in water for several days before construction, to make it pliable enough to be manipulated.

The Tohono O'odham people of the Sonoran Desert in the southwestern United States make miniature baskets of horsehair, frequently a mere 1½ to 3½ inches (3.8–9 cm) across and exquisite, regardless of whether they carry simple or complex patterning.

To use horsehair, rinse it to clean it, pick out the straw, and lay it flat to dry. Sort by color and quality (length, stiffness, fiber diameter). Pick it up directly and go to work, or organize it first with a coarse comb or a hackle.

Historic horsehair chachvan, or veil, from Afghanistan via eBay

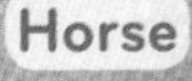

clean

raw

2-ply, from picked fiber

Most horsehair is mane or tail fiber, braided. This is Curly Horse brushings, spun. Because of the fiber's natural curl, spinning the singles S worked best.

Musk Ox

THE COMMON ENGLISH name of the musk ox (*Ovibos moschatus*) refers to the strong, musky odor that males of this species give off, but it is misleading: The males have a distinct smell, yet they don't have musk glands, and the term *ox* generally refers to a castrated male of the cattle family, yet these animals are more closely related to sheep and goats than to cows. The Inuit name for musk ox is Oomingmak, meaning "the bearded one." Both sexes have distinct beards hanging below their massive jaws.

Why, might you ask, would biologists place the musk ox, which admittedly looks sort of cowlike, closer to the sheep and goats than to the cows and bison? The answer lies in *sexual dimorphism*, a fancy term for big differences between the sexes. Sexual dimorphism can show up in traits like size or coloration, or in features like horns. It is very obvious to us in the world of birds, where there are amazing color differences between males and females, but the concept applies to all kinds of creatures. In this case, there is much greater sexual dimorphism among members of the sheep and goat family (the Caprinae subfamily for any biology geeks in the crowd) than there is in the cattle family (Bovinae subfamily).

One of the key areas where sexual dimorphism shows up is in horns. Both male and female cattle have horns of a similar size and structure, relative to the different breeds (for example, Texas Longhorn bulls and cows have long horns, whereas Shorthorn bulls and cows have short horns). By contrast, sheep and goats, and also our friend the musk ox, show extreme sexual dimorphism when it comes to their horns. Females' horns are significantly smaller than those of their male counterparts. In fact, the male musk ox's horns are so large that the base from which they grow covers the entire forehead, creating a crazy-looking helmet of horn bone (called a boss) at the top of his forehead.

Although musk oxen were native to parts of Alaska, they were hunted to extinction there in the 1850s. Thankfully, populations survived in Greenland and parts of Canada. The animals were reintroduced to Alaska from Greenland in 1931, when 34 animals were delivered to what is now the University of Alaska in Fairbanks in order to study their suitability for domestication. Only five years later, budget constraints led the university to release the animals on Nunivak Island in the Bering Sea, where they again became wild. In 1964 anthropologist John J. Teal, Jr., captured 33 calves from the Nunivak Island herd and began to develop the husbandry techniques that have allowed musk oxen to be raised in farming conditions.

Musk oxen grow several types of fiber, one of which is the animal's famous underdown, also known as qiviut or qiviuq (as well as a few other spellings). Qiviut deserves its legendary status. This rare fiber has not been readily available for spinning or in yarn form until recently, but now qiviut is making its way into our consciousness (ooh!) and then, if we're lucky, into our hands (aaah!). Now you can easily find qiviut in both spinning fluff and yarns.

Qiviut's exceptional combination of fineness, softness, lightness, and warmth make it a delight to work with and wear. It's also quite sturdy. About half an ounce (14 g) of qiviut in the form of a neck warmer — a tiny amount! — transformed our experience of winter.

Finest? Warmest? Who Says?

Let's start with a few statements about qiviut: It's the finest readily available fiber in the world, and the warmest. True, or not? Does it really matter which fiber is (speaking in averages) "finest" or "warmest"? From experience working with qiviut, we know that it is exceptionally fine and soft. From experience wearing it in bitter cold, we know that it is almost unbelievably warm: An ethereal lace scarf knitted from qiviut creates a sense of comfort and safety when the first thought on leaving the house is "frostbite." This fiber protects musk oxen to –100°F (–73°C). Yet worked up into a scarf at a light gauge and with lacy holes, it makes a fabric that is easy to breathe through and — this is so nifty — moisture doesn't condense on the inside with every breath.

Yet people often want to know which fiber is warmest, or finest, or most expensive. "Warmest" resides best in the category of subjective assessment because there are so many variables. Scientific measurement can help us think about fineness, although for qiviut there still isn't enough research data to make categorical statements about how fine it is. Part of the answer to our question depends on what we define as qiviut. Does it include all fibers with diameters below 30 microns, which is a standard cutoff point for testing fine fibers? Does it consist only of the finest down, with the intermediate fibers removed, as handspinners might be able to prepare it but commercial equipment cannot?

These types of measurement quandaries apply to all the data we use in talking about fibers. Which population are we testing? What are we measuring? What do our results tell us?

Here's what we found out when we tried to understand just how fine qiviut is: Two sets of widely quoted data are based on very small numbers of animals. In the first, from a 1931 study of three animals, measurements were taken near the root of the fiber, which the tester realized would be 1 to 2 microns finer than the middle. Those results were 11.3 to 15.3 microns. In the second set, more than 40 years later and involving 14 musk oxen, measurements were 13.0 to 17.3 microns. Both sets were grown by captive musk oxen from the group that had been moved from Greenland to Alaska.

In 1997, more than two decades after the second set, a large number (299) of animals' fiber was sampled. These samples were shaved from the hides of wild Canadian musk oxen. The samples were consistently taken from the same location on the hides and were consistently analyzed. This represents a major piece of work. Yet there was just one collection site per animal and because samples were shaved — in other words, the undercoat was not simply combed out — there were potentially more coarse fibers present than there would have been with a different source and method. Even with all these constraints on testing, the raw fiber averaged 19.5 to 21.5 microns, and the down, for which the samples were prepared in a way intended to approximate the results of commercial dehairing, came in at 16.5 to 18.2 microns. The researchers noted that this may not represent the animals' finest fiber.

In sum? Qiviut is exceptionally fine. It's right there in the cashmere department — possibly, with the intermediate hairs removed, a good deal finer than all but the very finest cashmere. It's in the realm of vicuña and guanaco (about which we also don't have enough data yet).

If you want to know how warm qiviut is, either get yourself a hand-knitted treat from the native Alaskan knitters of Oomingmak Musk Ox Producers' Co-operative (www.qiviut.com) or obtain some yarn or fiber and make something yourself. When it comes right down to it, this stuff is bliss.

Qiviut grows as a soft silver gray, which shifts with weather and age to a light gray-brown that is beautiful in its natural state and can also be modified through dyeing. Bleaching weakens the fiber, so there are no attempts to lighten it as there are for yak. Qiviut has no scales, so it will not shrink or felt (although it will full some), and it has no crimp, so it has no fiber memory; a heavy fabric will be overcome by gravity, drooping in a way that obscures its essential gossamer nature. It does have some tendency to pill with abrasion, because of the many short, fine fiber ends, so it makes sense to use it to construct items that have minimal exposure to repeated rubbing.

You can get the benefits of qiviut while also extending the fiber by blending it with other compatible, less-expensive fibers, such as cashmere. (Have you ever thought of cashmere as an economy move?) Other options are very fine wool, silk, alpaca, fine llama, or angora rabbit.

If you spin qiviut, two other fibers grown by the musk ox will slide into your awareness: the guard hairs (not the same as the long "skirt" hairs that hang almost to the ground) and an intermediate fine hair that remains mixed with the qiviut once the guard hairs have been removed. For a long time, musk oxen were thought to have only two coats: guard hairs and qiviut. Then some spinners, including Helen Griffiths Howard (who taught workshops that helped the knitters of Alaska's Oomingmak Musk Ox Producers Co-operative start their innovative business), Marguerite Cornwall, and Diane Olthuis, discovered another set of fine hairs mixed with the down. It's comparatively easy to see and pull away the heavy guard hairs. Those intermediate hairs, straighter and stiffer than the down, but quite fine, can also be removed by the patient artisan who wants to work with only the softest and most luxurious materials. (The intermediate hairs can be saved and spun on their own or added to a blend with other fibers.)

Dehairing is a big deal, no matter whether it's being done commercially or by hand. Commercial equipment does not even aim to remove the intermediate hairs. If you are going to dehair fiber, plan on spending a very long time, and don't try to do it all at once.

Musk oxen live in the wild in a few places in the far northern latitudes; there are small groups in Alaska and far larger herds in Canada. They also currently live in three captive herds under farming conditions of partial domestication at two locations in Alaska and one in Alberta. Consequently, qiviut fiber makes its way to fiber folk primarily through two channels: combed from living musk oxen as they shed at the farms (green light for vegans), or shaved from the hides of wild animals killed in regulated harvests in Canada, where food is the primary object but all parts of the animals — meat, horns, and fiber — are put to use. The shed fiber, which consists almost entirely of qiviut, requires less sorting and processing than the shaved fiber, which includes many more coarse hairs. Qiviut from both channels is available as spinning fiber and as ready-to-use yarns.

Qiviut yarn from a sock kit by Lorraine Weston

Musk Ox Facts

▶ **FLEECE WEIGHT**

5–8 pounds (2.3–3.6 kg); yield from combed animals 80–88 percent, from pelts 50 percent or less

▶ **STAPLE LENGTH**

Qiviut down ½–6 inches (1.3–15 cm)
Intermediate straight hairs Not determined; similar to lengths of down
Guard hairs ¾–6½ inches (2–16.5 cm)
Skirt hairs Up to 24 inches (61 cm)
There are also different-textured fibers in beard, mane, and blanket (or saddle area of the back) hair.

▶ **FIBER DIAMETERS**

Best estimates suggest between 11 and 19 microns for the qiviut (down), possibly as low as 10 microns and as high as 20. Qiviut becomes slightly coarser as the animal ages. Farmed animals seem to grow slightly finer fiber than wild musk oxen, averaging 14–16.5 microns rather than 17–18 microns.

▶ **LOCK CHARACTERISTICS**

Qiviut can be combed out of a shedding musk ox as a single blanketlike mass, because it all sheds at the same time; there are no locks to speak of.

▶ **NATURAL COLORS**

Has a characteristic soft gray-brown color.

Using Musk Ox

Dyeing. The gray-brown can be overdyed.

Fiber preparation and spinning tips. If your qiviut has been dehaired, proceed to either removing intermediate hairs or spinning. If your qiviut comes direct from shedding and has dandruff at the bases of the fibers, snip off those ends with sharp, small scissors. Remove the obvious guard hairs, and if you want, the intermediate hairs — a painstaking job that you may consider worthwhile. Spin the down directly from the fluff in your hand or card it, using fine-toothed (cotton) carders. Spin fine; qiviut is both a short fiber and a very effective insulator.

Knitting, crocheting, and weaving. On its own, qiviut is lofty and has some bounce, but has no fiber memory. Fabrics made from it may sag: Plan this into your design process. Most people knit with qiviut, but it's also great for weaving. It's fluffy and will tend to obscure texture patterns but works beautifully with laces, blooming a bit to fill the openings — so keep your patterning simple.

Best known for. Rare, soft, luxurious fiber with extreme warmth and lightness, as well as a distinctive gray-brown color.

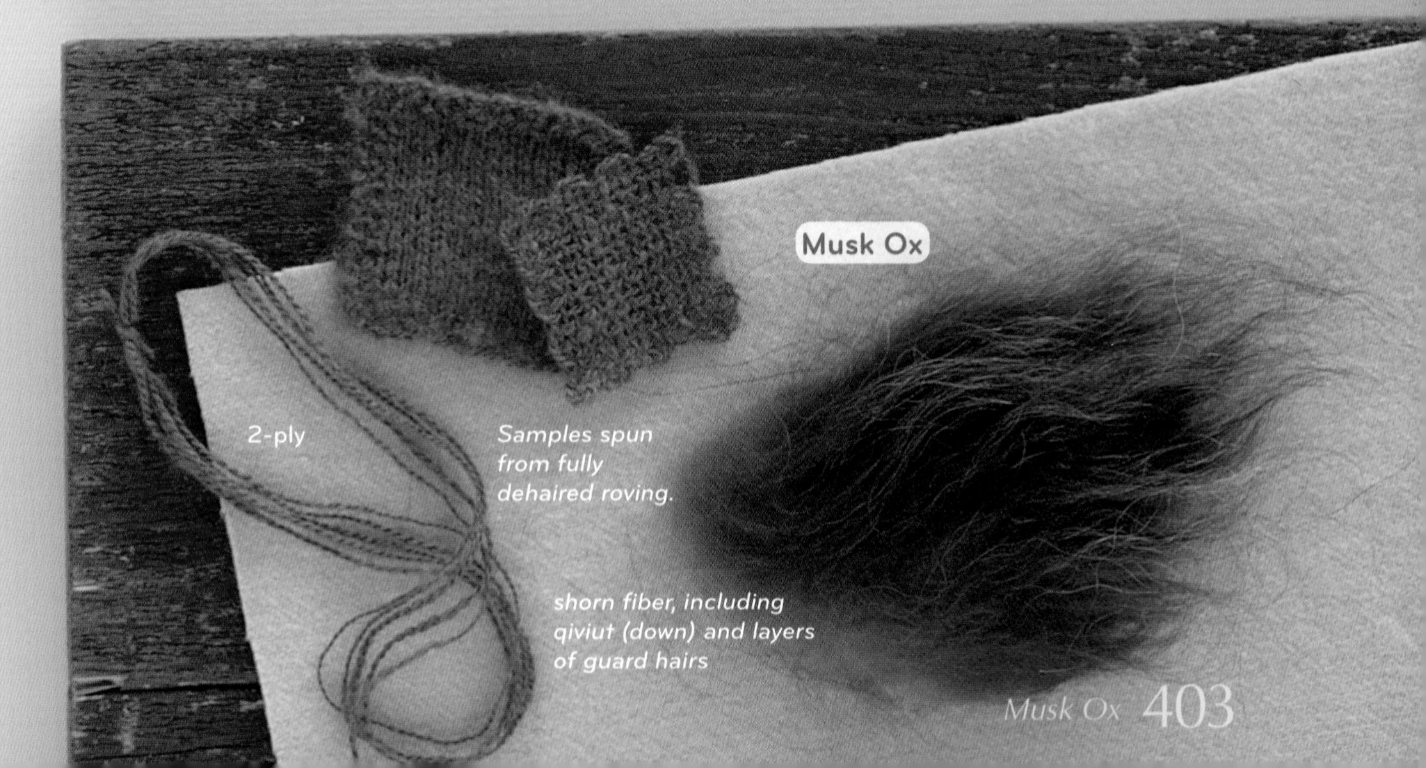

Rabbit

Rabbits are thought to be among the oldest mammals, dating back at least 30 million years. Nonetheless, rabbits were probably the last animals to be fully domesticated, starting around 1,400 years ago when Catholic monks in France began raising them for meat, supposedly because in about 600 CE Pope Gregory decided that rabbit wasn't considered meat and the monks could eat it during Lent.

Although there are approximately 50 species of wild rabbits in the Leporidae family, including numerous jackrabbits, hares, and cottontails, that family's *Oryctolagus* genus contains only a single species: *Oryctolagus cuniculus,* which includes the European wild rabbit and all of the many types of domesticated rabbits. The European wild rabbit, from which more than a hundred breeds of tame rabbits (worldwide) originated, is native to southwestern Europe — primarily Spain and Portugal — though humans have spread these rabbits across other areas of Europe, including the British Isles.

The practice of keeping rabbits for show and as pets developed in Victorian times. Interestingly, today rabbits are gaining in popularity as house pets, in part because they can be trained to use a litter box!

When it comes to rabbits, the Angora breeds are the primary fiber producers, though Fuzzy Lops, Long-Haired Holland Lops, and Jersey Woolies also grow spinnable wool. These animals' small size, however, means they are rarely raised commercially for fiber production.

In her book *Angora: A Handbook for Spinners,* Erica Lynne observed that "While learning the past of the angora rabbit . . . I have learned that history is not a truth carved in stone, but a puzzle with many missing pieces — an interpretive guessing game. The tantalizing clues are bits of bone and snippets of writing." This is, of course, true of many animals we are discussing in these pages: We know some things, while others remain mysterious.

No one knows exactly where Angora rabbits first appeared. As Lynne recounts, the name implies that they came from the area now known as Turkey (see Mohair, page 337). It may be that they sprang up in Britain, where eighteenth-century prohibitions on exporting these long-haired variants of the domesticated rabbit led to fibbing about their origins by traders. In 1823, traders are said to have taken the first Angoras to France (which then dominated angora production for a couple of centuries). Or perhaps, as a United Nations publication puts forth, they originated through a mutation in wild rabbits in France.

Although there are a number of Angora breeds around the globe, we highlight here the five found in North America: English, French, German, Giant, and Satin Angoras. The range of fleece characteristics we describe should give you an idea of what to expect overall from these rabbits, wherever they are found and whatever they are called. The American Rabbit Breeders Association (and its specialty group, the National Angora Rabbit Breeders Club) recognizes four of the breeds we cover — English, French, Giant, and Satin — while the German Angora, often confused with the Giant, has its own organization, the International Association of German Angora Rabbit Breeders.

Why do we love angora fiber? Warmth! Fluff! Delicacy and softness! Angora rabbit fur excels in all of these qualities, and the fiber often shows up in commercial yarns blended with one or more other fibers, because a small amount of angora has a profound effect on the finished product. Yet yarns can also be spun as pure angora, pushing the boundaries on all of the benefits and revealing a few additional attributes that need to be taken into account: almost no elasticity (provided in blends by other fibers), little resistance to abrasion, and, with some breeds' fluff, a tendency to mat and felt.

Angora yarn has also acquired a reputation for shedding, which is sometimes — but by no means always — warranted. Sharon Kilfoyle and Leslie Samson, in *Completely Angora,* speculate that shedding in a newly prepared angora fabric may be the result of the tips breaking off one type of fiber in the mix (the awn hairs). Shedding may not be an issue at all if the fiber has been grown, harvested, prepared, and spun appropriately. The fibers are indisputably slippery. If they are long enough and similar in length, and have been spun into a fine, orderly strand with adequate twist, they should stay put in the completed fabric — except for their tips, which will fluff out into either a lovely fuzz or a full-fledged halo, depending on the fiber mix and its processing.

Angora rabbits actually produce four types of fibers in their coats: two kinds of guard hairs, called bristles (or simply guard hairs) and awn hairs, and two kinds of wool, awn wool and down (or the undercoat). Each of the awn-type fibers has a distinctive profile along its length, narrowest toward the body and at the very tip, with a wider portion toward the outer end near the tip. While all these fiber types exist in all Angoras' coats, they vary dramatically in their proportions. The guard hairs are much more strongly medullated than the wool fibers. Unlike the fleeces of other multicoated animals, these combined fibers are ordinarily spun into yarn together, although some people separate out guard hairs in order to achieve a more consistently soft result — a goal that may be reached more easily by choosing a different Angora breed that grows fewer guard hairs. As is true for other animals, fiber qualities vary from one part of the rabbit's body to another, and health and nutrition play important roles in the amount and value of the harvest.

English, French, and Satin Angoras molt about every three months, and their fiber can be combed, plucked (when it loosens on its own), or shorn. Giant and German Angoras don't shed, and their fleeces must be shorn. Plucked or combed fiber is most likely to meet the non-shedding criterion of adequate and consistent length, as can shorn fleeces if they are collected at the right time, skillfully clipped, and carefully processed to remove any short, new growth that may end up in the mix.

An allergy to angora fibers may result from a reaction to the rabbit's saliva (deposited on the coat during grooming) or physical irritation caused by inhaling fine fibers broken up in mechanical processing. The former can be resolved through washing and the second by selecting long, custom-harvested fibers and carefully managing them through preparation and spinning.

Did You Know?

Spain's name actually derives from its native wild rabbits! Approximately three thousand years ago, Phoenician sailors landed on the coast of Spain. They called the area i-shephan-im ("the land of the rabbit"). The Romans converted the Phoenician term to a Latin form, Hispania, which has come to be Spain in modern English.

English Angora

The English Angora is a very fancy rabbit. It's relatively small, at 5 to 7½ pounds (2.3-3.4 kg), and covered with lots of wool — not only on its body, but also its face, ears, and feet (called furnishings). It can be hard to locate an English Angora's eyes for the pouf of wool sprouting from its forehead. The breed comes in 30 different colors, but due to their slightly smaller size compared to other Angoras, English Angoras are more often raised as pets or show animals and less often kept by serious fiber producers. Nonetheless, an English Angora can produce about ⅞ pound (400 g) of fiber annually, and it grows the finest fiber produced by any of the Angora breeds. English Angoras are sometimes plucked and sometimes shorn. The harvesting method has an effect on the quality of the wool, the presence of guard hair, and the fiber's tendency to mat.

English Angora wool felts especially easily. It will felt even when unspun and in storage, if anything compresses the fiber for long. Compared to the other breeds, English Angoras have relatively small amounts of guard hair, although the adults have enough guard hair to fend off felting while the coat is still on the animal, as long as it's groomed regularly. If you're inclined to pull the guard hairs out of the other types of Angoras, you might want to just start with the English breed.

Angora yarn from Local Harvest/Bloomingdale Farm Angoras

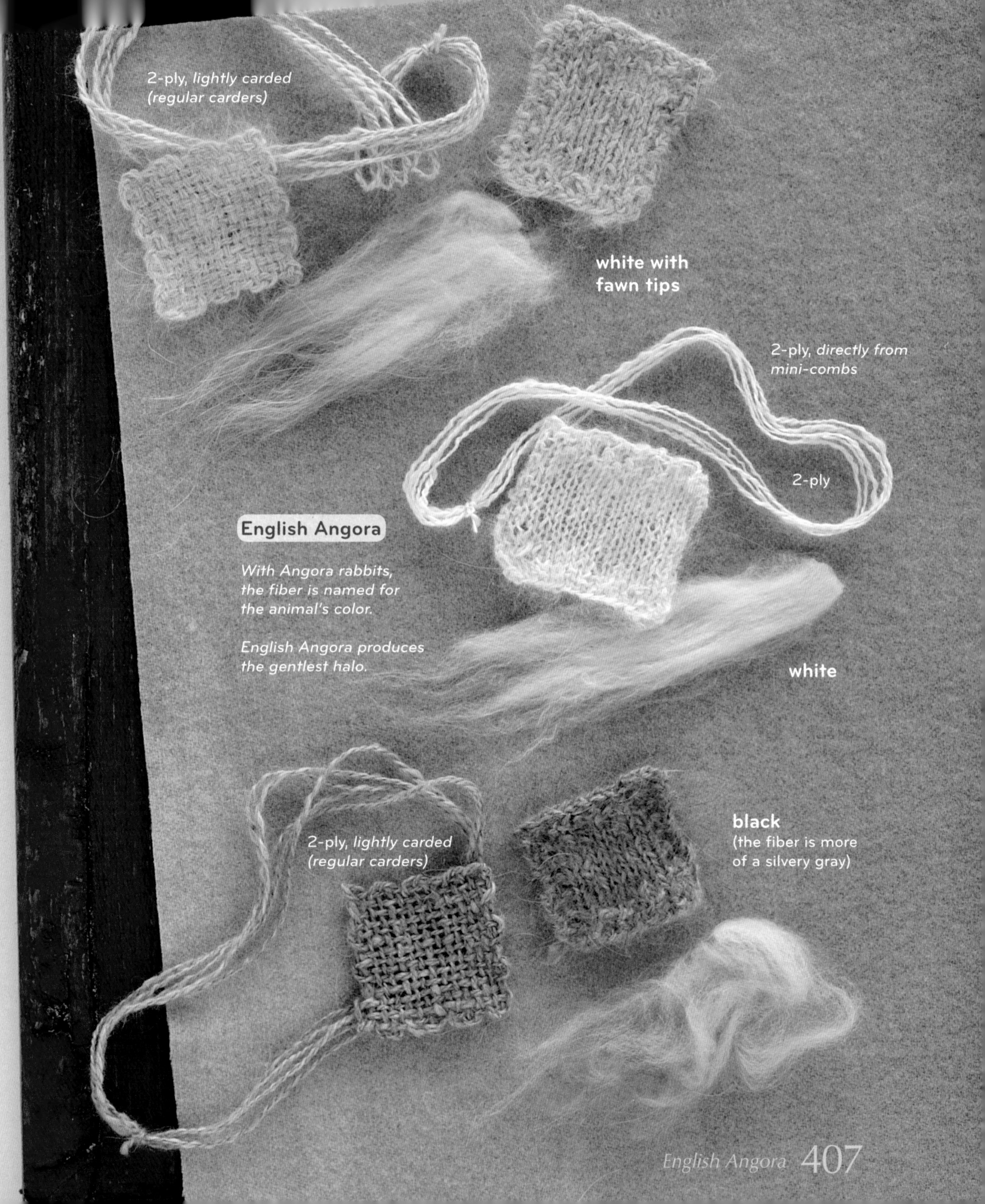

2-ply, *lightly carded (regular carders)*

white with fawn tips

2-ply, *directly from mini-combs*

2-ply

English Angora

With Angora rabbits, the fiber is named for the animal's color.

English Angora produces the gentlest halo.

white

2-ply, *lightly carded (regular carders)*

black
(the fiber is more of a silvery gray)

French Angora

French Angoras weigh between 7½ and 10½ pounds (3.4–4.8 kg). They were bred by the French specifically for wool production, and they come in a wide array of colors. Because of their wool-growth patterns, plucking has been the method of choice for harvesting their fiber, although they can also be shorn. A single rabbit may produce as little as ¼ pound (113 g) of fiber a year, or as much as 1 pound (454 g).

French Angoras grow wool that's a bit more substantial than that of their English cousins. As a result, it's easier to spin. Thanks to the larger percentage of guard hairs, it's also the slowest of all the angoras to felt. Those guard hairs — often called spiky, in reference to their visual effect — have a surprisingly soft texture, and the French Angora's fluff can generate the most halo.

2-ply, from loose fiber

sable (plucked)

2-ply, from loose fiber

chestnut (brushed)

2-ply, from loose fiber

black (brushed)

French Angora

2-ply, *directly from peasant combs*

broken red
(plucked)

French Angora produces the spikiest halo.

2-ply, *locks, fed to wheel base-first*

fawn

German Angora

In the middle of the twentieth century, German breeders began to develop a type of Angora rabbit that grew large quantities of wool and had a breadloaf-like body shape that suited efficient shearing. Rabbits of this type were imported into North America. The Germans now range from 5 to 11½ pounds (2.3-5.2 kg) and are similar to the English in that they are "furnished," or have wool on their face, ears, and feet; however, their furnishings are a bit lighter than those of the English. Annually, a high-producing German Angora can yield up to 5 pounds (2.3 kg) of well-crimped wool, which is gathered by shearing.

The breeders' association for German Angoras has helped start a yarn co-op, so finding breed-specific yarn isn't as tough as it is for some of the other Angoras.

Giant Angora

The combination of *giant* and *rabbit* may seem to form a bit of an oxymoron, but as bunnies go, these guys are big. They were developed in the United States from the German Angoras, with infusions of other breeds. Mature bucks have to weigh at least 9½ pounds (4.3 kg), and mature does at least 10 pounds (4.5 kg). But what's really amazing is that they produce between 1 and 2 pounds (450-900 g) of fiber per animal per year, sometimes even more. Giant Angoras' dense coats are shorn. Although the only recognized color for these rabbits is white, some breeders are working on developing other colors.

Satin Angora

Satin Angora rabbits weigh in between 6½ and 10 pounds (3-4.5 kg). Their origins are unclear; some say they came about through a mutation in a non-Angora breed simply known as the Satin, while others say they developed through crossing French Angoras with Satins. The original Satin rabbit is known for its fur's outstanding shine, due to the shape and surface quality of the fibers' scales, and that trait also characterizes the Satin Angora. The Satin Angora's fiber is somewhat finer yet slightly stronger than the French Angora's. As these smaller animals yield just up to ½ pound (228 g) of fiber per year, they are rarely raised by serious fiber producers.

Angora Facts

▶ FLEECE WEIGHT

Depending on breed, breeding, and husbandry techniques, from ¼ pound (113 g) to as much as 3 to 5 pounds (1.4–2.3 kg) a year, in portions gathered every 3 to 12 months (usually every 13 weeks). Yield is 90 to 99 percent of harvested weight.

▶ STAPLE LENGTH

Per harvesting cycle, depending on breed, breeding, husbandry, and processing techniques: commercially gathered for the industrial and global market, 1½–2¾ inches (3.8–7 cm); from small, custom suppliers, 2½–6 inches (6.5–15 cm), with the down on the shorter end and the guard hairs on the longer end. Note that there are longer fibers in the hand-processing market; this extra length can be beneficial in both yarn construction and the durability of finished textiles.

▶ FIBER DIAMETERS

For the down, from 5.5 to 20 microns, mostly 8–15 microns. Awn wool is similar to the down, although the fibers have the characteristic awn tip (described on page 405). Guard hairs average 44–48 microns. Awn hairs are 10–30 microns where they're narrow and larger than 40 microns where broad.

▶ **LOCK CHARACTERISTICS**

The down has crimp, although it is not like the crimp characteristic of fine sheep's wool. Some rabbits may grow wool with open, wavy crimp, good for making yarns that drape well, while others have tighter crimp that results in a more elastic yarn. The awns may have a little bit of crimp, and the guard hairs have none. If carefully handled, plucked or shorn fiber can remain parallel in what might be called locks, although the butt ends of shorn fibers may stick together and need to be teased apart. Combed fiber may tangle, having no identifiable form at all, and therefore spin into a textured yarn, regardless of the designer's intention.

▶ **NATURAL COLORS**

The color name of the rabbit is determined by its face, and the hair may be called by this name even when the name wouldn't be applied if the fiber were viewed on its own. This is most obvious with so-called black fiber, which looks gray to fiber folk. Regardless of the plethora of names, the working fiber colors include white, silvery gray (often called black), reddish shades, and fawns.

Using Angora Fiber

Dyeing. Unless you have experience and a good reason for doing so, dye angora in yarn form rather than as unspun fiber. Dyes that work for other protein fibers (anything in this book) work for angora. The different types of angora fibers take dye differently, with the down taking color more readily than the guard hairs, although the luster of the guard hairs gives any added hues brilliance. You will need to work especially hard to get the fiber thoroughly wet (use soap or detergent as a wetting agent; increase soaking time; and perhaps apply gentle, persistent massage without agitation or rubbing that will felt the angora). Because the fibers are delicate, keep heat, dyeing time, and chemical additives on the gentle side, although more dyestuff may be required than for an equivalent quantity of wool.

Fiber preparation and spinning tips

- Angora can (and almost always should) be spun without being washed first. Ideally, it was combed on the rabbit, to eliminate tangles, before it was plucked, brushed, or shorn. Washed unspun wool clumps together and is hard to manage.
- Use the longest fibers to spin pure angora yarn. Blend or felt the shorter fibers. Spin the fibers loose, or if necessary, card very lightly. Drum-card only if you are experienced with your equipment and have a fine-fiber drum. The purpose of carding may be to open clumps in combed or shorn angora, to mix the various types of hair so the halo and distribution of color will be even throughout, or to blend shorter angora fibers with another type of fiber.
- The fiber will tend to escape while you are working with it, so define a working area away from food-related spaces, and ideally spread a piece of cloth in a contrasting color across your lap when you are spinning to contain the fiber and help you see the drafting zone.
- Spin fine: The fibers need to be controlled by the twist, and the yarn will fluff up after it's finished. The fibers need enough twist to hold securely together but not so much that the yarn becomes wiry.
- If you are spinning plucked or shorn angora, spin with the cut or root ends toward the wheel, so they are securely caught in the twist. Colored angora will most dramatically display its unique qualities when spun this way, because the pigment is concentrated in the fiber tips. The halo develops fully later, when the yarn is used and the tips fluff out. French and Satin Angora produce more dramatic halos than German, Giant, or English Angora, which develop more of a soft haze.

Knitting, crocheting, and weaving. Pure angora yarn has no elasticity, fluffs up, and is incredibly warm; these factors will affect how you use it. It's sensitive to abrasion, especially when moisture is involved, so that, too, may need to be considered when you think of what to do with it. Any construction technique is appropriate, as long as there's room between the strands of yarn for the halo or fuzziness to come forth. That same halo or fuzziness will obscure intricate stitch or weave patterns. What's coming to mind for us is a warm cowl, a lacy vest, or a super-cozy beret, with carefully constructed ribbing (smaller needles and decreased stitch count, worked in the most-elastic option, a 2×2 rib) or a channel for elastic.

Best known for. Lightweight and extreme warmth, fine softness, and haloed or furry finish.

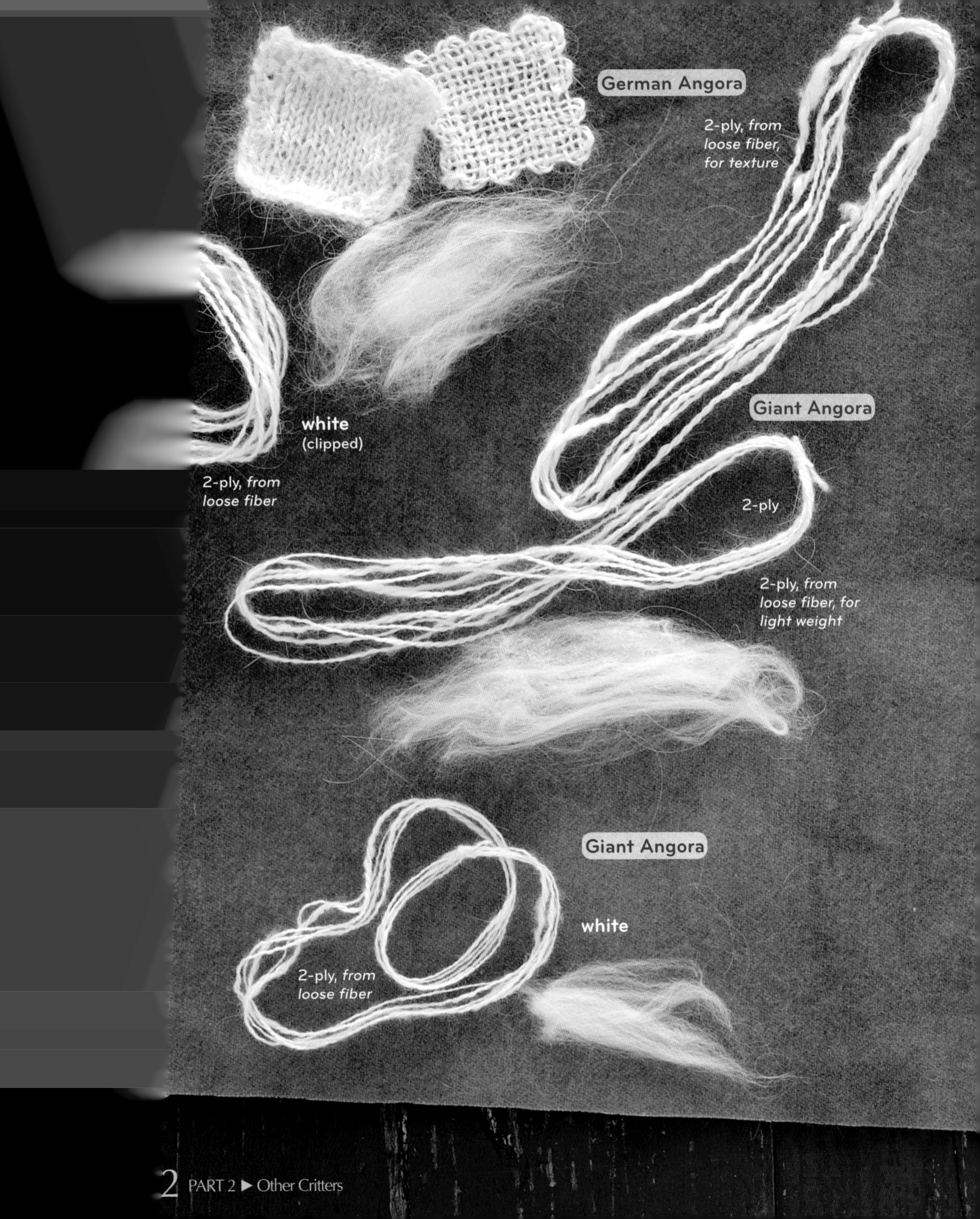
German Angora
2-ply, from
loose fiber,
for texture
white
(clipped)
2-ply, from
loose fiber
Giant Angora
2-ply
2-ply, from
loose fiber, for
light weight
Giant Angora
white
2-ply, from
loose fiber

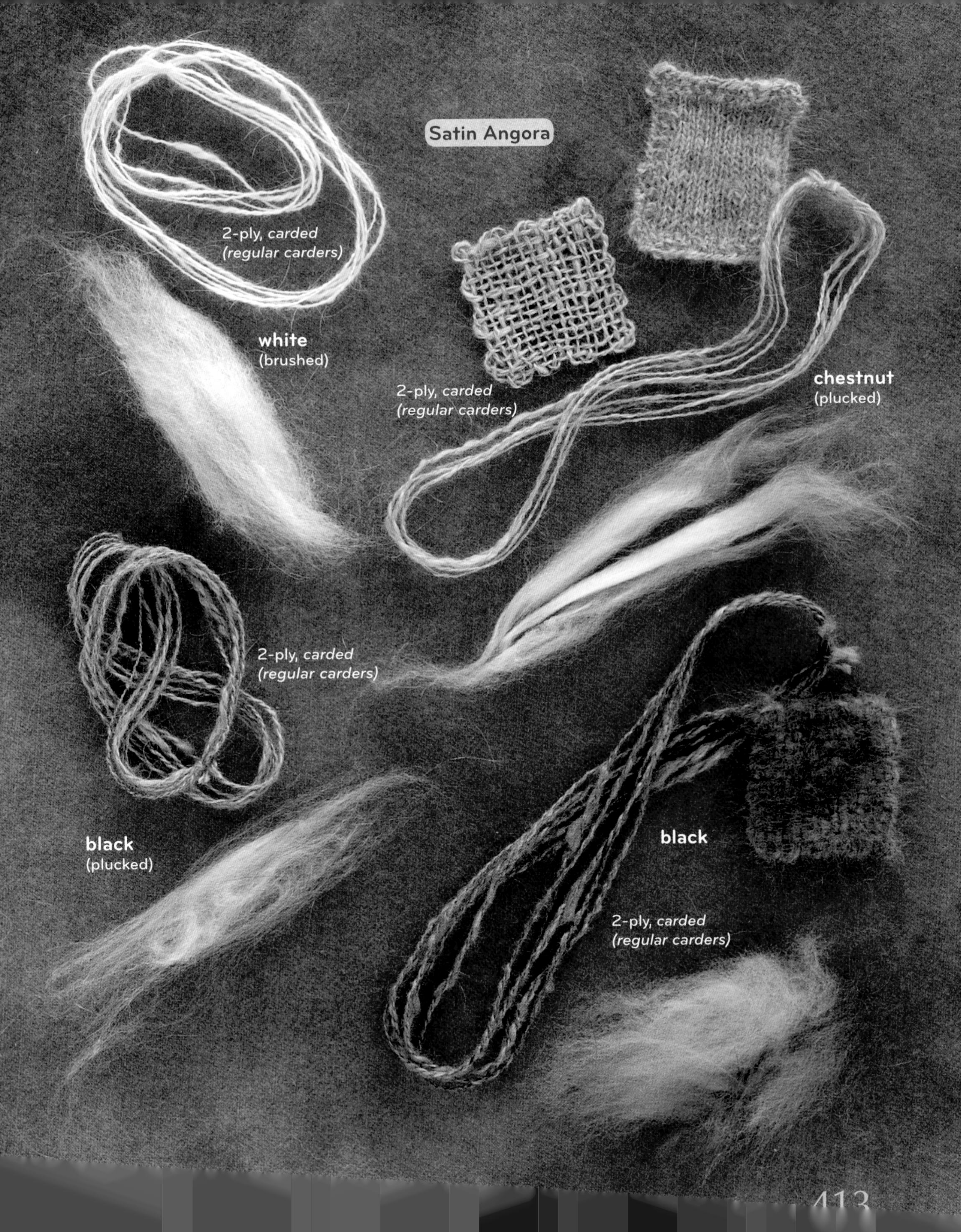
Satin Angora
2-ply, carded
(regular carders)
white
(brushed)
2-ply, carded
(regular carders)
chestnut
(plucked)
2-ply, carded
(regular carders)
black
(plucked)
black
2-ply, carded
(regular carders)

Yak

Conservation Species (in the Wild)

YAKETY YAK, don't talk back! Unless of course you are yakking about yaks! These hairy cowlike creatures are found throughout the Himalaya mountains, which cover parts of Afghanistan, Bhutan, China, India, Mongolia, Nepal, and Pakistan, and on the Tibetan Plateau, an area that is four times the size of Texas. The highest mountains in the world, K2 and Mount Everest, are in the Himalayas, but even the plateau is in the low-oxygen zone at an elevation of more than 13,000 feet above sea level (3,962 m).

Yaks are all-around economic engines for the people who inhabit this extreme part of the planet. Their down is used for wool; the long, outer hairs are used for rope and rugs and in the production of wigs, often for theatrical costuming. Yak milk is used as a beverage, as readers of this book may commonly drink cow's milk, and also fermented to make liquor. Yak meat tastes somewhat like beef, although it is sweeter and lower in fat. Yaks are used as plow animals, and they can be regularly saddled and ridden. In fact, yak racing is a traditional sport in Tibet, Mongolia, and parts of Pakistan, with top animals sprinting to the finish line at 10 miles (16 km) per hour. Tibetans also play a version of polo, called *sarlagan polo,* from the backs of their yaks.

The yak family tree has deep roots, and we need to shift to another geologic time to locate them, using fossils as our guides. Widely distributed fossil evidence links yaks to the Pleistocene epoch, more than 10,000 years ago, and yak ancestry may extend back to the Pliocene period (formerly called the late Tertiary), more than 2.5 million years ago. Domestication is thought to have occurred over a span of time between 10,000 and 4,500 years ago, in the vicinity of what is now known as Tibet. Yaks were domesticated later than sheep and goats, but not by long.

Two species of yaks exist today, wild (*Bos mutus*) and domestic (*Bos grunniens,* or sometimes *Poephagus grunniens*). Both grow similar multilayered coats, designed to help them survive at very high altitudes. The wild version, larger and solid black or very dark brown, is vulnerable to extinction, according to the International Union for the Conservation of Nature (IUCN), the organization that helps protect endangered animals around the globe. There are only about 10,000 *Bos mutus* still left in the wild. The smaller domestic species comes in many colors — black, white, red, brown, and golden, as well as spotted patterns. The lines between wild and domestic yaks are not clearly drawn, and the species do interbreed. In addition, a small group of feral yaks now exists in Inner Mongolia; although these animals live independently, they sport domestic-yak coloration.

Among domestic yaks, there are both breeds (as defined here) and types (local variations that may or may not be genetically distinct). In China and Tibet, there are eleven recognized breeds: Jiulong, Maiwa, Tianzhu White, Gannan, Pali, Jiali (or Alpine), Sibu, Huanhu, Plateau, Bazhou, and Ziongdian. There may

Yak by the Ton

A mature male yak is over 10 feet (3 m) long, and over 6½ feet (2 m) high at the withers (shoulders). He weighs more than a ton (or tonne)!

be other breeds, and there are definitely other types. There is a North American type, developed from yaks that were imported to Canada in 1909. No one knows if there were additional importations, but today animals from this source also live in the United States. Though still fairly rare, breeders estimate there are about two thousand animals in North America.

Although all yaks produce usable fiber, a strain of the Chinese Jiulong breed has been cultivated for particularly high fiber production — between 3 and 5 times more than other types, although some are said to grow 10 times as much, with half of that consisting of down. Record-setting Jiulong individuals are said to produce a staggeringly large mixed fleece of up to 55 pounds (25 kg).

White yaks produce more fiber than the darker ones, but because there are mostly dark yaks, white fiber is still far less common. With this range of options, it's not surprising that the fiber — even if you're just talking about the down — cannot be succinctly described. Some is crimpy, and some is more sleek, closer to alpaca than cashmere or qiviut.

Like bison and musk oxen, yaks grow several interleaved types of fiber. There's the long, shiny, coarse, and inelastic outercoat, used to make exceptionally durable ropes — some of which incorporate decorative combinations of

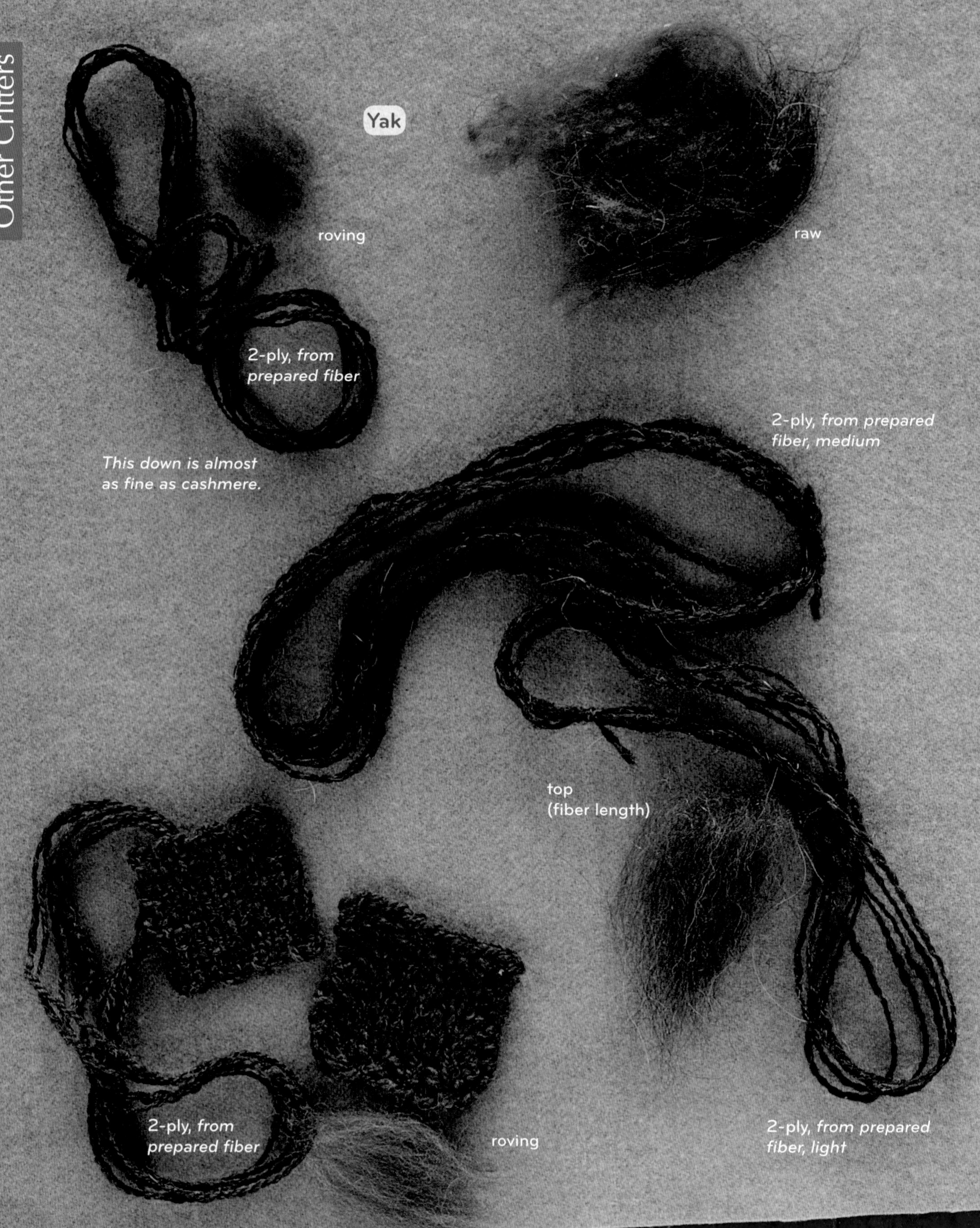
Yak
roving
raw
2-ply, from prepared fiber
This down is almost as fine as cashmere.
2-ply, from prepared fiber, medium
top (fiber length)
2-ply, from prepared fiber
roving
2-ply, from prepared fiber, light

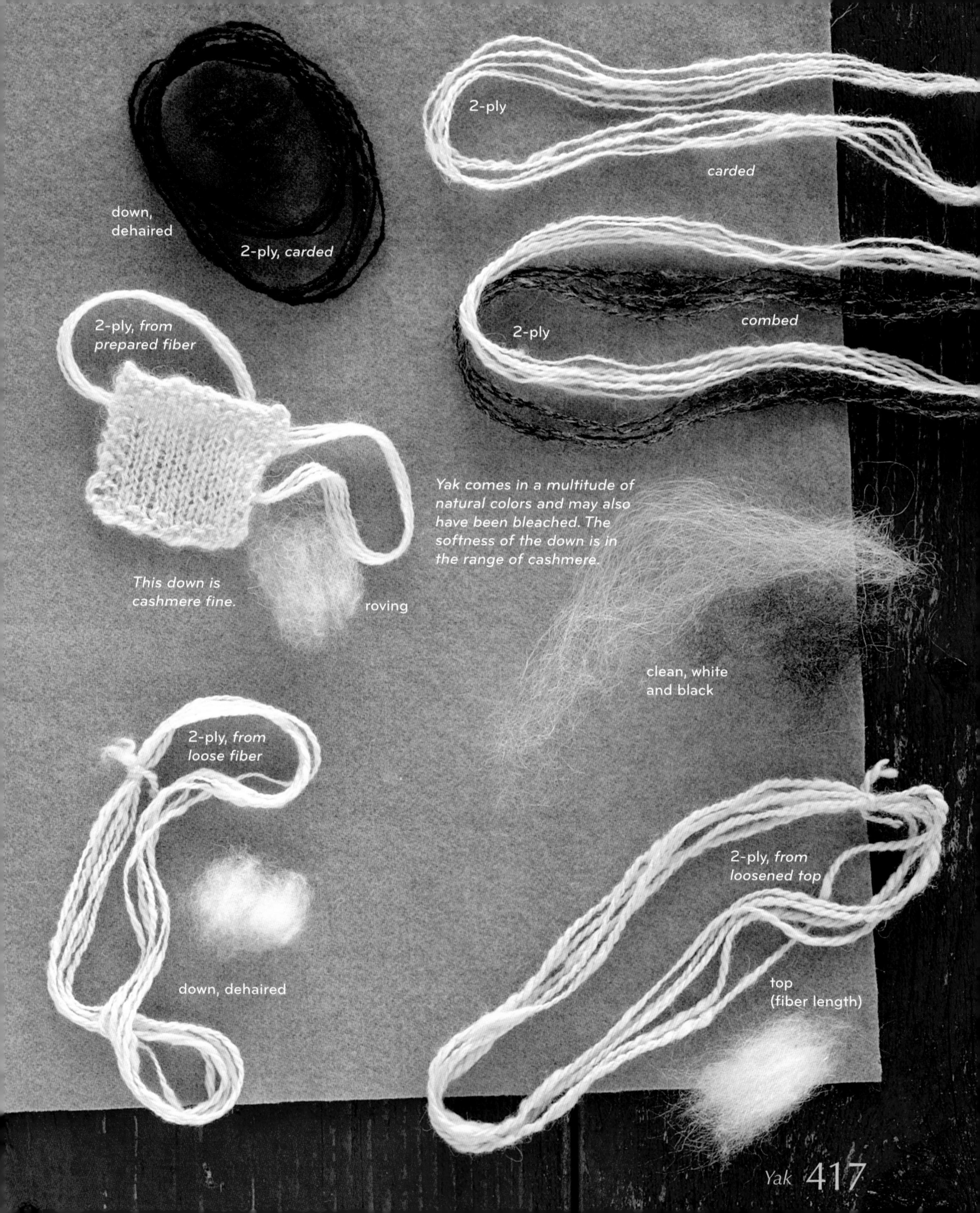

down,
dehaired

2-ply, *carded*

2-ply

carded

2-ply, *from prepared fiber*

2-ply

combed

Yak comes in a multitude of natural colors and may also have been bleached. The softness of the down is in the range of cashmere.

This down is cashmere fine.

roving

clean, white
and black

2-ply, *from loose fiber*

down, dehaired

2-ply, *from loosened top*

top
(fiber length)

the black and white hair colors — as well as tents, bags, rugs, and some very sturdy clothing. Then there's a midrange type of hair. Finally, there's the irregularly crimpy down undercoat, which usually has elasticity and bounce, as well as varying degrees of luster that can vanish unless it's spun into a yarn with the fibers as parallel as possible. (There's also tail hair, which is cut every other year instead of annually.) All of the fibers are very strong. Although the hair fibers have scales, their surfaces are smooth and they don't felt well; however, the down fibers do felt readily.

In traditional cultures, the midrange hair and down most often go through the textile processes without sorting. From this mixture, traditional herdsmen make garments that are waterproof and warm. Fiber folk elsewhere in the world focus on the down, which is about 1 to 2 inches long (2.5-5 cm) and — not surprising considering the diverse sources — varied in fiber diameters. One reason yak down has only recently been coming onto the market in yarn form is that the multiple coats are unusually hard to separate thoroughly. Because some of the guard hairs are very fine, a few hairs left in a prepared top or roving may not compromise the finished yarn's softness, as would guard hairs from many other animals. Some amazingly effective processing is giving us essentially hair-free yarns that can go straight to needles, hook, or loom: Kudos to the stubborn and visionary people who are making this happen!

Yak down yarn from Bijou Basin Ranch

The fiber diameters of the down may average an exquisitely slim 14 to 16 microns when the animals have been bred and the fibers have been processed to emphasize fineness and softness, and a respectably luxurious 18 to 22 microns otherwise. The finest yak down inhabits the same realm as the best cashmere and qiviut. Some of the down samples we obtained behaved more like alpaca (spinning up sleek, compact, and dense), and others had a fluffier feel, more like qiviut, camel down, or cashmere.

With increasing demand for the fine fiber, which is a function of global market influence, gathering and processing techniques are being adjusted to improve the quality of the down harvest and to keep it separate from the hair. The down can be combed out before shearing, or the whole coat can be shorn and then sorted. (In Tibet, the breeding bulls are not shorn, to help them maintain their macho images.) In most, but not all, industrial production, the three coats are generally split into only two qualities, with the dividing line set at 35 microns. Different yaks produce varying amounts of down, although it's not yet clear whether yield depends on breed, geographic conditions, or both. Yaks definitely produce more fiber when they are in their normal, cold environments.

Yak Facts

▶ **FLEECE WEIGHT**
The amounts of fiber vary greatly. Down yields for adult animals can be as low as 7 ounces (198 g) or as high as 2 pounds (907 g), although the Jiulong yaks bred for fiber produce substantially more. Measurements (from Mongolian yaks) suggest a total of all fiber types (down and hair) of 2¾–3⅞ pounds (1.3–1.8 kg).

▶ **STAPLE LENGTH**
Plateau breed of Chinese yak:
Outercoat 4½–8½ inches (11.5–21.6 cm)
Midcoat 2¼–4¾ inches (5.7–12 cm)
Down 1¼–2¼ inches (3.2–5.7 cm)
Bhutanese yaks:
Outercoat has been measured at 7¾–15¾ inches (20–40 cm)

▶ **FIBER DIAMETERS**
Coarse, long, crimp-free, shiny hair fibers with diameters that exceed 52 microns, going up to as much as 140 microns; midrange hair, also lustrous, with a small amount of wavy crimp, 25–52 microns; and irregularly crimpy down, with fiber diameters less than 25 microns. Measurements taken from Plateau breed of Chinese yak: outercoat, 30–52 microns; midcoat, 30–52 microns; down, 13–22 microns. Some processed North American down, unusually well dehaired, averages 14–16 microns, which is in the same range as very fine cashmere and qiviut.

▶ **LOCK CHARACTERISTICS**
Whether or not any locks are visible and distinct depends on the type and age of the animal, as well as where within its coat the fiber comes from. Generally, the three defined coats come off together, whether shed or shorn.

▶ **NATURAL COLORS**
Black, dark gray, brown, white, reddish, and, more rarely, a beige or golden color. Dark brown is by far the most common. Light-colored fiber may be natural or bleached.

Using Yak Fiber

Dyeing. Yes, it can be dyed; color clarity will depend on the underlying natural color. Unlike many similar fibers, yak down's strength permits it to be bleached, increasing its color range and dyeability.

Fiber preparation and spinning tips

- The down can be short, shiny, and slippery, or short, matte, and crimpy. In either case, it needs to be spun fine and with enough twist to hold the yarn together securely.
- The down can be spun from a loose mass, delicately carded (use fine carders and a light touch, or you'll make neps), or from commercially processed roving.
- Make sure there's enough twist in your singles to get the strand off the bobbin or spindle, as well as onto it: With some yak fiber (but not all) you'll need more twist than you think, but not so much that the yarn becomes wiry or stiff. Build up any more substantial yarns you want by plying.
- The hair, should you have some, will feel ornery and opinionated but can be spun into interesting, heavily textured, strong yarns; it's tempting to call them cords instead.

Knitting, crocheting, and weaving. There are strong traditions of weaving and felting with the mixed fibers, taking advantage of the water-repellent powers of the guard hairs and the cohesive contributions of the down. Most Western contemporary fiber folk knit with yak-down yarn, in part because of its luxury feel and status — you wouldn't want to lose any to loom waste — although well-spun yak also makes a great weaving fiber. The yarns will bloom when they're finished and in use.

Best known for. Often dark brown, extraordinarily soft, down fibers that are delightfully warm and lightweight and have some elasticity, loft, and bounce.

Bibliography

Note: This is a representative, not exhaustive, list of sources consulted. We accessed many more breed societies' and breeders' websites, as well as more books and articles, than we have space to include.

Albright, Barbara. *The Natural Knitter: How to Choose, Use, and Knit Natural Fibers from Alpaca to Yak.* New York: Potter Craft, 2007.

Alderson, Lawrence. "Breeds of Coloured Sheep in Great Britain and Their Potential Value." In Erskine, *Colored Sheep and Wool*, 19-23.

———. *The Chance to Survive.* Northamptonshire, UK: Pilkington Press, 1994.

———. "Shetland as Laboratory: The Genetics of Island Populations." In Kohlberg, *Shetland Breeds*, 135-150.

Alpaca Owners and Breeders Association. www.alpacainfo.com (accessed 3/08/08).

American Livestock Breeds Conservancy. "Breed Information — ALBC Conservation Priority List." www.albc-usa.org/cpl/wtchlist.html#sheep (accessed repeatedly 2008-2011).

American Romney Breeders Association. *American Romney.* 2005 ed. American Romney Breeders Association, 1995.

American Sheep Industry Association. *The SID Sheep Production Handbook*, 7th ed. Centennial, CO: American Sheep Industry Association, 2003.

Amos, Alden. *The Alden Amos Big Book of Handspinning.* Loveland, CO: Interweave Press, 2001.

Armstrong, Wayne P. *Wayne's Word: An On-Line Textbook of Natural History.* http://waynesword.palomar.edu (accessed repeatedly between February 2008 and May 2010).

Australian Association of Stud Merino Breeders. "Evolution of the Australian Merino." www.merinos.com.au/merinos.asp?pageid=16 (accessed June 1, 2009).

Barber, E. J. W. *Prehistoric Textiles: The Development of Cloth in the Neolithic and Bronze Ages with Special Reference to the Aegean.* Princeton, NJ: Princeton University Press, 1991.

———. *Women's Work, the First 20,000 Years: Women, Cloth and Society in Early Times.* New York: Norton, 1994.

Benedict, George L. "The Sheep of Shetland: A Historical Perspective." North American Shetland Sheepbreeders Association. www.shetland-sheep.org/pdf/the_sheep_of_shetland.pdf (accessed October 26, 2009).

Bergen, Werner von. *Wool Handbook: Volume 1.* 3rd enlarged ed. New York: John Wiley, 1963.

———. *Wool Handbook: Volume 2, Part 1.* 3rd enlarged ed. New York: John Wiley, 1969.

———. *Wool Handbook: Volume 2, Part 2.* 3rd enlarged ed. New York: John Wiley, 1970.

Bergen, Werner von, and Walter Krauss. *Textile Fiber Atlas: A Collection of Photomicrographs of Common Textile Fibers.* New York: American Wool Handbook Company, 1949.

Birutta, Gail. *Storey's Guide to Raising Llamas.* North Adams, MA: Storey Publishing, 1997.

Bixby, Donald, Carolyn J. Christman, Cynthia J. Ehrman, and D. Phillip Sponenberg. *Taking Stock: The North American Livestock Census.* Blacksburg, VA: The American Livestock Breeds Conservancy, 1994.

Black and Coloured Sheep Breeders' Association of New Zealand. www.colouredsheep.org.nz (accessed May 4, 2009).

Bowen, Godfrey. *Wool Away: The Art and Technique of Shearing.* New York: Van Nostrand Reinhold, 1974.

Bowie, G. G. S. "New Sheep for Old — Changes in Sheep Farming in Hampshire, 1792-1879." *Agricultural History Review*, 35: 15-24.

Briggs, Hilton Marshall. *Modern Breeds of Livestock.* New York: Macmillan, 1958.

Broudy, Eric. *The Book of Looms: A History of the Handloom from Ancient Times to the Present.* New York: Van Nostrand Reinhold, 1979.

Brown, Geoff. "Rough Fell and Herdwick Sheep in Cumbria: A Proposal for Re-Building Flocks and the Breeds in Cumbria." January 2002. Cabinet Office. http://archive.cabinetoffice.gov.uk/fmd/fmd_report/documents/a-submissions/ref%20568.pdf (accessed September 15, 2009).

Brown, Mike. *Aunt Millie's Guide to Llama Fiber: A Primer on Llama Fleece Preparation and Its Use.* El Prado, NM: Mike Brown, 1997.

Buchanan, Rita. "American Romney." *Spin-Off*, Winter 2001, 34-43.

———. "Spinner's Question: on Wet-Spinning Flax." *Spin-Off*, Summer 1999, 36-37.

———. *The Weaver's Garden.* Loveland, CO: Interweave Press, 1988.

———, coordinator. "Wool Combs and Combing: A Roundtable Discussion with Bill Benham, Iris Dozer, Patricia Emerick, Priscilla Gibson-Roberts, Lee Raven, Sharon Reese, Paula Shull, and Noel Thurner." *Spin-Off*, Summer 1991, 50-59.

Bureau of Land Management, U.S. Department of the Interior. "History of Public Land Livestock Grazing." www.blm.gov/nv/st/en/prog/grazing/history_of_public.html (accessed October 24, 2009).

Camelid ID Working Group. "Status Report." http://animalid.aphis.usda.gov/nais/naislibrary/documents/plans_reports/CamelidWG_2006_Status_Report.pdf (accessed February 28, 2008).

Cardamone, Jeannette M., William C. Damert, John G. Phillips, and William N. Marmer. "Digital Image Analysis for Fabric Assessment." *Textile Research Journal* 72, 10 (October 2002): 906-916. http://findarticles.com/p/articles/mi_qa4025/is_200210/ai_n9143245/tell (accessed May 15, 2008).

Carr, Margaret. "Famous Flock Led to Lecturer's New Life on North Ronaldsay." *The Orcadian.* www.orcadian.co.uk/features/articles/northronsheep.htm (accessed May 4, 2009).

Carson, Amanda, Matt Elliott, Julian Groom, Agnes Winter, and Dianna Bowles. "Geographic Isolation of Native Sheep Breeds in the U.K. — Evidence of Endemism as a Risk Factor to Genetic Resources." *Livestock Science* 123, no. 2-3 (August 2009): 288-299.

Christman, Carolyn J., D. Phillip Sponenberg, and Donald E. Bixby. *A Rare Breeds Album of American Livestock.* Pittsboro, NC: The American Livestock Breeds Conservancy, 1997.

Clutton-Brock, Juliet. *A Natural History of Domesticated Animals.* Austin: University of Texas Press, 1987.

Cortright, Linda. "Brown Sheep Keeps Spinning in America's Heartland." *Wild Fibers,* Winter 2008, 38-42.

———. "Eviction at Sea: Why the St. Kildans Vanished." *Wild Fibers,* Winter 2009, 5th anniversary ed: 44-54.

———. "Going Home to a Strange Land." (Topic: Cashmere.) *Wild Fibers,* Summer 2008, 8-17, 19.

———. "Have You Taken a Walk on the Wild Side?" *Wild Fibers,* Spring 2005, 32-37.

———. "Living on the Wooly Side of the Wild West." *Wild Fibers,* Winter 2008, 8-14, 16-17, 19.

———. "Raising Your First Kids at 90: Walter Pack's Cashmere Goats!" *Wild Fibers,* Summer 2004, 10-13.

———. "Threatened by the 'Throat of Fire.'" *Wild Fibers,* Winter 2008, 20-23, 25-26.

———. "What Nomad Is This?" *Wild Fibers,* Winter 2007, 16-24.

CSIRO Materials Science and Engineering. "Achievement: OPTIM Fine and OPTIM Max." www.csiro.au/science/OPTIMfineOPTIMmax.html (accessed November 28, 2009).

Darwin, Charles. *Voyage of the Beagle.* London: P. F. Collier & Son, 1937.

Davenport, Elsie. *Your Handspinning.* Pacific Grove, CA: Select Books, 1964.

Dohner, Janet Vorwald. *The Encyclopedia of Historic and Endangered Livestock and Poultry Breeds.* New Haven, CT: Yale University Press, 2001.

Domestic Animal Diversity Information System (DAD-IS). http://dad.fao.org (accessed repeatedly between 2007 and 2010).

Dowling, Robert, Lawrence Alderson, and Roger A. Caras. *Rare Breeds.* Boston: Little, Brown, 1994.

Dun, Kathryn, and Paul Farnham. *Beautiful Sheep: Portraits of Champion Breeds.* New York: Thomas Dunn Books, 2008.

Ekarius, Carol. *Storey's Illustrated Breed Guide to Sheep, Goats, Cattle, and Pigs.* North Adams, MA: Storey Publishing, 2008.

Enzlin, George J. *Colour and Island Sheep of the World.* Haarlem, Netherlands: Ram Press, 1995.

Erskine, Kent, ed. *Colored Sheep and Wool: Exploring Their Beauty and Function.* The Proceedings of the World Congress on Coloured Sheep, U.S.A., 1989. Ashland, OR: Black Sheep Press, 1989.

Evans, Nicholas, and Richard Yarwood. "The Politicization of Livestock: Rare Breeds and Countryside Conservation." *Sociologia Ruralis* 40, no. 2 (April 2000): 228-248.

Evitt, Gisela. "Gotland Afghan." *Spin-Off,* Fall 1990, 82-83.

Farlam, Gill. "The Manx Loghtan." *Spin-Off,* Spring 1990, 56-57.

Fegan, John M. *Merino Wool: A Study of Its Characteristics and Classing.* Sydney: Grahame Book Company, 1947.

"Fiber Review: Musk Ox (Qiviut), Spelsau, Gotland." *Spin-Off,* Spring 1990, 36, 41-42.

Field, Anne. *The Ashford Book of Spinning.* 2nd rev. ed. Christchurch, NZ: Shoal Bay Press, 1995.

———. *Spinning Wool: Beyond the Basics.* Christchurch, NZ: Shoal Bay Press, 1995.

Fournier, Jane. "Fiber Basics: Bouncy and Lightweight — Clun Forest." *Spin-Off,* Fall 1993, 50-53.

———. "Fiber Basics: Elasticity Characterizes the Targhee." *Spin-Off,* Spring 1992, 23-25.

———. "Fiber Basics: Rambouillet." *Spin-Off,* Fall 1994, 40-44.

———. "Fiber Basics: Shetland." *Spin-Off,* Winter 1992, 25-29.

Fournier, Nola. "The Stansborough Gotland Sheep." *Spin-Off,* Spring 2002, 34-36.

Fournier, Nola, and Jane Fournier. *In Sheep's Clothing: A Handspinner's Guide to Wool.* Loveland, CO: Interweave Press, 1995.

Frank, Robert, ed. *Silk, Mohair, Cashmere, and Other Luxury Fibres.* New York: CRC Publishing, 2001.

Franquemont, Abby. "The Effects of Twist in Plying." *Spin-Off,* Spring 2008, 42-44.

———. "What are Roving, Top, and Sliver?" *Spin-Off,* Winter 2007, 34-35.

Gauthier-Pilters, Hilde, and Anne Innis Dagg. *The Camel: Its Evolution, Ecology, Behavior, and Relationship to Man.* Chicago: University of Chicago Press, 1981.

Gomez-Pompa, Arturo. *The Lowland Maya Area: Three Millennia at the Human-Wildland Interface.* Riverside, CA: Hawthorne Press, 2003.

Gordon, Iain J., ed. *The Vicuña: The Theory and Practice of Community-Based Wildlife Management.* New York: Springer Science and Business Media, 2009.

Hall, Steven J. G., and Juliet Clutton-Brock. *Two Hundred Years of British Farm Livestock.* London: British Museum of Natural History, 1989.

Handwerk, Brian. "Wild Bactrian Camels Critically Endangered, Group Says." *National Geographic News,* December 3, 2002. http://news.nationalgeographic.com/news/pf/56517764.html (accessed December 13, 2009).

Hansford, K. A. "Real-Time Measurement On-Farm of Mean Fibre Diameter, Fibre Diameter Variability and Fibre Curvature Using Sirolan Laserscan." *Wool Technology and Sheep Breeding* 47, no. 1 (1999): 19-33.

Harris, Jennifer. *Textiles, 5000 Years: An International History and Illustrated Survey.* New York: Harry N. Abrams, 1993.

Heathman, Margaret. "Fiber Basics: Cormo." *Spin-Off,* Spring 2008, 60-64.

Heise, Laurie, and Carolyn Christman. *American Minor Breeds Notebook.* Pittsboro, NC: The American Minor Breeds Conservancy, 1989.

Henry, Oliver. "Shetland Sheep: A Brief Description." Jamieson and Smith. www.shetlandwoolbrokers.co.uk/epages/BT2741.sf/en_GB/?ObjectPath=/Shops/BT2741/Categories/Shetland_Wool_and_Sheep (accessed July 20, 2009).

Herriott, Jody Nankivell. "Angora Blends." *Spin-Off,* Fall 1993, 41-43.

Hochberg, Bette. *Fibre Facts.* Santa Cruz, CA: Bette Hochberg, 1981.

———. *Handspinner's Handbook.* Santa Cruz, CA: Bette and Bernard Hochberg, 1978.

———. *Reprints of Bette Hochberg's Textile Articles.* Santa Cruz, CA: Bette Hochberg, 1982.

Hoffman, Eric. "The Comeback Camelid." *Wild Fibers,* Winter 2008, 46-53, 55-56.

———. "Fiber as a Transitory Medium: Factors Affecting a Histogram." Bonny Doon Alpacas, 1998. www.bonnydoonalpacas.org/fiberart.html (accessed March 2, 2008).

———. "The Kaleidoscope & Fiber Evaluation." Bonny Doon Alpacas. www.bonnydoonalpacas.org/kaleido.html (accessed March 2, 2008).

———. "The South American Camelidae Family." *Wild Fibers,* Winter 2008, 58-59.

———. "The World Fiber Market and Why the Vicuna May Be the Key to Improving the Alpaca." Bonny Doon Alpacas, 2004. www.bonnydoonalpacas.org/worldfibermarket.html (accessed March 2, 2008).

Horne, Beverley. *Fleece in Your Hands — Spinning with a Purpose: Notes and Projects.* Loveland, CO: Interweave Press, 1979.

Ingram, Gwen, and Jim Krowka. "Llama Fiber." Lost Creek Llamas. http://lostcreekllamas.com/fiber.htm (accessed February 29, 2008).

International Lama Registry. *Llama Fiber.* Educational Brochure 9. Kalispell, MT: International Lama Registry, 2004.

International Year of Natural Fibres, 2009. "Natural Fibres: Ancient Fabrics, High-Tech Geotextiles." www.naturalfibres2009.org/en/fibres/index.html (accessed May 6, 2009).

Irwin, Bobbie. *Spin-Off Magazine Presents the Spinner's Companion.* Loveland, CO: Interweave Press, 2001.

Irwin, John, and P. R. Schwartz. *Studies in Indo-European Textile History.* Ahmedabad, India: Calico Museum of Textiles, 1966.

Isle of Man Government. "The Manx Loaghtan Sheep." www.gov.im/tourism/trade/members/secure/documents/press/loaghtan.doc (accessed May 20, 2008).

Isle of Man Guide. "Loghtan Sheep." www.iomguide.com/loghtansheep.php (accessed May 20, 2008).

Johnston, J. Laughton. *A Naturalist's Shetland.* London: T. & A. D. Poyser, 1999.

Kadwell, Miranda, Matilde Fernandez, Helen F. Stanley, Ricardo Baldi, Jane C. Wheeler, Raul Rosadio, and Michael W. Bruford. "Genetic Analysis Reveals the Wild Ancestors of the Llama and the Alpaca." *Proceedings of the Royal Society B: Biological Sciences,* 268 (2001): 2575-2584. www.pubmedcentral.nih.gov/picrender.fcgi?artid=1088918&blobtype=pdf (accessed March 2, 2008).

Katzenbach, William, ed. *Threads of History.* New York: American Federation of Arts, 1965.

Keeler, Patricia, and Francis X. McCall, Jr. *Unraveling Fibers.* New York: Atheneum Books for Young Readers, 1995.

Keyser, Willet. "A Study of the Brown Color in Columbia Sheep." M.S. thesis, University of Wyoming, July 1957.

Kinsman, David. *Black Sheep of Windermere: A History of the St Kilda or Hebridean Sheep.* Windermere, UK: Windy Hall Publications, 2001.

Kirple, Kenneth, and Kriemhild Coneè Ornelas, eds. *The Cambridge World History of Food.* Cambridge: Cambridge University Press, 2000.

Klein, Julius. *The Mesta: A Study in Spanish Economic History, 1273-1836.* Cambridge: Harvard University Press, 1920.

Klein, Lettie. "Judging Karakul Sheep: Their Conformation and Fiber." Karakul Judging Workshop, Boonville, MO, June 22, 2003. American Karakul Sheep Registry. www.karakulsheep.com/judging_karakul_sheep.htm (accessed November 2, 2009).

Kohlberg, Nancy, and Philip Kopper. *Shetland Breeds, 'Little Animals . . . Very Full of Spirit': Ancient, Endangered and Adaptable.* Chevy Chase, MD: Posterity Press, 2003.

Lawrence, Carl A. *Fundamentals of Spun Yarn Technology.* Boca Raton, FL: CRC Press, 2003.

Leask, Agnes. "Island View: An Older Breed of Sheep in Shetland?" In Kohlberg, *Shetland Breeds*, 152-155.

Lent, Peter C. *Muskoxen and Their Hunters: A History.* Oklahoma City, OK: University of Oklahoma Press, 1999.

Linklater, Andro. "Eden with Weather: A Short History of This Valiant Place." In Kohlberg, *Shetland Breeds*, 1-35.

Lovick, Elizabeth. *A North Ronaldsay Yarn: The Sheep, Their Wool and Their Island.* Orkney, Scotland: Elizabeth Lovick, 2004.

Lutwyche, Richard H. L. "The Missing Pig: Beyond the Brink." In Kohlberg, *Shetland Breeds*, 157-164.

Lynne, Erica. *Angora: A Handbook for Spinners.* Loveland, CO: Interweave Press, 1992.

Manx Loaghtan Breeders' Group. www.manxloaghtansheep.org (accessed May 20, 2008).

Manx Loaghtan Produce Company Ltd. www.manxloaghtan.com (accessed May 20, 2008).

McColl, Angus. "Methods for Measuring Microns." Northwest Alpacas. www.alpacas.com/PDFs/Library/MeasuringMicrons.pdf (accessed March 2, 2008).

———. "Understanding Micron Reports: Integrity of Sampling and Use of Micron Results." Yocom-McColl. www.ymccoll.com/micron_reports.html (accessed March 2, 2008).

McCorkle, Constance M., ed. *Improving Andean Sheep and Alpaca Production: Recommendations from a Decade of Research in Peru.* Small Ruminant Collaborative Research Support Project, University of California/Davis. Columbia, MO: University of Missouri-Columbia, 1990.

McCuin, Judith MacKenzie. *Teach Yourself Visually Handspinning.* Hoboken, NJ: Wiley, 2007.

———. "Women Who Spin with the Wolves." *Spin-Off,* Fall 2009, 58-62.

McGregor, Bruce. "Variation in and Sampling of Alpaca Fleeces." State Government of Victoria, Department of Primary Industries, 2007. www.dse.vic.gov.au/DPI/nreninf.nsf/childdocs/89E7A8DAFEA417624A2568B30004C26A20402CDDBB6CE5FCCA256BC700811E708CC631FC461B1804A256DEA00274044D17D841537B7716ECA256C1A0014A4BD?open (accessed March 2, 2008).

McNeal, Nancy Wilkie. "Navajo Sheep Project: A Survivor!" *Spin-Off,* Summer 1993, 35, 37.

Mercer, John. *The Spinner's Workshop: A Social History and Practical Guide.* Dorset, UK: Prism Press, 1978.

Mulholland, Diane. "Home of the Polwarth: A Visit to Tarndwarncoort." *Spin-Off,* Fall 2008, 62-63.

Muller, Donna. "Sarah Natani: A Year with Navajo Sheep." *Spin-Off,* Winter 1992, 40-45.

———. "Spinning on a Navajo Spindle: A Visit with Sarah Natani."*Spin-Off,* Spring 1995, 88-95.

Murray, Roberta. "Shetland Sheep — Beyond the Islands." *Spin-Off,* Summer 2000, 56-60.

Natural Tasmanian Wool. www.naturaltasmanianwool.com (accessed November 20, 2009).

New Zealand Romney Sheep Breeders Association. "The Typical NZ Romney Sheep." www.romneysheep.org/typical.htm (accessed February 27, 2008).

Nicolson, James R. "Shetland Sheep: A Wealth of Wool and Words." In Kohlberg, *Shetland Breeds*, 70-87.

Nordic Gene Bank, Animals. "NorthSheD: Origin and Diversity of North European Sheep Breeds." www.rala.is/beta (accessed repeatedly during July 2008).

Ogilvie, Malcolm. "Choughs and Cowpats." *Birds of Britain.* May 2008. www.birdsofbritain.co.uk/features/mao-march-00.asp (accessed June 12, 2008).

Oklahoma State University, Department of Animal Science. Breeds of Livestock. www.ansi.okstate.edu/breeds (accessed repeatedly 2008-2011).

Olberding, Susan Deaver. "Churro Sheep: A Southwest Legacy." *Spin-Off,* Summer 1993, 34, 36-37.

Olson, Torie. "Chasing Skirts in Guatemala." *Wild Fibers,* Winter 2008, 28-36.

Paco-Vicuña Registry, The. "Grade Chart." www.paco-vicunaregistry.com/grade_chart.htm (accessed March 12, 2008).

Parkes, Clara. *The Knitter's Book of Yarn: The Ultimate Guide to Choosing, Using, and Enjoying Yarn.* New York: Potter Craft, 2007.

Piroch, Sigrid. "Handspinning in Slovakia: A Closer Look at a Folk Culture from Eastern Europe." *Spin-Off,* Spring 1992, 60-65.

Pohle, Elroy M. "Grading and Production of Wool." In Bergen, *Wool Handbook: Volume 1,* 547-615.

Porter, Valerie, and I. L. Mason. *Mason's World Dictionary of Livestock Breeds, Types and Varieties.* Oxon, UK: CABI Publishing, 2002.

Pulliam, Deborah. "Controlled Substances for Spinners." *Spin-Off,* Summer 1995, 103.

———. "Fiber Basics: Cotswold." *Spin-Off,* Fall 2007, 42-49.

Quine, David. *St Kilda.* Colin Baxter Island Guides. Grantown-on-Spey, Scotland: Colin Baxter Photography, 1995.

Rainsford, Francis. "Concern Over Peru's Coarsening Alpaca Fibre." Northwest Alpacas. www.alpacas.com/PDFs/Library/concern.pdf (accessed March 2, 2008).

Rare Breeds Conservation Society of New Zealand. "About Feral Sheep in New Zealand." New Zealand Rare Breeds. www.rarebreeds.co.nz/ferals.html (accessed May 9, 2008).

———. "Arapawa Sheep: A Rare Breed of New Zealand Origin." New Zealand Rare Breeds. www.rarebreeds.co.nz/arapawa.html (accessed May 9, 2008).

———. "New Zealand: What and Where Are We?" New Zealand Rare Breeds. www.rarebreeds.co.nz/newzealand.html (accessed May 9, 2008)

Rare Breeds Survival Trust. "Watchlist: Sheep." www.rbst.org.uk/watch-list/sheep/ (accessed repeatedly 2008-2011).

Reeve, Eric, and Isobel Black, eds. *Encyclopedia of Genetics.* London: Routledge, 2001.

Rhoades, Carol Huebscher. "Fiber Basics: Bactrian Camel." *Spin-Off,* Fall 2007, 50-52, 54, 56-57.

———. "Fiber Basics: Black Welsh Mountain." *Spin-Off,* Spring 2005, 66-70.

———. "Fiber Basics: Clun Forest." *Spin-Off,* Spring 2009, 72-76.

———. "Scandinavian Wools: Unique Fleeces for Special Spinning and Knitting Techniques." *Spin-Off,* Winter 2000, 78-88.

Robbins, Christopher. *Apples Are from Kazakhstan: The Land That Disappeared.* New York: Atlas & Co., 2008.

Robertson, B. T., ed. *Fine Wool: Proceedings of the 1982 Merino and Halfbred Producers' Seminar,* special publication no. 24. n.p. New Zealand: Tussock Grasslands and Mountain Lands, 1982.

Robson, Deborah. "Diné bé 'Iiná: The Navajo Lifeway." *Spin-Off,* Winter 1996, 82-84

———, ed. *Spin-Off Magazine Presents Handspun Treasures from Rare Wools: Collected Works from the Save the Sheep Project.* Loveland, CO: Interweave Press, 2000.

Romney Sheep Breeders Society. http://romney.webeden.co.uk (accessed February 27, 2008).

Ross, Mabel. *The Encyclopedia of Handspinning.* Loveland, CO: Interweave Press, 1988.

———. *The Essentials of Yarn Design for Handspinners.* Kinross, Scotland: Mabel Ross, 1983.

Russell, Margaret B. "Rare Thoughts: Cotswold." *Wild Fibers,* Summer 2008, 54-55.

———. "Rare Thoughts: Forgotten?" *Wild Fibers,* Winter 2009, 56-60.

———. "Rare Thoughts: The Tale of the Little Gray Sheep." *Wild Fibers,* Summer 2009, 54-58.

Ryder, M. L. "Genetics of Wool Production." In Reeve, *Encyclopedia of Genetics,* 374-379.

———. "The History of Sheep Breeds in Britain," *The Agricultural History Review* 12, 1 (1964): 1-12.

———. *Sheep and Man.* London: Gerald Duckworth and Co., 2007.

Safley, Mike. "The Case for Crimp." Northwest Alpacas. www.alpacas.com/AlpacaLibrary/CaseforCrimp.aspx (accessed March 2, 2008).

———. "Some Views on Evaluating Suri Fiber." Northwest Alpacas. www.alpacas.com/PDFs/Library/ViewsOnEvaluatSuriFiber.pdf (accessed March 2, 2008).

———. "The Wool Industry Faces a Prickly Question: Are People Allergic to Wool?" Northwest Alpacas. www.alpacas.com/PDFs/Library/WoolIndustry.pdf (accessed March 2, 2008).

Saitone, Tina L., and Richard J. Sexton. "Alpaca Lies? Do Alpacas Represent the Latest Speculative Bubble in Agriculture?" *Applied Economic Perspectives and Policy* 29, 2 (2007): 286-305.

Schmid, Sarah. "The Value Chain of Alpaca Fiber in Peru, An Economic Analysis." M.S. thesis, Swiss Federal Institute of Technology, Zurich, 2006. http://tarwi.lamolina.edu.pe/~cgomez/Diplomarbeit%20_Endfassung_Tesis%20_2006.pdf (accessed March 6, 2008).

Schoeser, Mary. *World Textiles: A Concise History.* London: Thames & Hudson, 2003.

Schorsch, Anita, ed. *The Art of the Weaver.* New York: Universe Books, 1978.

Sheep 101. www.sheep101.info (various dates in 2009).

"Sheep Is Life: A Celebration of Navajo-Churro Shepherds and Weavers." *Spin-Off,* Winter 1996, 95.

Shull, Paula. "Fiber Basics: Polwarth, A Personal Experience." *Spin-Off,* Winter 1995, 42-45.

———. "Washing Wool for Combing." *Spin-Off,* Fall 1993, 109.

———. "Worsted Yarns: The Classic Way." *Spin-Off,* Fall 1993, 98-108.

Simmons, Paula, and Carol Ekarius. *Storey's Guide to Raising Sheep: Breeding, Care, Facilities.* North Adams, MA: Storey Publishing, 2001.

Sjöberg, Gunilla Paetau. *Felt: New Directions for an Ancient Craft.* Loveland, CO: Interweave Press, 1996.

Slater, Keith. *Environmental Impact of Textiles: Production, Processes and Protection.* Cambridge: Woodhead Publishing, 2003.

Smith, William Clarence. "A Comparison of Wool from the Down Breeds of Sheep." M.S. thesis, University of Wyoming, 1929.

Soay Sheep Society. "Boreray Sheep." www.soaysheepsociety.org.uk (accessed February 13, 2008).

———. "Soay Sheep." www.soaysheepsociety.org.uk (accessed February 13, 2008).

Spark, Pat. "Chart of Some Felting Wools." www.peak.org/~spark/feltingwools.html (accessed February 25, 2008).

Sponenberg, D. P. "Color Genetics in Coopworth Sheep." Deer Run Sheep Farm. www.deerrunsheepfarm.com/genetics.html (accessed September 1, 2009).

Sponenberg, D. Phillip, and Donald E. Bixby. *Managing Breeds for a Secure Future: Strategies for Breeders and Breed Associations.* Pittsboro, NC: American Livestock Breeds Conservancy, 2007.

Stove, Margaret. *Handspinning, Dyeing and Working with Merino and Superfine Wools.* Loveland, CO: Interweave Press, 1991.

Strick, Candace Eisner. *Beyond Wool: 25 Knitted Projects Using Natural Fibers.* Woodinville, WA: Martingale, 2004.

Suri Llama Association and Registry. "Suri Llama Breed Standards: Fiber Standards." www.surillama.com/fiber_standards.php (accessed March 2, 2008).

Switzer, Chris. *Spinning Llama and Alpaca.* Second edition. Estes Park, CO: Chris Switzer, 1998.

Tasmanian Wool. www.tasmanianwool.com (accessed November 20, 2009).

Teal, Peter. *Hand Woolcombing and Spinning: A Guide to Worsteds from the Spinning-Wheel.* Dorset, UK: Blandford Press, 1976.

Thurner, Noel A. "Viking Wool Combs: A Primitive Tool for Modern Handspinners." *Spin-Off,* Spring 1992, 44-49.

Trow-Smith, Robert. *A History of British Livestock Husbandry, 1700-1900.* Reprint, London: Routledge, 2006.

University of Michigan Museum of Zoology. "Animal Diversity Web." http://animaldiversity.ummz.umich.edu (accessed repeatedly from February 2008 to June 2010).

Vickrey, Anne Einset. *Felting by Hand.* Geneva, NY: Craft Works Publishing, 1987.

Wade-Martins, Peter. *Black Faces: A History of East Anglian Sheep Breeds.* Kent, UK: Geerings of Ashford and Norfolk Museum Services, 1993.

———. *The Manx Loghtan Story: The Decline and Revival of a Primitive Breed.* Kent, UK: Geerings of Ashford and The Rare Breeds Survival Trust, 1990.

Walker, Earnest P., ed. *Mammals of the World.* Baltimore, MD: Johns Hopkins University Press, 1964.

Walker, Linda Berry. "Know Your Sheep Breeds: Corriedale." *Spin-Off,* Summer 1989, 24.

———. "Rambouillet." *Spin-Off,* Spring 1989, 8.

Weaver, Sue. "Ancient Treasure: Soay Sheep." *Hobby Farms,* January/February 2008: 36-41. www.hobbyfarms.com/livestock-and-pets/soay-sheep.aspx (accessed October 31, 2009).

Wendelboe, Linda, and Kathy Baker. Shetland Sheep Information. www.shetlandsheepinfo.com (accessed October 2009, repeatedly).

Wensleydale Longwool Sheep Breeders' Association. www.wensleydale-sheep.com (accessed February 13, 2008).

Wheeler, Jane C. "A Brief History of Camelids in the Western Hemisphere." www.icidirectory.com/PDFs/AncientHistory.pdf (accessed March 2, 2008).

Wheeler, J. C., A. J. F. Russel, and H. F. Stanley. "A Measure of Loss: Prehispanic Llama and Alpaca Breeds." *Archivos de Zootecnia* 41, no. 154 (1992): 467-475. www.surillama.com/wheeler_467_475.pdf (accessed March 2, 2008).

Whiteface Dartmoor Sheep Breeders Association. www.whitefacedartmoorsheep.co.uk (accessed November 15, 2009).

Wilkinson, Rosemary. "The View from the Rookery." *Wild Fibers,* Winter 2008, 60-65.

Williams-Davies, John. *Welsh Sheep and Their Wool.* Dyfed, Wales: Gomer Press, 1981.

Wilson, John M. *The Rural Cyclopedia, or A General Dictionary of Agriculture.* Edinburgh: A. Fullerton & Co., 1851.

Wulfhorst, Burkhard, Thomas Gries, and Dieter Veit. *Textile Technology.* Munich: Hanser Publishers, 2006.

Yocom-McColl Laboratories. "Fiber Testing Terminology." Northwest Alpacas. www.alpacas.com/AlpacaLibrary/Html/FiberTesting.htm (accessed March 2, 2008).

Acknowledgments

The nearly four years we spent working on this book involved the creation of an amazing project-centered social network. We have talked to fiber and animal people from around the world, and they have truly shared their love and knowledge, provided insights, and donated fleeces, products, and photos. We deeply appreciate their contributions. Any missteps that remain, or omissions of important names!, are our own.

Though hundreds of people helped along the way, a few went above and beyond the call of duty: Jennifer Heverly of Spirit Trail Fiberworks (www.spirit-trail.net), Melissa Laffin of Peace of Yarn (www.peaceofyarn.com), and Beth Smith of The Spinning Loft (www.thespinningloft.com) generously supplied abundant samples and helped hunt down myriad missing fleeces and fibers. Tim Booth of the British Wool Marketing Board (www.britishwool.org.uk) sent us raw fleece samples of many British breeds we hadn't been able to get our hands on — a herculean feat, considering that our request reached him months after the normal shearing season, when we had exhausted our other options.

For helping us in different ways, special thanks go to (in an idiosyncratic alphabetic order):

Lawrence Alderson, Sue Blacker, and Robert Padula for specialized fiber insights

Ellen Bendtsen, Birgit Boberg, and Ólafur R. Dýrmundsson for helping us understand Nordic breeds

Carol Christiansen, of the Shetland Museum, and Linda Wendelboe, of Fibre Works Farm, for helping us understand Shetland sheep

Janette Cooper, of the Welsh government, for helping us to understand some of the breeds from Wales

Jessica Derksen and Jonathan Derksen, for making space for Deb to work

June Hall, for too many reasons to list

Keith Kendrick, for sharing his research on sheep cognition

Steve Messam, the artist who created the CLAD building (page 211)

Clara Parkes for her love of breed-specific wools, her fine books, and her encouraging words

Donna Puccini for sharing her poem (page 31)

Barb Ranson, for helping with research, and for her long and valued friendship with Carol

Joanne Seiff for filling gaps at the last minute

Olga Soffer for elaborating on fiber in prehistory

Sarah Swett, for promoting breed-specific wool through her art and tapestry

Unicorn Power Scour for significantly expediting and perfecting our fiber-cleaning

Priscilla and Steve Weaver and Kathie Miller for insight into Soays

Halcyon Blake, Kathy Elkins, Jeanne Giles, Mary Harris, Donna Herrick, Darcy Justine, Diane Minard, and Laura Zander, for helping us understand fiber users' needs from a shop owner's perspective

We both could not even imagine having produced this work without the exceptionally visionary and dedicated people associated with Storey Publishing, including but not limited to: Pam Art, Deb Balmuth, Deb Burns, Sarah Guare, John Polak, Jessica Richard, Gwen Steege, Mary Velgos, and Mars Vilaubi.

Deb would especially like to thank her mother, Allene Robson, and her daughter, Rebekah Robson-May, along with Judy Fort Brenneman, Virginia Cross, Pam DeVore, Donna Druchunas, Kit Dunsmore, Priscilla Gibson-Roberts, Linda Ligon, Margaret Mahoney, Carrie Manley, Kris Paige, Kristi Schueler, David and Rae Smith, Jonathan and Suzanne Taylor, Susan Tweit and Richard Cabe, and the Monday night knitters for all kinds of essential and much appreciated support. And, of course, Carol Ekarius.

Our donors and supporters (many have websites, and a Google search will take you to them) include:

Ginny & Jeff Adams, Walnut Hill Farm at Elm Springs
Barry Allen, Renwick Mill
American Livestock Breeds Conservancy
Kay Applebee, Dorcas Fields
Ashland Bay Trading Company
David Barnes, Bonnyrigg Hall Farm
Birkeland Bros. Wool, Ltd.
Sue Blacker, Blacker Designs
Ellen Bloomfield
Birgit Boberg, Northern Shorttail Sheep Working Group
Patti Bobonich, Sweet Grass Wool
Margaret Boos, The Wool Barn
Cat Bordhi
Sarah Bowley, SVF Foundation
British Wool Marketing Board
Juli Budde, Little Country Acres
Hilde Buer, Northern Shorttail Sheep Working Group
Nancy Bush
Anna Carner, Unicorn Fibre
Gregory Carrier
Kathi Cascio, Apple Hollow Fiber Arts
Colorado Art Ranch
Linda Cortright, Wild Fibers
Sandy Dass, Bryn Hollow Farm
Greg Deakin, Deakin Family Farms
Diamond Valley Alpaca Ranch (Kathy & Dale Gilliland and Elaine Sipes)
Swithin Dick
Gillian Dixon, South Yeo Farm East
John Dodd, Belview Stud
Robbie Donovan, Bovey Tracey
Dean Drake, Drake Farm
Ólafur R. Dýrmundsson, Northern Shorttail Sheep Working Group
Jacqueline Ericson, Northridge Farms
Estes Park Wool Market
Elizabeth Ferraro, Apple Rose Farm
Diane Fisher, The Murmuring Wheel
Philip Fisher, Singing Shearer
Lori Flood, Lori Flood Felted Fibers
Abby Franquemont
Joanna Gleason, Gleason's Fine Woolies
Diana Hachenberger, Three Bags Full and Castle Crags Ranch
Elizabeth Hails
June Hall, The Wool Clip
Maurice Hall, Old Bank
Karen Hallett, Crossfire Hill Corriedales
Elsa Hallowell, Elsawool Company
Lorraine & Bennett Hammond
Bill & Helen Hardman, Uncompahgre Polypay Farm
Bill Hauer, Large Animal Research Station, University of Alaska
Barbro Heikinmatti
Sue Hepworth, Bush Park
Jennifer Heverly, Spirit Trail Fiberworks
Roy Hill
Kathy Huebner, Oldhaus Rabbits
Sarah Jane Humke-Mengel
Shirley Inman, Oak Lodge
Alison Jackson-Bass, Eco Eco, Ltd.
Bill Julien, High Meadows Farm
Pam Keaton, Divine Sheep Farm
Debbie Kingsley, South Yeo Farm West
Barbara Kingsolver and Steven Hopp Family Farm
Elaine & Dennis Kist, WindKist Farm
Carl & Eileen Koop, Bijou Basin Ranch
Kay Kreutzer, Kreutzer Farms
Beate Kubitz, MakePiece
Melissa Laffin, Peace of Yarn
Richard & Donna Larson, Old Gjerpen Farm
Carol Leonard, Spindler
Linda Ligon, Interweave Press
Louet North America
Elizabeth Lovick, Northern Lace
Judith MacKenzie McCuin
Martha McGrath, Deer Run Sheep Farm
Barb McRoberts, McRoberts Game Farm
Laurie & Michael Martin, Martin Farms
Maryland Sheep and Wool Festival
Terry Mattison, Rainbow Yarns Northwest
Barbara Merickel, Mañanica Farm
Jim & Kathy Mills
Ron Miskin, Buffalo Gold
Amy Clarke Moore, *Spin-Off* Magazine
June Morris, North Ronaldsay
William Morrow, Whitmore Farm
Mount Vernon Farm (George Washington's Mount Vernon)
Marilyn Murphy, Cloth Roads
Wayne Myers, Beau Chemin Farm
Lucy Neatby
Kris Paige, Sunflower Ridge Ranch
Christiane Payson, North Valley Farm
Cindy & Jeff Phillips, Reindeer Games
Graham Phillipson, Littledale Farm
Susan Prechtl, Rainbow Yarns Northwest
Deborah Pulliam
Rare Breeds Survival Trust
Gina Rawle
Jeni Reid
Carol Huebscher Rhoades
Kay Richardson, Richardson Texels
Patrick Roll, West Elm Farm
Ronan Country Fibers
Norma Sanders
Marie Schmidt, Durakai Sheep and Fiber Arts
Don Schrider, Schrider Farm
Joanne Seiff
Charles and Karen Sigler, Benjamin Farms
Linda Singley, Bearlin Acres
Beth Pink Smith, The Spinning Loft
Carolyn Smith, Viva Yarns
Colin & Leslie Smith, Bowscale
Jenny Smith, Underhill Farm
Kathy Soder, K Bar K Farm
Gwen Steege
Sonja Straub, Legacy Farm
Marta Sullivan, Currow Hill Ranch
Monty Sullivan, Two Boy Farm
Chris Switzer, SwitzerLand Alpacas
Connie Taylor, Bayeta Classic Sheep and Wool
Kate Tetlow
Mick Venters, Aston Rowant National Nature Refuge
Priscilla & Steve Weaver, Saltmarsh Ranch
Barbara Webb, Jager Farm
Becky Weed, Thirteen Mile Wool Company
Linda Wendelboe, Fibre Works Farm
Lisa Westervelt, Cranberry Moon Farm
W.O.L.F. Rescue
Woodland Woolworks
Sarah Wroot

Credits (Interior Photography)

Interior photography by © John Polak except as noted below:

© Accent Alaska/Alamy: 388 (Musk Ox)
© Agripicture Images/Alamy: 78 top
© Alwiyn J. Roberts/FLPA/Minden Pictures: 294
© American Livestock Breeds Conservancy: Jeanette Beranger 5 bottom right, John and Sue McCummins 5 top right and 132
© Andre Maslennikov/agefotostock: 314 (Gute)
© AnimalsAnimals/Superstock: 122 (Arapawa Island Sheep)
© Anthony Hope/Alamy: 52 (Cheviot)
© Anthony Mayatt/iStockphoto.com: 355 left
© ARCO/J. De Cuveland/agefotostock: 162
© Ashley Cooper/agefotostock: 195
© Bernard Gagnon/Wikimedia Commons: 4
© British Wool Marketing Board: 40, 43, 72, 100, 154, 203, 207, 210, 215 left, 217, 219, 240, 257, 263, 311, 314 (Welsh Mule), 329
© Catharina van den Dikkenberg/iStockphoto.com: 360 (Bactrian Camel)
© cfgphoto.com: 76, 245
© Chris Pole/iStockphoto.com: 62
© Christine Wilson: 122 (Gulf Coast Native)
© Csaba Petrik: 320
© Darrell and Freda Pilkington/Highergills Farm: 116
© Dave Porter/Alamy: 107
© 2009 David C. Phillips/Garden Photo World: ii (Alpaca), 360 (Alpaca), 367, 375, 378, 381
© David Dalton/FLPA/Minden Pictures: 302
© David Platt/Alamy: 96
© David Ridley/Alamy: 275
© Design Pics/agefotostock: 271
© Dmitry Dedyukhin/iStockphoto.com: 383
© Douglas Houghton/agefotostock: 175
© Elizabeth Ferraro/Apple Rose Farm: 233
© Ercan Sucek/iStockphoto.com: 388 (Angora Rabbit), 406 top
© Evans Dan/iStockfoto: 224 (Herdwick)
© Fabian von Poser/agefotostock: 387 top
© FLPA/Alamy: 84
© FLPA Angela Hampton/agefotostock: 30 left
© FLPA John Eveson/agefotostock: 66 (Hampshire Down), 74, 104, 232, 306 top
© FLPA Primrose Peacock/agefotostock: 230, 261
© FLPA Tony Wharton/agefotostock: 102
© FLPA Wayne Hutchinson/agefotostock: 36 (Rough Fell), 38, 44
© FMZET/Dreamstime.com: 148
© Geoff Trinder/ardea.com: 398 left
© 2009 Georgianna Lane/Garden Photo World: 365
© Gerald Lacz FLPA/Minden Pictures: 270
© Gillian and Ian Dixon/South Yeo Farm East: 66 (Shropshire)
© Grace Smith: v, 88 left
© Graham Taylor/iStockphoto: 266
© Gregory Carrier: 124 top
© Gretchen Frederick: viii
© Holger Burmeister/Alamy: 36 (Scottish Blackface)
© Jackson M. Dzakuma, PhD: 334 (Spanish Goat)
© Jason Houston: vii (Pygmy Goats)
© Jean G. Green: 292
© Jeff Adams/Walnut Hill Farm at Elm Springs: 128
© Joanna Gleason: 256
© Johan Rieneke/Dreamstime.com: 60 (Dorper)
© John Daniels/ardea.com: 66 (Suffolk), 80 top
© John Eveson FPLA/Minden Pictures: 198 (Badger-faced Welsh), 224 (Zwartbles)
© John James/Alamy: 82 (Cotswold)
© John Van Decker/Alamy: 226
© John W. Banagan/Getty Images: 144
© Joost De Raeymaeker: 379
© Juan Carlos Munoz/bluegreenpictures.com: 185
© Juniors Bildarchiv/Alamy: 168 top
© Ken Chalmers/www.backacresangora.com: ii (Angora Rabbit), 408 left
© Kenneth W. Fink/ardea.com: 335
© Kirk Anderson/Alamy:115
© Kseniya Abramova/iStockphoto.com: 389
© Lawrence Wright/www.middlecampscott.co.uk: 213 left
© Lewis White: 296
© Linda Davis, Davis Farmland: 5 middle right
© Lynn Stone: 395
© Mark Bouton/Alamy: 328
Mars Vilaubi: 10, 18, 19, 28-29, 68
© Mary & Gene Langhus/Langhus Columbia Sheep: 242
© Mary Swindell/Bellwether Farm: 234
© Mike Lane/Alamy: 152 (Hebridean)
© Mornay Van Vuuren: 334 (Angora Goat), 337

© National Geographic Image Collection/Alamy: 251
© Nature Picture Library: 153
© Naturepix/Alamy: 206
© Niall Benvie/Alamy: 317
© Nigel Catlin/Alamy: 30 right, 52 (North Country Cheviot), 59, 63 top
© Nigel Catlin/FLPA/Minden Pictures: 236 top, 289
© Nndemidchick/Dreamstime.com: 276
© Pablo Caridad/iStockphoto: 363
© Patric Lyster/Coyote Acres: 60 (Dorset Horn)
© Paul Collins/Alamy: ii (Angora Goat)
© Paul Tessier/iStockphoto.com: 336 top
© PCL/Alamy: 111 top
© Pekka Sakki/agefotostock: 159
© Petur Farkas: 314 (Racka)
© Philip Perry/FLPA/Minden Pictures: 172
© R Hemming/Spered Breizh Ouessants: 182
© Radius Images/Alamy: ii (Musk Oxen), 400
© Richard Osbourne/Alamy: 286 top
© Robert Dowling: 69, 82 (Leicester Longwool), 92, 120, 152 (Soay), 180, 224 (Jacob), 347
© RS-foto/Wikimedia Commons: 309
© Sarah Roland/FLPA/Minden Pictures: 157
© Scott Mann: ii (Debouillet Sheep), 134 (Debouillet), 142
© Simon Whaley/Alamy: 211 top
© Stephen Dalton/Minden Pictures: 237
© Steve Lovegrove/123RF: 134 (Saxon Merino)
© Steve Messam: 211 bottom
© Sue Weaver: 56
© Tanya Charter and Greg Shore/McKenzie Creek Ranch: 282
© tbk media.de/Alamy: 260
© Terry Hankins/Egypt Creek Ranch: 334 (Kiko Goat)
© Top-Pics TBK/Alamy: 189
© Trinity Mirror/Mirrorpix/Alamy: 202
© Steve Messam: 211 bottom
USDA/ARS/Jack Dykinga: 390
© Wayne Hutchinson/Alamy: 47 top, 55 top
© Wayne Hutchinson FLPA/Minden Pictures: 31, 36 (Swaledale), 49, 198 (Beulah Speckled Face)
© Wild Fibers Magazine: vii (Yak), 415
© Yin Yang/iStockphoto.com: 5 top left

Index

Page numbers in *italics* indicate illustrations and photographs; page numbers in **bold** indicate charts and tables.

C

D

E

M

N

O

T

U

V

W

Y

Z

A Closer Look at Selected Sheep Breeds
Where We Think (and Sometimes Know) that Certain Sheep Breeds Originated
A star (★) indicates that this breed originated at a narrowly defined location. A highlight indicates a general area of origin.
*The origin of Jacob sheep is uncertain, but the breed was first identified by that name in this location.
N
W
E
S
1 inch = 75 mi/120 km
Shetland
★North Ronaldsay
North Country Cheviot
Soay★
★Boreray
Hebridean
British Isles
Scotland
★Border Leicester
Scottish Blackface
Cheviot
British Milk Sheep
★Castlemilk Moorit
★Bluefaced Leicester
Northern Ireland
Rough Fell★
★Teeswater
Herdwick
Dalesbred★
★Swaledale
★Manx Loaghtan
★Wensleydale
Lonk
Isle of Man
★Whitefaced Woodland
★Jacob*
Lincoln Longwool
★Derbyshire Gritstone
★Galway
Republic of Ireland
Welsh Mountain & Badger Face Welsh Mountain
Lleyn★
★Leicester Longwool
★Shropshire
Norfolk Horn
Beulah Speckled Face★
★Kerry Hill
★Clun Forest
★Welsh Hill Speckled Face
★Suffolk
★Hill Radnor
England
Wales
Ryeland
Llanwenog★
Black Welsh Mountain & South Wales Mountain
★Colbred
Balwen★
Cotswold
★Oxford Down
★Brecknock Hill Cheviot
★Exmoor Horn
British Milk Sheep
★Hampshire Down
★Romney
★Dorset Down
Southdown
Devon Closewool
Dorset Horn
Devon & Cornwall Longwool
★Greyface Dartmoor & Whiteface Dartmoor
★Portland